Advanced Sciences and Technologies for Security Applications

Indexed by SCOPUS

The series Advanced Sciences and Technologies for Security Applications comprises interdisciplinary research covering the theory, foundations and domain-specific topics pertaining to security. Publications within the series are peer-reviewed monographs and edited works in the areas of:

- biological and chemical threat recognition and detection (e.g., biosensors, aerosols, forensics)
- crisis and disaster management
- terrorism
- cyber security and secure information systems (e.g., encryption, optical and photonic systems)
- traditional and non-traditional security
- energy, food and resource security
- economic security and securitization (including associated infrastructures)
- transnational crime
- human security and health security
- social, political and psychological aspects of security
- recognition and identification (e.g., optical imaging, biometrics, authentication and verification)
- smart surveillance systems
- applications of theoretical frameworks and methodologies (e.g., grounded theory, complexity, network sciences, modelling and simulation)

Together, the high-quality contributions to this series provide a cross-disciplinary overview of forefront research endeavours aiming to make the world a safer place.

The editors encourage prospective authors to correspond with them in advance of submitting a manuscript. Submission of manuscripts should be made to the Editor-in-Chief or one of the Editors.

More information about this series at http://www.springer.com/series/5540

Hamid Jahankhani · Arshad Jamal · Shaun Lawson
Editors

Cybersecurity, Privacy and Freedom Protection in the Connected World

Proceedings of the 13th International Conference on Global Security, Safety and Sustainability, London, January 2021

 Springer

Editors
Hamid Jahankhani
London Campus
Northumbria University
London, UK

Arshad Jamal
Northumbria University
London, UK

Shaun Lawson
Department of Computer and Information
Sciences
Northumbria University
Newcastle upon Tyne, UK

ISSN 1613-5113   ISSN 2363-9466 (electronic)
Advanced Sciences and Technologies for Security Applications
ISBN 978-3-030-68536-2   ISBN 978-3-030-68534-8 (eBook)
https://doi.org/10.1007/978-3-030-68534-8

This Springer imprint is published by the registered company Springer Nature Switzerland AG
The registered company address is: Gewerbestrasse 11, 6330 Cham, Switzerland

Committee

General Chair:

Prof. Hamid Jahankhani, UK
hamid.jahankhani@northumbria.ac.uk

Steering Committee

- Prof. Babak Akhgar, Director of CENTRIC, Sheffield Hallam University, UK
- Prof. Giovanni Bottazzi, Luiss University, Italy
- Dr. Brij Bhooshan Gupta, National Institute of Technology Kurukshetra, India
- Mrs. Sukhvinder Hara, Middlesex University, UK
- Prof. Ali Hessami, Director Vega Systems, UK
- Dr. Amin Hosseinian-Far, University of Northampton, UK
- Dr. Nilakshi Jain, Shah and Anchor Kutchhi Engineering College, Mumbai, India
- Dr. Arshad Jamal, QAHE and Northumbria University London, UK
- Prof. Gianluigi Me, Luiss University, Italy
- Dr. John McCarthy, Director, Oxford Systems, UK
- Dr. Ian Mitchell, Middlesex University, UK
- Dr. Reza Montasari, University of Huddersfield, UK
- Dr. Sina Pournouri, Sheffield Hallam University, UK
- Prof. Bobby Tait, University of South Africa, South Africa
- Prof. Willy Susilo, Director—Institute of Cybersecurity and Cryptology, University of Wollongong, Australia
- Dr. Guomin Yang, University of Wollongong, Australia
- Prof. Shaun Lawson, Northumbria University, UK
- Dr. Neil Eliot, Northumbria University, UK
- Kudrat-E-Khuda (Babu), Department of Law, Daffodil International University, Bangladesh
- Prof. Sergio Tenreiro de Magalhaes, Chair of Cyber Security, Champlain College Online, USA

Programme Committee

- Dr. Fiona Carroll, Cardiff Metropolitan University, UK
- Dr. Christos Douligeris, University of Piraeus, Greece
- Dr. Omar S. Arabiat, Balqa Applied University, Jordan
- Mr. Konstantinos Kardaras, Technical Consultant, Greece
- Dr. Chafika Benzaid, Computer Security Division, CERIST, University of Sciences and Technology Houari Boumediene, Algeria
- Prof. Bobby Tait, University of South Africa, South Africa
- Dr. Fouad Khelifi, Northumbria University, UK
- Dr. Kenneth Revett, University of Maryland College Park, Massachusetts, USA
- Dr. Ken Dick, Nebraska University Centre for Information Assurance, USA
- Dr. Elias Pimenidis, University of West of England, UK
- Prof. Henrique M. D. Santos, University of Minho, Portugal
- Prof. Alexander Sideridis, Agricultural University of Athens, Greece
- Mr. Ray Brown, QAHE and Northumbria University, UK
- Prof. Murdoch Watney, University of Johannesburg, South Africa
- Dr. Mamoun Alazab, Macquarie University, Australia
- Dr. Gregory Epiphaniou, Commercial Director, Wolverhampton Cyber Research Institute (WCRI), University of Wolverhampton, UK
- Dr. Haider Alkhateeb, Wolverhampton Cyber Research Institute (WCRI), University of Wolverhampton, UK
- Dr. Carlisle George, Middlesex University, UK
- Prof. Gianluigi Me, University of Rome LUISS "Guido Carli", Italy
- Prof. Antonio Mauro, Head of Security and Compliance Assurance, Octo Telematics S.p.A., Italy
- Dr. John McCarthy, Director, Oxford Systems, UK
- Dr. Ameer Al-Nemrat, University of East London, UK
- Mrs. Sukhvinder Hara, Middlesex University, UK
- Dr. Ian Mitchell, Middlesex University, UK
- Prof. Ahmed Bouridane, Northumbria University, UK
- Dr. Amin Hosseinian-Far, University of Northampton, UK
- Dr. Reza Montasari, Birmingham City University, UK
- Mr. Ayman El Hajjar, QAHE and Northumbria University, UK
- Dr. George Weir, University of Strathclyde, UK
- Dr. Sufian Yousef, Anglia Ruskin University, UK
- Dr. Alireza Daneshkhah, Coventry University, UK
- Dr. Maryam Farsi, Cranfield University, UK
- Mr. Usman Javed Butt, QAHE and Northumbria University, UK
- Asim Majeed, QAHE and University of Roehampton, UK
- Dr. Rose Fong, QAHE and Northumbria University London, UK
- Dr. Athar Nouman, QAHE and Northumbria University London, UK
- Mr. Umair B. Chaudhry, Queen Mary College, University of London, UK
- Mr. Nishan Chelvachandran, Cyberreu, Finland
- Mr. Stefan Kendzierskyj, Cyfortis, UK

13th International Conference on Security, Safety and Sustainability, ICGS3-21

Cybersecurity Privacy and Freedom Protection in the Connected World 14th and 15th January 21

About the Conference

Following the successful 12th ICGS3, we have much pleasure in announcing the 13th International Conference on Global Security, Safety & Sustainability.

Year 2020 will be remembered as a year of COVID-19 pandemic year with a profound impact on our lives and challenging times for governments around the world with technology as a backbone of tools to assist us to work remotely while in self-isolation and in total lockdown. Every individual and organisation are faced with the challenges of the adoption of technological change and understand that nothing will be the same for some time after and even when normality resumes. AI, blockchain, machine learning, IoT, IoMT, IIoT all are tested and applied at a very rapid turnaround as we go through the pandemic day after day. However, all these transformations can bring many associated risks regarding data, such as privacy, transparency, exploitation and ownership.

Since the COVID-19 pandemic in January 2020, governments and organisations; small or large have seen an increase in cyber-attacks. Hackers, through reconnaissance and some network intrusions, targeting nation states medical research facilities, healthcare organisations conducting research into the virus, financial sectors and so on. Early in April 2020, the UK and the US security agencies sound COVID-19 threat alert "…Cybersecurity Infrastructure and Security Agency issued a joint warning that hacking groups associated with nation-state governments are exploiting the COVID-19 pandemic as part of their cyber operations…".

The advancement of Artificial Intelligence (AI), coupled with the prolificacy of the Internet of Things (IoT) devices is creating smart societies that are interconnected. The expansion of Big Data and volumetric metadata generated and collated by these mechanisms not only give greater depth of understanding but as change is coming at an unprecedented pace COVID-19 driving the cultural changes and the public sector innovations to a new height of responsibilities.

Further development and analysis of defence tactics and countermeasures are needed fast to understand security vulnerabilities which will be exploited by the cybercriminals by identifying newer technologies that can aid in protecting and securing data with a range of supporting processes to provide a higher level of cyber resilience.

The ethical implications of connecting the physical and digital worlds, and presenting the reality of a truly interconnected society, presents the realisation of the concept of smart societies in reality.

This Annual International Conference is an established platform in which security, safety and sustainability issues can be examined from several global perspectives through dialogue between academics, students, government representatives, chief executives, security professionals and research scientists from the United Kingdom and from around the globe.

The TWO-day Virtual conference will focus on the challenges of complexity, the rapid pace of change and risk/opportunity issues associated with the twenty-first-century lifestyle, systems and infrastructures.

October 2020   Prof. Hamid Jahankhani
                                            General Chair of the Conference

Contents

Artificial Intelligence and the International Information and Psychological Security

Konstantin A. Pantserev⊙ and Konstantin A. Golubev⊙

Abstract The paper considers the problem of malicious use of technologies that are based on artificial intelligence (AI). The authors presume that the need to develop advanced technologies is seen by states as essential to ensuring their global leadership and technological sovereignty. Particular focus is on AI-based technologies whose capabilities are growing at unprecedented rates. AI has already become part and parcel of intelligent machine translation and transport systems. AI-based technologies are widely used in medical diagnostics, e-Commerce, online training and even in the production of news and information. Meanwhile, world's top search engines have offered their users voice assistants that significantly simplify and accelerate search for relevant information. Yet, evidently most technological innovations that are meant to make our lives easier could potentially be used for malicious purposes. Therefore, the rapid growth of our dependency on hybrid computer intelligent systems renders national critical infrastructure extremely vulnerable to attacks by those who would like to use AI-based technologies to cause significant harm to a nation, which in turn poses a serious challenge to psychological and information security of people around the world. The paper discusses ways of malicious use of AI and offers possible instruments of mitigating the threat that advanced technologies are posing.

Keywords Artificial intelligence · Strategic communication · International information and psychological security · Information and psychological warfare

1 Introduction

Developing advanced technologies is deemed a sine-qua-non to ensure one's global leadership in today's world. Particular focus is on technologies that are based on artificial intelligence (AI). The latter's capabilities have been growing at unprecedented rates. Nowadays, AI-algorithms are widely used in intelligent machine translation systems, medical diagnostics, electronic commerce, on-line education, intelligent

K. A. Pantserev (✉) · K. A. Golubev
Saint-Petersburg State University, 7/9 Universitetskaya Emb, 199034 Saint-Petersburg, Russia
e-mail: k.pantserev@spbu.ru

© The Author(s), under exclusive license to Springer Nature Switzerland AG 2021
H. Jahankhani et al. (eds.), *Cybersecurity, Privacy and Freedom Protection in the Connected World*, Advanced Sciences and Technologies for Security Applications, https://doi.org/10.1007/978-3-030-68534-8_1

1

transport systems and even in the production of news and information. Meanwhile, world's top search engines have offered their users voice assistants that significantly simplify and accelerate search for relevant information.

Often a time, developers of AI-based solutions receive financial support from their national governments. According to official data, over 30 countries have adopted national strategies and roadmaps related to AI. Those include the USA, China, France, Japan, Russia, and UAE to name just a few [1].

Yet, evidently most technological innovations that are meant to make our lives easier could potentially be used for malicious purposes. Therefore, the rapid growth of our dependency on hybrid computer intelligent systems renders national critical infrastructure extremely vulnerable to attacks by those who would like to use AI-based technologies to cause significant harm to a nation, which in turn poses a serious challenge to psychological and information security of people around the world.

2   Artificial Intelligence: From Imaginary Threats to Real Ones

As a result of the advances in computer science, complex algorithms are becoming an integral part of people's daily routines. Yet, one has to be conscious that those technologies in fact could present a real threat to individual and collective security at both national and international levels. Thus, researchers who are involved in the development of AI are divided over the issue of assessing the threat to the future of humanity in the age of smart societies. Some (e.g. Bill Gates, Mark Zuckerberg) view technological breakthroughs in AI as an opportunity for people to leave behind monotonous work and engage in self-actualization. Yet, others (e.g. Moshe Vardi, Steven Hocking) admonish that a rapid development of such technologies could trigger mass unemployment or even World War III. Therefore, we should get ready to deal with the risks coming from advanced technologies, trying to mitigate those before they turn into a real threat. That is why it seems crucial to elaborate effective mechanisms of legislature at both national and international levels to regulate the application of AI-technologies.

Particularly daunting are the rapid rates at which contemporary information technologies develop, so clearly neither national legislature, nor the system of international law, let alone the existing mechanisms of control, can keep up with those. This is the main challenge of the digital age. Most technological innovations developed over the recent years are supposed to make people's lives easier, yet the lack of effective control mechanisms, as well as that of a proper regulatory framework, dramatically increases the risk of malicious use of such technologies.

As D. Bazarkina and E. Pashentsev argue, possible malicious use of artificial intelligence "can cause serious destabilizing effects on the social and political development of the country and the system of international relations" [2]. However, given that AI-based technologies are a relatively recent phenomenon, there are few if any real cases of malicious use of artificial intelligence either by actual actors of international psychological warfare or by terrorists. Yet, it seems almost inevitable that destructive elements of all kinds, when they get familiar with such technologies, will put the latter to malicious use before long. Therefore, we need to be prescient about the ways of possible malicious use of AI in order to elaborate pre-emptive measures of neutralizing those threats. Nonetheless, some threats have already materialized.

Thus we suggest dividing all threats connected to artificial intelligence into two types, namely latent challenges that threaten international psychological security and actual ones that threaten individual psychological security. Latent ways of malicious use of artificial intelligence have been studied by professors Bazarkina and Pashentsev in their work "Artificial Intelligence and New Threats to International Psychological Security." In it they highlight the following potential threats that have to do with an active use of advanced technologies:

- Malicious takeover of integrated, all-encompassing systems that either actively or primarily use AI;
- Delivery of explosives by or causing crashes with commercial AI systems such as drones or autonomous vehicles [2].

To that list should be added such threats as:

- Autonomous weapon systems programmed for killing. Nowadays, most leading nations are racing to develop various intelligent armament systems. One can speculate that in the nearest future the nuclear arms race will give way to that of developing military hybrid intelligent systems. One can only imagine what would happen if a national army lost control over such intelligent systems or if such systems fell into the hands of terrorists.
- Social manipulative practices. Currently, social media aided by AI-algorithms can effectively target prospective consumers. Likewise, using access to personal data of millions of people, knowing their needs, strengths and weaknesses, AI-algorithms could be used maliciously also to engage in mind manipulation and direct propaganda at specific audiences.
- Invasion of privacy. This threat is already with us. Now ill-minded individuals can track every step of online users, as well as calculate the exact times of their doing daily chores. In addition, surveillance cameras are being installed almost everywhere, taking advantage of facial recognition algorithms that easily identify each and every one.
- Mistakes by operators. The human element will remain crucial even if there appear smart machines capable of learning on their own. AI is valuable to us primarily because of its high productivity and efficiency. However, if we fail to define the objectives clearly for an AI-system, their optimal attainment could have unintended consequences.
- Lack of data. As is well known, AI-algorithms are based on processing data and information. The more data is fed into the system, the more accurate the result will be. Yet, if there is insufficient data to perform a particular task or if the data

has been corrupted by perpetrators, it could cause fatal glitches within the entire AI-system, which could cause unpredictable outcomes.

Finally, we should pinpoint yet another risk factor that might be the most significant one—our increased dependency on advanced technologies that are integrated into every aspect of daily life of each individual, overseeing the functioning of numerous applications and even most critical infrastructure. Clearly such technologies will be a magnet to ill-minded people and terrorists of all sorts.

In particular, nowadays one can observe a rapid growth of complex intelligent automated systems. Such systems span a wide range of useful applications such as the organisation and optimisation of traffic or the management of large facilities and infrastructure. In the meantime, one should bear in mind that all those intelligent systems can become an easy target to high-tech terrorist attacks. One can only imagine the consequences of an interception of control over the transport management system of a major city such as New York, Moscow, Paris, Sidney, Beijing, Shanghai, Tokyo, Seoul or over intelligent navigation systems that are used by vessels and aircraft. Undoubtedly such incidents could result in numerous casualties or cause panic and create a climate that from information and psychological perspectives would be conducive to further hostile actions.

Another plausible scenario is for future high-tech AI-based systems that manage energy grids of large industrial regions to be hacked by an adversary or terrorists, resulting in unfathomable damage.

The aforementioned examples are only imaginary threats; nevertheless, society must come to grips with such challenges. In our opinion, as of now, the real threat comes not from hypothetic terrorists trying to gain control over hybrid intelligent automated systems but from relatively experienced perpetrators who possess sufficient knowledge of AI-algorithms. For example, they can devise intelligent bots whose mission is to misinform ordinary people, create unfavourable public opinion or manipulate news agenda [3]. Undoubtedly such bots can be widely used in defamation campaigns against individuals, groups and entire nations.

The greatest danger still, as of now, is posed by those fake videos created with the help of AI, whose fabricated yet realistic footage is capable of confusing the general public regarding what in fact is happening around the world.

In and of itself, the technology (Generative Adversarial Networks) of making fake videos, also known as deepfakes, applies certain algorithms to synthesise facial movements of humans based on AI. Most importantly, one needs neither fancy skills nor knowledge in AI or machine learning to create such videos. All tasks are performed by a computer application which can be freely downloaded from the Internet.

One of such applications is FakeApp. It can superimpose a person's face onto someone else's. It has a rather intuitive interface and a detailed help file about how to produce deepfake videos. What is required is a sufficient amount of authentic video footage of the person one intends to recreate so that the application could render it to be as realistic as possible.

The application allows users with the most basic skills in computers to produce replicas of any person that walk and talk like their prototype and pronounce any nonsense one can only think of.

This latter threat has already become part of our reality. Starting in December 2017, there appeared on the Web a number of fake porno video clips starring well-known Hollywood actresses such as Gal Gadot, Chloë Moretz, Jessica Alba and Scarlett Johansson. Those films by no means pose a threat to international psychological security but they demonstrate a possibility to threaten personal psychological security because few people would want to be cast in such films. Yet, what is worse, there are already fake videos out there featuring such salient political figures as Vladimir Putin, Donald Trump and Barak Obama.

So far, fake footage featuring world leaders has been put out either for fun or to demonstrate the capability of this new technology. However, some experts caution that in the future deepfakes could become so realistic that this could have a detrimental impact on world politics as a whole [4].

One of the most extraordinary examples of deepfake videos was the one created by American filmmaker Jordan Peel back in April 2018. It featured the former US President Barak Obama badmouthing the incumbent President Donald Trump. The idea behind the film was not to misinform the audience but to draw attention to the potential danger of this new advanced technology [5].

The Russian President Vladimir Putin has also been a target of such deepfakes. One of such video clips featured Mr. Putin boasting that it was him who put Donald Trump into the Oval Office [6].

Last but not least, there was a phony video that showed U.S. President Donald Trump speaking at the White House, declaring that the United States was going to withdraw from the Paris climate agreement and calling on Belgium to follow suit. That video was made and distributed by the Flemish Socialist Party. So far, this has been the first and only real case of using deepfake technology to pursue political aims. Still, it was a relatively innocuous endeavour since its creators did not intend to misinform their audiences. On the contrary, at the end of it, the fictitious character of Donald Trump revealed that the video was fake. In fact, the Flemish Socialist Party simply wanted to draw public attention once again to the problem of climate change by calling on the people to sign a petition to invest more money into alternative sources of energy and electric cars and to shut down the nuclear power plant Doel in Flanders [7].

Still, the above cases clearly demonstrate the harmful potential of this new technology that allows any person with basic computer skills, using a desktop application available to be downloaded from the Web, to create a deepfake video of any person and to put any words in his or her mouth. It seems highly likely that terrorist groups will shortly take advantage of this promising technology and start using it in order to incite ethnic hatred or to recruit new acolytes to their ranks. This last point brings us to a conclusion that in the current age of fake news and disinformation society will have to grapple with some serious challenges to national and international security.

We are already suffering from information overload due to huge volumes of information created not only by professional journalists but ordinary social media users who very often disseminate unwittingly or on purpose unverified stories or even outright lies. The advent of technology capable of creating deepfake videos presents yet another challenge—we can no longer trust what we see or hear for any video that one comes across on the Web, no matter how veritable it may seem, may in fact be fake, contributing further to "widespread uncertainty" which "may enable deceitful politicians to deflect accusations of lying by claiming that nothing can be proved and believed" [8].

Worse still, according to some experts, in the nearest future, as deepfake technology progresses, one simply will not be able to tell a real video from a fake one [9]. The future, then, may turn out rather terrifying. One can imagine the chaos that would result if a fake video came out with the US President announcing an "impending nuclear missile attack on North Korea" [10]. This could absolutely perplex audiences as to what is happening to their world. And if people are not able to trust what they see or hear, they will choose for themselves what they want to believe [11]. As Chesney and Citron argue:

"Imagine a video depicting the Israeli prime minister in private conversation with a colleague, seemingly revealing a plan to carry out a series of political assassinations in Tehran. Or an audio clip of Iranian officials planning a covert operation to kill Sunni leaders in a particular province of Iraq. Or a video showing an American general in Afghanistan burning a Koran. In a world already primed for violence, such recordings would have a powerful potential for incitement. Now imagine that these recordings could be faked using tools available to almost anyone with a laptop and access to the Internet—and that the resulting fakes are so convincing that they are impossible to distinguish from the real thing. Advances in digital technologies could soon make this nightmare a reality. Thanks to the rise of "deepfakes"—highly realistic and difficult-to-detect digital manipulations of audio or video—it is becoming easier than ever to portray someone saying or doing something he or she never said or did" [12].

The cases cited above lead one to ponder about how we should respond to those challenges. What clearly comes out is that it is necessary to work out effective mechanisms that could put a lid on uncontrolled spread of deepfakes of this kind so as to neutralize their toxic content. In the meantime, computer science experts are working hard to come up with specific algorithms that would be able to detect deepfakes. Once in effect, such algorithms could be used by social media platforms such as Facebook and Twitter to inspect and mark up uploaded videos before they could be viewed by other users.

Yet, as of now, there are no efficient algorithms that could detect deepfakes with 100% certainty. Moreover, it should be noted that deepfake technology is getting better with each detection cycle. And this is just one side of the coin. The other one is that there exists no law to regulate the process of creation and distribution of deepfakes.

When working on the appropriate legislation, aiming to stop the spread of toxic deepfakes, we would run into a serious problem since any unjustified prohibition to create or distribute fake videos would be interpreted as a violation of the basic principle of freedom of speech and expression. Therefore, we believe it important that the legislature draw a distinction between malicious use of deepfakes that are aimed to create some toxic content and the sort of satire, creative effort and self-expression. Until this conundrum is solved, there will be no law to regulate the process of creation and distribution of deepfakes. This means that in the short term one should expect a flurry of highly realistic and difficult-to-detect deepfakes to come out, raising the risk of destabilisation of international system and threatening the global psychological security.

3 Conclusion

Nowadays, most advanced nations increasingly focus on developing AI-based systems. This trend in science and technology becomes a key priority for the national development of every country to ensure technological sovereignty in the contemporary digital age. Yet, the rapid integration of AI into our daily life increases the risk of malicious use of such technologies, which can put personal, national and international psychological security in jeopardy.

In our opinion the real and present danger, as of now, lies with deepfakes. In the contemporary information environment, it has become the rule of thumb to double-check any information published on the Web. However, the emergence of deepfakes puts disinformation at a qualitatively new level where it could be used maliciously not only by cyber prankers but also by various perpetrators, terrorists and other destructive elements. The numerous cases of deepfake videos out there has shown clearly that this new technology could be used to blackmail and terrify people who in fact have done nothing wrong.

Thus it becomes crucial for the international community to elaborate effective mechanisms to control the spread of deepfakes with toxic content. Nowadays, computer science experts are trying to elaborate appropriate algorithms to detect deepfakes. The problem, however, is that there currently exists no methodology to ensure 100 percent detection rate. Worse still, the practical implementation of such algorithms would be stumbling over some serious legislative obstacles for as long as this field remains free of legislative regulation.

Given all that, it seems extremely important to elaborate an appropriate legal base as quickly as possible to deal with deepfakes so as to distinguish between malicious use of this technology and the innocuous one—that of satire, creativity and self-expression. Until then, we can expect some toxic content created with the help of AI-based deepfake technology to threaten the stability of political systems around the world.

References

1. Dutton T (2016) An overview of national AI strategies. Politics + AI, https://medium.com/pol itics-ai/an-overview-of-national-ai-strategies-2a70ec6edfd. Last accessed 15 Aug 2016
2. Bazarkina D, Pashentsev E (2019) Artificial intelligence and new threats to international psychological security. Russia in Global Affairs 1 (2019). https://eng.globalaffairs.ru/articles/ artificial-intelligence-and-new-threats-to-international-psychological-security. Last accessed 17 Aug 2016
3. Horowitz MC, Scharre P, Allen GC, Frederick K, Cho A, Saravalle E (2018) Artificial intelligence and international security. Washington: center for a new American security (2018). http://www.cnas.org/publications/reports/artificial-intelligence-and-international-security. Last accessed 17 Aug 2016
4. Palmer A (2018) Experts warn digitally-altered 'deepfakes' videos of Donald Trump, Vladimir Putin, and other world leaders could be used to manipulate global politics by 2020. Mail Online. http://www.dailymail.co.uk/sciencetech/article-5492713/Experts-warn-deepfakes-vid eos-politicians-manipulated.htm. Last accessed 19 Aug 2016
5. Pantserev KA (2020) The malicious use of AI-based deepfake technology as the new threat to psychological security and political stability. In: Jahankhani H, Kendzierskyj S, Chelvachandran N, Jimenez JI (eds) Cyber defence in the age of AI, smart societies and augmented humanity. Springer Nature Switzerland AG, pp 37–55
6. Putin: Face Replacement. https://www.youtube.com/watch?time_continue=3&v=hKxFqx CaQcM&feature=emb_logo. Last accessed 20 Aug 2016
7. Von der Burchard H (201) Belgian socialist party circulates 'deep fake' Donald Trump video (2018). https://www.politico.eu/article/spa-donald-trump-belgium-paris-climate-agreement-belgian-socialist-party-circulates-deep-fake-trump-video. Last accessed 20 Aug 2016
8. Vaccari C, Chadwick A (2020) Deepfakes and disinformation: exploring the impact of synthetic political video on deception, uncertainty, and trust in news. Social Media + Society
9. Browne R (2018) Anti-election meddling group makes A.I.-powered Trump impersonator to warn about 'deepfakes'. https://www.cnbc.com/2018/12/07/deepfake-ai-trump-impersonator-highlights-election-fake-news-threat.html. Last accessed 23 Aug 2016
10. Harris D (2018) Deepfakes: false pornography is here and the law cannot protect you. Duke Law Technol Rev 17(1)
11. Dack S (2019) Deep fakes, fake news, and what comes next. https://jsis.washington.edu/news/ deep-fakes-fake-news-and-what-comes-next. Last accessed 23 Aug 2016
12. Chesney R, Citron D (2019) Deepfakes and the new disinformation war: the coming age of post truth geopolitics. Foreign Affairs

Blockchain Medicine Administration Records (BMAR): Reflections and Modelling Blockchain with UML

Ian Mitchell

Abstract Modelling blockchain should be easy? There are many figures and diagrams that are intuitive, however they are often unorthodox and do not comply with standard modelling techniques, e.g., UML. The enterprise blockchain system designed is for Medicines Administration Records (MARs) and requires audits for accreditation, that exposes vulnerabilities to unauthorised access and alteration. These alterations are made in order to pass inspection of the auditors and often completed to correct human errors unforeseen until the audit. The design stage has *five* stages and includes *four* stages from UML. The initial stage looks at how effective blockchain is as a solution for the problem domain, and then followed by Use Case, Sequence, State Machine and Deployment. The final design was implemented and the overall approach is evaluated.

Keywords Blockchain modelling · Permissioned blockchain · Blockchain development · Medication administration records · Medication administration errors

1 Introduction

The ability to model before building is strategic to any engineering domain. Blockchain engineering is no different, and blockchain modelling has produced many illustrations that are intuitive and bespoke. The diagrams are different and unorthodox to other software engineering practices. The approach taken here utilises and tries to adapt existing models to blockchain engineering project in healthcare, Blockchain Medicines Administration Records, or BMARs [1] for short. So, the problem is twofold: (i) develop a blockchain consortium for BMARs; and, (ii) understand the issues with modelling blockchain with UML. The next section introduces the problem domain and the rest of the paper looks at how to model this as a blockchain using UML notation.

I. Mitchell (✉)
Middlesex University, London, UK
e-mail: i.mitchell@mdx.ac.uk

9

2 Problem Domain

Falsification of documentation or records is something that healthcare professionals deny publicly, but privately there are issues. Healthcare professionals have to complete equipment checks, document the administration of medication and its reactions to patients, and many other processes that all require auditing by some central governing quality assurance agency. The evidence is anecdotal that falsification occurs, and only comes to light when there are whistle-blowers or the patient makes a litigation. The temptation when a lawsuit is issued, is to amend, edit or destroy the documentation and to retrospectively contaminate the evidence in favour of the defendant. Whilst rare, there are cases that indicate that this occurs and how tragic the consequences are [2]. In this case the U.S. supreme court decided that the destruction and alteration of medical records by a paediatrician was to evade the potential medical malpractice that could have followed.

It is not only paediatricians or General Practitioners, that can spoliate evidence or make mistakes. All healthcare professionals have opportunities to do this. Removing this opportunity, promotes trust between all parties, and is the ambition of BMARs [1]. Briefly, BMARs uses blockchain to automate and audit the completion of Medicines Administration Records, see [3] for further information on advice on MARs completion.

2.1 Blockchain

Permissioned blockchain development is inspired by the work of [4], and relies on append-only decentralised ledgers, but this is where the similarities stop. Permissioned blockchain development does not require Practical Byzantine Fault Tolerance, PBFT [5, 6], since all the parties involved in the blockchain development channel are to be trusted and therefore requires only fault tolerance. This difference allows faster and more efficient consensus algorithms since they do not have to ascertain trust between each party. This in turn sees increased throughput in updating the blockchain and reduces the energy expedited. The other differences are: (i) that permissioned blockchains are written in Turing complete languages, e.g., Java, and is therefore deterministic; and (ii) tokens or crypto-currency is optional.

So far, so good but what about patient privacy and confidentiality, surely this can be compromised on a network? This final point is significant and important to all healthcare professionals, if we are going to use blockchain technology to overcome this, the question of security and service-user (SU) privacy and confidentiality has to arise. This will be looked at in more detail, but essentially the same technology that keeps transactions anonymous on a permissionless blockchain, e.g., bitcoin, is used to keep information secure and private on a permissioned blockchain.

2.2 Medicines Administration Records, MARs

The motivation is to prevent falsification of documentation. Falsification can include, the destruction, editing, overwriting, and any other means of altering the medical data. The solution is to use permissioned blockchain technology to medical data. The problem domain is still required, one of the simple tasks that healthcare professionals have to complete when administering drugs to service users (SU), is the use of a MARs sheet. The MARs sheet records when an SU receives medication and their reaction to it. It can also record the refusal to take medication and the method in which it was applied. Guidance on completion of MARs sheets includes the 6 R's [7], which are:

1. right resident,
2. right medicine,
3. right route,
4. right dosage,
5. right time; and,
6. right to refuse.

The MARs sheet is preserved and included in quality assurance inspections. For the UK this is completed by the Care Quality Commission, CQC, which is the accrediting body for all healthcare providers, including care homes. So, the problem definition is to design and model a permissioned blockchain application that implements MARs.

The rest of this paper is divided up into, Sect. 3 the method and approach used to model blockchain, Sect. 4 the evaluation of the system and Sect. 5 discusses the issues and summarises the approached used.

3   Method

For reasons aforementioned, permissioned blockchain is conducive to implement applications that deal with medical data or records. There are many blockchain medical applications, e.g., see [8–10] that handle medical data. From these, and other applications it is known that auditability is apt for blockchain; and, e-Health Records (EHR) can be stored on blockchain, without contravening any of the Caldicott principles [11, 12]. However, the first stage of any design is to ask if blockchain is a suitable solution? This is not a feasibility study but a quick review to decide if blockchain technology is conducive to solving the given problem. In technology, buzzwords and trends sometimes overrule common sense, in the 1990s everything was Web 1.0 and everyone wanted a website. Web 2.0 saw everything shift to user-content generation, or/and social media, again the applications of this had huge success but also saw many instances of Web 2.0 backfire on organisations. Web 3.0 [13] is blockchain or decentralised ledgers, and the advantage is that the mistake of using gratuitous technical solutions to problems that are solved has been done in Web 1.0 and 2.0—it is the intention of Web 3.0 not to make the same mistakes.

3.1 Blockchain Efficacy Analysis

So, when designing a decentralised application, be it permissioned or permissionless, there are some guidelines to see if your problem domain is conducive to blockchain. This simple set of guidelines can be found as a flowchart in [24, p. 57]. There are six questions, adapted from [14], that are going to be answered below for BMARs:

Need a shared consistent data store? According to Oscar Research [15] there are over 20,000 care homes registered in the UK. Whilst not all of these may have facilities or trained personnel to administer medicines, many of them will, e.g., 5028 are Nursing Homes. There is not so much a need to share data across a community of care homes, however there is a need to share data with the auditors. So, yes there is a need to share consistent data with auditor.

More than one participant to contribute data? In each care home there are multiple SUs, across a single country there are multiple care homes. There is not an exact number available in the UK, but if only 10% of the care homes used MAR sheets, then everyday there would be at least 2,000 records produced. All could be subject to falsification or errors. So, yes more than one participant would contribute data.

Data records are never updated or deleted? These records should never be edited or deleted. Even after the SU's death these anonymous records would remain for relevant investigations. MAR sheets themselves should never be altered or edited in anyway. The information about participants, Care Home Providers (CPH), prescriptions and medication are stored on a database—these already exist. The blockchain, via a secure channel, would only store information when a specific event occurs and store a record of that as a transaction. The transaction is evidence of a specific instance of an event occurring. So, yes data records are never updated or deleted.

Sensitive identifiers are not written to the data store? This is the biggest challenge and all data would be anonymised. The auditors are not interested in the personal details of the SUs, they are only interested the quality of care the SU receives. This issue has been discussed for this application in [12] and all the Caldicott principles are adhered to. So, yes there are no sensitive identifiers written to the blockchain.

Who would control the data? The blockchain would be permissioned and central control be given to the auditing agency, e.g. CQC. So, yes the auditing agency would have control of nodes and therefore the data.

A tamperproof log of all the writes to the data store? This is absolutely necessary to ensure that the medication is administered correctly and therefore, yes a tamperproof log is required.

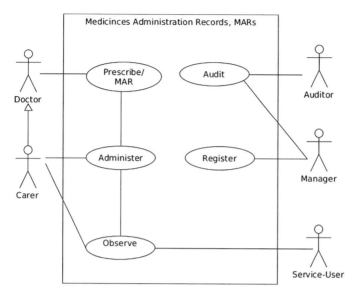

Fig. 1 Use case diagram for medication administration records, MARs

In essence, the above questions have all affirmative answers and therefore Medication Administration Records is conducive to blockchain technology. Any negative answers to the above would require further research and proceed with caution. This stage is probably the most important and decides whether to progress to the following subsections.

3.2 Use Case

Figure 1 illustrates the Use Case Diagram for the system. The use case shows the interaction of participants with the system and each use case is described below:

Register: The registration and deregistration of an SU to an instance of a CHP is completed by a manager. This registration would trigger smart contracts, that may re-order prescriptions from a medical professional and more. Smart contracts would invalidate any further MARs entries and prevent re-ordering of prescriptions to the same Care Home provider when an SU has been deregistered. All transactions are recorded anonymously on the blockchain.

Prescribe: The Doctor participant in our use case diagram is responsible for generating the prescription, which in turn generates the MAR sheet. This procedure may differ from country to country, in the UK the prescription is forwarded to the pharmacy, who dispense the medication and generate the corresponding MARs sheet. A prepared MAR sheet will include the dosages, timings, medicine administer code

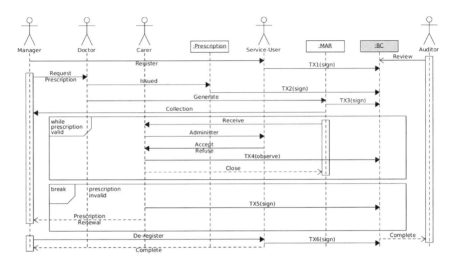

Fig. 2 Sequence diagram for medicine administration records blockchain application

(e.g. oral), and other information for the duration of the prescription, which can be 28 days or 1 week.

Administer: The carer will administer the medication overtly and thus allow the SU to refuse or reject the dosage. This would then be recorded; such actions could involve the smart contract contacting for further medical assistance or advice.

Observe: Any reactions that are abnormal would be recorded and observed and then the Carer would provide further support by looking at the SU's personal care plan and request further professional advice of how serious the consequences are.

Audit: Auditors can audit the system at any time, as can managers. The Auditors cannot identify individuals from querying the blockchain. Running smart contracts will have sufficient privileges to access generated MARs and determine from the blockchain if any transgressions should be bought to the attention of the Auditor. The system is not concerned with attribution and this would be the responsibility of the Auditing agency and the Manager.

Once agreed and completed, the modelling moves to the next stage, the sequence diagram. Normally, a class diagram would be produced for data, however, there is probably an adequate database with the information installed and therefore this would not be required. If there is no database then a class diagram would be required, we have skipped this part since the database already existed and the blockchain complements the database, not replaces it.

3.3 Sequence Diagram

The sequence diagram will help identify the transactions explicitly and independently. The traditional approach to sequence diagrams is to have a sequence diagram per use case, however, a holistic approach has been taken and consider the overall system as a sequence diagram and in it identify the transactions.

Figure 2 illustrates the sequence diagram and the key issue in this diagram is the addition of the blockchain, represented by the class BC. This would not appear in the class diagram and at first would appear as an error. This is deliberate and represents the information store on the blockchain. There are *six* different types of transactions (TX) represented in the lifecycle of a MAR sheet, and they are described below:

TX1: is to ensure that an immutable record is kept of the registration of an SU to a care provider by a manager and prevent medication being administered to unregister SUs. Information included: manager signature, SU primary key, care home primary key and timestamp.

TX2: is to ensure that an immutable record of the prescription by a qualified medical professional. Other systems will record this and it is not the intention of this system to control the ordering and re-ordering of prescriptions. This system is just to record the prescription has been completed. The identity of the individual is kept anonymous and has to be completed before a MAR sheet can be generated. Information included: SU primary key, digital signature of the Doctor/Participant, and timestamp.

TX3: is to ensure that an immutable record of the generation of the MAR sheet. Documents stored on blockchain is frowned upon, since they would require constant updates and contradict one of the questions on never changing information, so it is not the MARs sheet that is stored on the blockchain, but a record of it being generated. Information included: SU primary key, digital signature of the generator and timestamp.

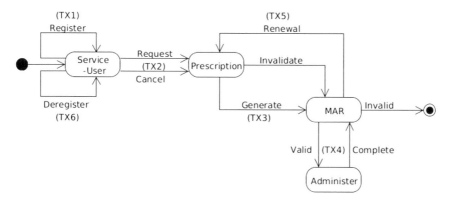

Fig. 3 State machine diagram for medicine administration records blockchain application

TX4: is to ensure that an immutable record of the administration of a medicine to an individual. This record is repetitive and continues until the prescription becomes invalid. It would include: carer signature; SU primary key; MARS primary key; timestamp; and, observation notes.

TX5: is to ensure that an immutable record is kept of when the prescription becomes invalid and a renewal process is requested. Essentially, this is when the MARs sheet runs out of rows to complete and a new one is reordered, however, there could be other reasons, e.g., patient is showing a bad reaction to medication. This also ends the combined fragment in Fig. 3. Information included: carer signature, SU primary key and timestamp.

TX6: is to ensure that an immutable record is kept of the deregistration of an SU from a home. This would be completed by a manager and invalidate any MARs entries. There would be a period of grace for transition and agreed by the manager and dependent on the SU's needs. Information included: care home primary key, manager signature, SU primary key and timestamp.

The use case, along with knowledge of the MARs procedure has informed the identification of the transactions above. The next stage is to think of each of these transactions and what further events they can trigger. The next section considers smart contracts modelling with state machine diagrams.

3.4   Smart Contracts—State Machine Diagram

In [16] smart contracts, SCs, are defined as, "… business rules and state…". SCs automatically update the ledger when certain pre-conditions are agreed. The design of these has been covered in [17], despite the excellent description and case study, this primarily looked at individual smart contract design and was completed using unorthodox diagrams. Our challenge is to observe the state change for each of the transactions, and there is no better diagram than UMLs state machine diagrams. Figure 3 shows the state machine diagram for the case study and associates each of the transactions, TX1-6, identified in Sect. 3.3.

Then a detailed analysis, much like the one offered in [17], is required of the possible smart contract implications. For example, the smart contracts for the completion of a MARs entry could complete the automatic renewal request when its state is 5 working days, estimated time to renew prescription, from completion. During an audit of this nature it is what is missing, not present, that indicates errors. MARs entries left empty are crucial and indicate the omission and are not to be confused with SU's refusal or inability to accept medicines, of the administering of a prescribed medicine. Again, smart contracts would be responsible for highlighting these omissions to auditors, they could even be used to help prevent omissions due to forgetfulness and remind Healthcare professional of an imminent medicine requiring administering, reduction in omissions have been recorded in [18] as a result of sending reminders to healthcare professionals.

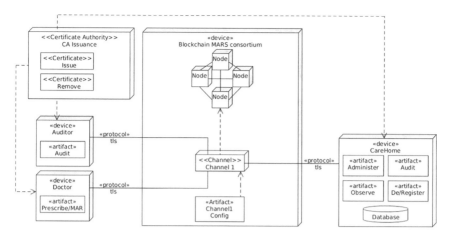

Fig. 4 Deployment diagram for medicine administration records blockchain application

Such detail is crucial to the development of a future Decentralised Autonomous Organisation, DAO. Until now everything is suggest in a sequence, however, the next sub-section can be completed in parallel with the detailed analysis of smart contracts, and concerns the physical architecture.

3.5 Architecture—Deployment Diagram

Figure 4 illustrates the UML deployment diagram. The Blockchain MARs consortium would be deployed on the auditing agency's server, in the UK this is the CQC. For auditing networks, the control is normally centralised to a governing body that can be trusted. Within this device access to several 'Nodes' or 'peers' is permitted, which are used to determine, via consensus, the blocks added to the blockchain. Each node would have its own copy of the blockchain, which would be accessible for querying for Audits.

Note that the CHP has its own database, and that the proposed blockchain system is not to replace the database, but complement it by providing an append-only ledger that is updated by secure protocol and can be used for auditing purposes. Such a database that could be reflected in the class diagram model for data and provide key information as to the attributes being passed.

The auditor would not have system access, but would be granted read access privileges by. Certificate authorities would issue various certificates to communicate and would be part of the channel configuration.

Normally, instances on a deployment diagram are not shown, however, it is important to consider the consequence of adding another instance of a CHP. In our case study, the instances of Doctor and Auditor have no consequence to the physical design. However, the introduction of a second CHP has a significant effect. This is

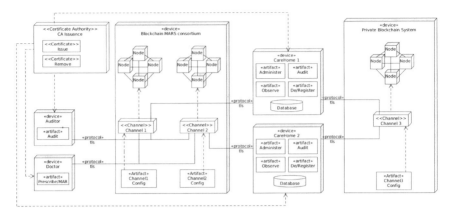

Fig. 5 Deployment diagram for multi-channel architecture for medicine administration records blockchain application with two Care Home Providers (CHP)

something worthwhile considering during the design stage, to determine if a single channel is sufficient for communication as the number of instances increase. This is discussed in the next sub-section.

Multi-Channel Fig. 5 shows two organisations, CHP 1 & 2. These devices represent managers and healthcare professionals, essentially all devices working in the Care Homes under a single organisation. GDPR [19] and National data protection laws, see [20], permits sharing data within an organisation and therefore multiple care homes managed by a single organisation operate under a single channel. Naturally, there would be restrictions in place to prevent unauthorised access and this can be completed using ABAC [21], that is supported by Hyperledger.

This does require access by both Doctors and Auditors, so use cases for auditing and prescribing/MARs would have to reflect the correct channel and certificate to use - this would probably require a portal to access the individual CHPs.

The separation of a group of nodes indicate that there are two ledgers, one for each channel. Access control can be implemented to restrict access to only trans-actions from a particular care home. Therefore, there is a decision to implement a multi-channel blockchain, each channel having its own ledger. Terminating blockchains can also be difficult and the ledgers could easily be removed and would not persist elsewhere by closing the channel and therefore removing the associated nodes, e.g., where a CHP fails accreditation and is removed.

Throughput is difficult to estimate, however the blockchain system can be optimised based on: blocksize; number of channels; StateDB used; and, peer resources. Optimising these for your Hyperledger consortium is discussed in [22] and should be considered when designing your architecture. Depending on the settings [22] shows that these can exceed 2000 transaction per second (TPS) for a single channel network.

Finally, something outside the realms of this paper, there is the opportunity for the two CHPs to communicate via their own private blockchain. This could be the hiring of staff, use of temporary staff or other resources these organisations may share.

4 Evaluation

The prototype was built using hyperledger composer [23]. Whilst deprecated, it is an excellent tool for building prototype permissioned blockchains before the construction on hyperledger fabric.

One issue that was soon apparent is that people make mistakes, or user errors. This did not stop the model, nor information being stored on a blockchain. It is not unusual for someone to push or click a wrong button on a form, whilst this would still be recorded on the blockchain, an additional functionality was added to include the reversal of this action and another TX. This would be followed by the corrected action and an appropriate TX made—users should be made aware that mistakes cannot be altered, only corrected and this does not overwrite the error, merely adds a new block demonstrating that the previous error is acknowledged and a further transaction is to be made to make the appropriate correction. This correction facility is time-bound and cannot be completed after a certain time has elapse from the original submission.

In summary, here there are five important stages of development for blockchain engineering:

1. **Blockchain Efficacy Analysis**: Is the dog wagging the tail, or the tail wagging the dog. Is blockchain generating problems, or being used to solve them? Is the blockchain system proposed the most effective technology to provide a solution to the given problem. Auditability is conducive to blockchain, but it does not mean every audit requires a blockchain.
2. **Use Case**: Identify the main updates required for the blockchain.
3. **Sequence Diagram**: Identify the key transactions to generate information on the blocks.
4. **Smart Contracts**: Use state machine diagrams to model state changes, these should help identify the transactions and where smart contracts can interact with the system.
5. **Deployment**: Most permissioned blockchains are going to have a similar deployment figure. You need to identify the use cases and how and where they are deployed.

The modelling of blockchain certainly helped with the design, but more with maintenance and explanations. Explaining with the aid of diagrams is always going to be easier than explaining code. The BMARs was implemented with success and exists as a prototype blockchain consortium, further results can be found in [1].

5 Conclusion

The main contribution is twofold: (i) application of decentralised systems to healthcare; and, (ii) a *five* stage to modelling process for blockchain applications. Specifically, this model was applied to an auditing network and may not adapt so easily to a different type of network, e.g., a supply-chain. In such cases different models may be used, such as BPMN, to accommodate the change in the blockchain application, see [24] for further information.

This offers a template for the design of a blockchain consortium for auditing. There may be differences between auditing services, but in general there are going to be several crucial decisions, that will revolve around the architecture and type of throughput required. If the throughput is high, say $\gg 1,000$ TPS, then optimisation of the network will need to be considered, this could be a single channel, or smaller blocks. If the throughput is low, say <100 TPS, then you can afford to have more peers and a greater number of channel without little consequence to the latency of each block submitted.

There is a long way to go in modelling blockchain. At the time of writing the technology supporting blockchain is [4] *twelve* years old. There have been many major developments, from Ethereum to Hyperledger, but there is a need for a common method to model blockchains and this *five* stage methodology is making a start.

Acknowledgements Author would like to thank Dr Harjinder Rahnu for consultation during the development of this paper and the reviewers for their feedback and general comments.

Competing Interests None identified.

References

1. Mitchell I, Hara S (2019) BMAR—blockchain for medication administration records. Blockchain and Clinical Trial—Securing Patient Data, pp 231–248
2. Cabatic GV (2018) 159 AD3d 62. New York, Supreme Court, App. Div, 2nd Dept
3. Justard J (2020) Filling out medication administration records. https://dphhs.mt.gov/Portals/85/dsd/documents/DDP/MedicalDirector/MARs-howtofillthemout.pdf. Accessed Aug 2020
4. Nakamoto S (2008) Bitcoin: A peer-to-peer electronic cash system
5. Castro Miguel, Liskov Barbara et al (1999) Practical byzantine fault tolerance. OSDI 99:173–186
6. Lamport L, Shostak R, Pease M (1982) The byzantine generals problem. ACM Trans Program Lang Syst (TOPLAS) 4(3):382–401
7. National Institute for Health and Care Excellence. Managing medicines in care homes, March 2014. nice.uk.org/guidance/sc1
8. Azaria A, Ekblaw A, Vieira T, Lippman A (2016) Medrec: using blockchain for medical data access and permission management. In: Open and big data (OBD), international conference on. IEEE, pp 25–30
9. Kumar R, Tripathi R (2019) Traceability of counterfeit medicine supply chain through blockchain. In: 2019 11th international conference on communication systems & networks (COMSNETS). IEEE, pp 568–570
10. Vithanwattana N, Mapp G, George C (2017) Developing a comprehensive information security framework for mhealth: a detailed analysis. J Reliable Intell Environ 3(1):21–39
11. Caldicott F (2013) Information: to share or not to share? The information governance review. Dept. of Health, UK
12. Mitchell I, Hara S (2019) Quality audits and blockchain for healthcare in the uk. In: Health IT workshop (HITWS)

13. Wood G (2020) Web 3.0 foundation. web3.foundation. Accessed: Aug 2020
14. Yaga D, Mell P, Roby N, Scarfone K (2018) Blockchain technology overview. Technical report, National Institute of Standards and Technology
15. Oscar Research Ltd (2020) Care homes in the uk. https://www.oscar-research.co.uk/datasheets/carehomes. Accessed Aug 2020
16. Bashir I (2018) Mastering blockchain. Packt, 2 edn
17. Karamitsos I, Papadaki M, Al Barghuthi NB (2018) Design of the blockchain smart contract: A use case for real estate. J Inform Secur 9(3):177–190
18. Bennett JW, Glasziou PP (2003) Computerised reminders and feedback in medication management: a systematic review of randomised controlled trials. Med J Aust 178(5):217–222
19. Council of European Union. Council regulation (EU) no 2016/679. https://eur-lex.europa.eu/legal-content/en/LSU/?uri=CELEX%3A32016R0679. Accessed Aug 2020
20. Carey P (2018) Data protection: a practical guide to UK and EU law. Oxford University Press, Inc.
21. Hu VC, Ferraiolo D, Kuhn R, Schitzer A, Sandlin K, Miller R, Scarfone K (2014) Guide to attribute based access control (ABAC) definition and considerations. SP 800-162
22. Thakkar P, Nathan S, Vishwanathan B (2018) Performance benchmarking and optimizing hyperledger fabric blockchain platform. arXiv preprint arXiv:1805.11390
23. Gaur N, Desrosiers L, Novotny P, Ramakrishna V, O'Dowd A, Baset SA (2018) Hands-on blockchain with hyperledger: building decentralised applications with hyperledger fabric and composer. Packt
24. Weber I, Xu X, Riveret R, Governatori G, Ponomarev A, Mendling J (2016) Untrusted business process monitoring and execution using blockchain. In: International conference on business process management. Springer, pp 329–347

An Investigation into an Approach to Updating the Governance of Satellite Communications to Enhance Cyber Security

Lesley-Anne Turner and Hamid Jahankhani

Abstract The UK is keen to be part of the Global Space market that has seen the recent activities of Virgin Galactic and SpaceX. This research will take into consideration the UKs operability parameters in relation to International Space agreements to investigate whether the Governance surrounding Satellite communications prepares for the new Spaceports that are being developed and the impact on UK Government departments. The research aim for A Qualitative Approach to Updating the Governance of Satellite Communications to Enhance Cyber Security is to produce a set of guidelines to be used for the Governance of Satellite communications through researching the level of Governance that exists for Satellite communications. The research includes which techniques and technology that satellites use, the impact of Governance on Satellite communications within Government departments, COBIT 2019 in Satellites, a review of literature to pinpoint current research of legislation, a critical review of current Government frameworks for satellite communications and what public organisations will be taken into consideration.

Keywords Virgin Galactic and SpaceX · Satellite communications · COVID-19 Pandemic · Cyber security · COBIT 19 · SOA · ITIL4

1 Introduction

Space, and what the future explorations of its sciences and travel might bring, is on the minds of the UK Government.

With the very recent activities of Virgin Galactic and SpaceX, 2020 is set to be the beginning of big things to come from the Space community. 2019 celebrated the 50th Anniversary of the Moon Landing and the UK's House of Commons published its Library Podcast on 1st April 2019 headlining, "When will UK spaceports be ready for lift-off?" The Podcast unveiled the UK's plans for Spaceports to be built at each end of the Country. The UK is perfectly situated for its Spaceports to deploy and be

L.-A. Turner · H. Jahankhani (✉)
Northumbria University, London, UK
e-mail: Hamid.jahankhani@northumbria.ac.uk

© The Author(s), under exclusive license to Springer Nature Switzerland AG 2021
H. Jahankhani et al. (eds.), *Cybersecurity, Privacy and Freedom Protection in the Connected World*, Advanced Sciences and Technologies for Security Applications, https://doi.org/10.1007/978-3-030-68534-8_3

capable of reaching the Polar orbits. The blog goes on to explain the requirements of the Country's interests in 10% of the Global Space market by 2030 so Governance of its Satellite communications will be a requirement.

Satellites have been deployed more and more frequently over the past decade and the United Kingdom are keen to be part of the Global Space market. Satellite numbers are also set to increase further with the building of International satellite deployment stations so appropriate cyber security Strategies and Policy requirements, across International States regarding Satellite communications for new and small satellites and for the existing satellites, will be a priority across the whole Space community.

Inter-Governmental 'asset pooling' between the stakeholders of the Space community is worth investigating because it could provide a welcomed addition to the UK's plans to bolster its initiatives and capabilities whilst deploying its own satellites and communication into the International arena. There has been little investigation into what Inter-Government agreements exist and for the UK's Space academia and business interests, it is an exciting prospect.

The synergy of science exploration and the integral use of cyber security must be realised though to bring the Treaty fully up to date and to cater for the future of Space exploration at pace. Users will be aware of the dangers of not updating the Treaty, as not doing so would leave the rules wide open to translation as, for example, Japan has recently demonstrated in its very basic Memorandum of Understanding with a private stakeholder.

The societal, ethical and political extraction of Space resources remains unregulated and as commercial private companies are taking up the exploration of the possibilities that Space offers, Memorandums of Understandings are the only regulations in existence, and those are imposed by the asset owners.

The research aim for A Qualitative Approach to Updating the Governance of Satellite Communications to Enhance Cyber Security is to produce a set of guidelines to be used for the Governance of Satellite communications through researching the level of Governance that exists for Satellite communications. In order to put forward proposals for guidelines, the research questions are:

- which techniques and technologies do Satellites use?
- what is the impact of Governance on Satellite communications within Government departments?
- does COBIT 2019 fit into Satellite communications?
- what is the current legislation for Satellite communications?
- what are the current Government frameworks for Satellite communications?
- what public organisations will be taken into consideration?

The scope of this investigation will take into consideration the UK's operability parameters in relation to International Space agreements, the Governance surrounding Satellite communications and the impact on UK Government departments. The use of secondary data collection to increase research validity and reliability throughout the research will aid an investigation into what Governance of

Satellite communications is currently in place, coming from available Space associations' and member States' agreements, strategies and policies and previous research materials.

The methodology adopted for this investigation will be a Qualitative approach in order to probe the diverse nature of Governance. A triage involving a critical Literature Review, to form a gap analysis of Governance in Space, case studies interviews and data method validation using SWOT (Strengths, Weaknesses, Opportunities and Threats) analyses will all be performed throughout the investigation to validate the research.

2 Literature Review

The inevitable death of WiFi will see all traffic bouncing through satellites and the Governments need to be ready to accommodate all of the advancements of 5G. Satellite services have evolved and Satellite techniques and technologies are set to evolve at pace with the emerging services that are in great demand Globally to be available anytime, anywhere. de la Plaza Ortega, [1] explains the four IBIS Satellite Communications Systems in the world which are Amerhis 1 and 2 in operation within Hispasat Amazonas Satellites and REDSAT and OVERHORIZON, in manufacturing at the time of writing. The IBIS is a Satellite Multimedia System, where the Broadcast Network part is integrated with the Interaction Network part, for supporting Interactive TV, Internet and Multimedia services. The up link is based on the DVB-RC standard and the down link is compliant with the DVB-S standard.

The information and data is transported according to the MPEG-2 and IP Standards, and the system is implemented by following the ETSI Reference Model for Interactive Systems. The development and new architecture technology of this system has allowed for advanced performance with lower costs when sharing and is easily obtained by the citizen. Digital visual Broadcasting Satellite (DVB-S2) has since been developed and extended to produce DVB-S2X which allows for the use of smaller antennas and broadband interaction and is covered by the ETSI standards.

The introduction of laser Satellite communications to send more data (1 terabit per second), with more efficiency, using light signals with less power has advanced Satellite communications performances exponentially. Inter-satellite links can also be performed to exchange data. Laser Satellite communication is now a necessity and radio frequency has become defunct. VSATs (Very Small Aperture Terminals) are also increasing in numbers because they are easier to launch.

The Governance which informs the Policies of all of the International advances into Space Satellite communications has not been kept up to date. The impact on the UK and the other Government departments from the need for a new Governance model for GOVSATCOM to engage with modern security needs will eventually and inevitably draw the Global State Members of the Treaty back to the table to sign up to agreements concerning Satellite communications and security and collaborations on the priorities of International missions to Space.

With International Governments already reaping the rewards of the early owner-ship of their own Outer Space satellites and data, namely America and Japan as examples, the recently reported statistics of success and the benefits derived from ownership are very attractive.

The costs of Satellite deployment is decreasing through lack of exclusivity due to the increase of 'hitchhiking' canisters and 'ridesharing' into Space. The UK would want to have ownership of their Satellites and data to be able to offer those services. Some questions raised though around 'going it alone' on such complex advances in Space are research and performance costs, usability—who will use them and why, resources to plan, build, launch monitor and repair, decision making, and the security and Governance of communications surrounding all of these factors.

Whilst COBIT 2019 is concerned only with what technology should be delivering or doing and focuses on the business of the overarching Governance, ITIL 4 gives an in depth guide to how that technology should be implemented and used by the business from within. Both COBIT 2019 and ITIL 4 frameworks are industry recognised Standards, encouraging a Best Practice approach and are used by industries Globally.

Combining these Global methodologies would be extremely beneficial to Satel-lite communications as implementing and embedding the COBIT 2019 framework, together with the ITIL 4 framework would ensure that every facet of Governance and Management are taken into consideration for the techniques and technology of Satellite communications.

The resulting lower Risk posture of Satellite communications is of paramount importance, not only from a GDPR (General Data Protection Regulation) assurance perspective across intended international stakeholders but also to ensure the confi-dentiality, Integrity, accessibility, accountability and non-repudiation (CIAAN) of the users themselves. However, concerns are emerging about whether GDPR holds sufficient Law clout throughout the EU (European Union) and beyond regarding the sharing of data. An example of this is Irelands Lawsuit against a large organisation that has its offices in Ireland in order to take advantage of lower taxes.

The IRGC (International Risk Governance Council) website also released the IRGC's 2015 Guidelines for Emerging Risk Governance Report—Guidance for the Governance of Unfamiliar Risks, which shows flexible guidelines supporting public and private sectors [2]. Of particular interest in the IRGC report, for use in Gover-nance of Satellite communications, are the categories that explain scientific knowl-edge and experience required for Governance activities, complex environments and technology interactions specific to communications and changes in evolving contexts that create a magnitude of impacts. Add these to the methodology pot of COBIT 2019 and ITIL 4 and the enterprise is set to future-proof its business.

From the UK's perspective, BSI 13,500 (British Standards Institute 13,500) Code of Practice for delivering effective Governance of organisations by way of offering a framework to continuously improve and assess development is the national Stan-dard for delivering effective Governance. This would benefit from an update to take into consideration the changing landscapes of technology, for example the Blackett Review in 2018 looked at the UK's vulnerabilities to over-reliance on Global Navigation Satellite Systems (GNSS).

The content of this investigation produces some insight into intergovernmental Governance leading to Policy setting which is also lacking in content. Producing a scalable Cloud enabled Mobile Governance Framework as citizen mobile phones utilise Satellites from land, air and sea (5G) is a further challenge.

Papadimitriou et al. [3], asks questions about how outer space exploration prompted an international community to co-operate with each other, leading to an age of Global Governance to include Space Orbit Governance and ensure the suitable use of Global 'commons'. This investigation produces further information on where Satellite communications sits in this regard.

Martinez et al. [4], explores the differences between security 'in' space and 'from' space. Is a new Governance model required? EC (European Commission) proposed a GOVSATCOM initiative and the EDA (European Defence Agency) released GOVSATCOM (Government Satellite Communications) was given a mandate to prepare the next generation of Satellite communications within a 2025 timeframe. The EU Global Strategy is investing in Satellite communications. This research details the UK's relationship with ESA (European Space Agency) and its plans from 2021.

Space Agencies should adopt procedure and processes appropriate to scale of projects (preserving open data policy) Space Agencies should collaborate on PRIORITIES for international missions for smaller satellites.

Recommendations to Policy Makers are to ensure adequate access to spectrum, ensure that report control guides are easier to understand so there's a balance between National Security and scientific interests and to provide education and training on Regulations to the access of small Satellites.

Recommendations to COSPAR (Committee on Space Research) who should facilitate (but not fund) the fostering of small Satellites' ground rules (Governance) CubeSat cannisters Standards came in and gave commercial opportunity for 'sharing'.

UK is launching in Sutherland, Scotland via Space Growth Partnership, which is well suited for Polar Orbits. This research will be a new and fundamental part of Space exploration for the UK launches (6 launches to be hosted):

- Sutherland = virticle launch
- Cornwall and Newquay = horizontal launch Glasgow Prestwick = horizontal launch.

Regulated by the UK Space Agency, CAA (Civil Aviation Authority) and BEIS. Licensing Framework for—Space Industry Act 2018

- need for Standard and recommended Practices
- Cyber security connections to the emerging Legal norms are not raised in the COPOUS Proposal on Space Governance
- nor does the April 2016 draft Report to Chair of Working Group on Status, Application and Enforcement of the Five Space treaties
- needs a Global Framework for multi-stakeholders.

To put a context onto the legislation for Satellite communications, and by way of clarifying the mixed messages and understandings of what the rules are regarding the use of resources in Space as a whole, a person hiking in a national park is classed as an "authorised entrant", whereas a fisherman with a fishing license in a lake is an "authorised user" thereby highlighting the lawful difference between 'going' or 'being' in Space and 'using things' in Space [5].

'The OST Article One provides that Outer Space "shall be free for exploration and use by all States", and it is therefore important to ensure that this freedom is not restricted based on misunderstanding or misuse of the notion of "commons". The article goes on to distinguish between the economic and legal meanings of "commons" and between resource systems and resource units. It concludes that rephrasing and relaunching the discourse must differentiate between the resource itself (and its economic features) and the legal regime applied to it. Also, each part of Space or at least categories of parts of Space, e.g. planets, stars, moons, asteroids, resource units (e.g. helium, platinum, water); and artificial objects such as satellites, all need to be rephrased. After having identified a specific part of Space or category of parts of Space and having established which kind of resource it is, we can associate the efficient Governance regime to it and then move forward to suggest the appropriate legal regime (what kind of legal regime do the Space Treaties provide, if any), thus crafting policy—with CPR's being best managed by their users.

The technologies currently used by owned and shared Satellites extend the requirements for Global Governance. The OST (Outer Space Treaty), which had its 50th anniversary in October 2017, is looked after by the only Globally focussed COPUOS United Nations body for a legal regime in outer space. Further questions are raised by the Papadimitriou et al. [3] highlight that, do we need today an update on the OST? Maybe a 'Peaceful Uses of Outer Space 2.0? Only the placement of weapons of mass destruction is explicitly prohibited? The other questions of the concept of ICOC (International Code of Conduct) by the EU in 2015 failed to be accepted. The IAA (International Academy of Astronautics) in 2006 carried out a study that put forward the idea of implementing a Space Traffic Management system. By 2020 a guidance document was agreed to be developed to encourage States to become a party to the 5 United Nations.

The UKSA (UK Space Agency) has in place the National Space Policy, the Space Innovation and Growth Strategy 2014–2030—Space Growth Action Plan, the National Space Security Policy, the National Security Strategy and the Strategic Defence and Security Review 2015. The UK also has a dedicated Government expenditure for defence Space programs from which the UK MOD procures under PFI (Private Finance Initiative) for satellite communications.

The security services that are in place today for border surveillance (through FRONTEX) and maritime surveillance (through EMSA), the recognition on 14th November 2016 of the EU Global strategy for the EU's foreign and security policy includes space and security objectives, the coordination of SECPOL3 (Security Policy and Space Policy Unit) with the Special Envoy for Space—the EEAS actions on the field of space and security, the adoption in 2015 of the international code of conduct by the EU Council. The European Defence Agency (EDA) is an inter-governmental Agency under the Common Security and Defence Policy. ESA and EDA signed an Administrative Arrangement in June 2011. This has all led to the EU (European Commission) proposing a GOVSATCOM initiative and strengthening of the security requirements when developing EU Space systems, alongside the two flagship programmes being Galileo and Copernicus [4].

The UK Space industry's Space Growth Partnership involves the UK Space Agency, Department for International Trade and the Department for Business, Energy and Industrial Strategy. On 11th May 2018 the Prosperity from Space strategy was published and set out four priorities, including security [6].

The UK Science Minister Amanda Solloway announced on 19th February 2020 the World's first National Timing Centre, which will ensure the UK economy and public services have additional resilience to the risk of satellite failure. Alongside the new centre, the Government is investing in a further 40 M Pounds in new research on Quantum Technologies for Fundamental Physics. Chief Executive Professor Sir Mark Walport from UK Research and Innovation said that our emergency services, energy network and economy rely on the precise time source that global satellite navigation systems provide—the failure of these systems has been identified as a major risk [7].

Whilst spending time researching at the UK Space Agency, the following information from other public organisations were collected:

- New public cyber security campaign launch: www.gov.uk/government/spe eches/baroness-morgan-speaking-on-how-we-can-make-technology-work-for-everyone
- BEIS at: www.gov.uk/government/consultations/future-frameworks-for-internati onal-collaboration-on-research-and-innovation-call-for-evidence
- Government Science Capability Review, publication on 5 Nov 2019 and Feb 2014—Government Office for Science organisation chart see: www.gov.uk/gov ernment/publications/government-science-capability-review
- 26 May 2016 publication—BEIS Quality Assurance Model methodologies used: www.gov.uk/government/publications/assumptions-log-template,

The research aim is to produce guidance by way of a conceptual Framework that consists of implementing 3 components into the business from Board level—the COBIT 2019 and ITIL 4 frameworks and importantly, taking into consideration the IRGC (International Risk Governance Council) Guidance for the Governance of Unfamiliar Risks.

3 New Governance Model for the International Space Communications

Creating a Roadmap to pull the 3 chosen main research area components together will help to determine whether they are suitable to use together in a simulated experiment.

SOA (Service Oriented Architecture) is the type of strong architecture that is Agile and also risk informed. SOA is a good top level model to begin the Roadmap with. The SOA concentrates on 5 areas of the business (or enterprise):

- Services
- Best Practice
- Process
- Users
- Platform.

The research into which frameworks are appropriate to use against the 5 areas of the SOA will look at COBIT 2019 (Overarching Governance Framework) and ITIL 4 (Aligning IT Services and Business Needs). Layering these 3 components across the SOA will plug all gaps for Governance and IT service management, bearing in mind the new 5G (wireless) capabilities now available to businesses that will increase attack surfaces.

The use of a security layered approach using a CSMM (Cyber Security Maturity Model) provides the business with a snapshot of present and required future levels of cyber security and shows the upward projection of 3 levels within the business:

- Protection Level
- Maturity Level
- Cost Level.

Each of the steps (or columns) across the line graph shows the level at which a particular security goal is achieved. Carrying out an initial Cyber Security Maturity Assessment is recommended and should ideally be based on the businesses security assessment framework.

For this method to be proactively iterative it must be underpinned with the evaluation of new technologies frequently.

The reason for doing things this way is due to recently unprecedented changes Globally caused by the Pandemic that have moved us into unknown territory. The physical interconnectivity of millions of citizens whilst they freely travel between countries has allowed for an 'unseen' and 'unfamiliar' catastrophic risk—COVID-19. An adequate transport network operating between countries has enabled the worst seen virus since 'Spanish Flu' to spread beyond borders and overseas to threaten citizens' lives. To prevent coming into contact with one another citizens have been asked to quarantine at home. This is a big problem for security resilience as it means that the new technology that has been implemented leaves citizens without a secure zone. Furthermore, where businesses are concerned, while citizens are in 'Lockdown' the risks of not being able to remove business vulnerabilities from Risk Management registers in order to increase the business resilience are increasing day by day. Delayed Pentesting, through the commissioning of IT Health Checks that can only now be conducted remotely, presents a huge challenge to the cyber security of businesses, especially when checking the vulnerabilities of the physical security of data centres and any parts of the business that requires people to be onsite. These are systemic risks that deal with physical and socio-ecological systems.

The UK has withdrawn from the pan-European Galileo—Satellite navigation project, which has 22 usable Satellites for SAR (Search and Rescue) and is also used in mobile phones which will still be available after 1 January 2021, leaving the UK with a gap in its access to Space. However, the UK is still a member of the ESA (European Space Agency) which is not an EU organisation [8]. The UK Space Agency leads on policy relating to satellites and the resilience of the UKs Space programme, as it gears up to launch its own Satellites from the UK by 2025, will be of paramount importance during the earliest stages that will go on to be completed by 2030. The National Space Council was set up in 2019 to oversee Space across Government and it is from there that Governance will be the most robust.

Expiration of the UK MODs SKYNET-5 in August 2022, under a (PFI) Private Finance Initiative, has intensified the UKs requirements where resilience is concerned and the UK are awaiting an audit of Britains military Space capabilities as of Feb 2019.

The question of whether International Space communications need a new Governance model to engage with modern security needs comes from the increased advances in technology, for example Drones are now being used commercially as well as in Military arenas and are connected directly to Satellites to use GPS (Global Positioning System) to determine their positions, along with GNSS Position Navigation Timing which generates automatically geocoded outputs. The GPS URE (User Range Error) comes from the signals transmitted in Space with 95% probability [9]. If the transmission were to be intercepted for example, the impact of the Drone receiving the wrong coordinates from a Satellite could be catastrophic. This means that the risks impact of a Satellite being attacked would far outweigh the benefits because of the nature and the use of GPS across the Globe.

Along with implementing ISO 27001 and 2 into an organisation, it is becoming increasingly important that the management of Unfamiliar Risks (unknown territory) need to be taken into consideration when preparing a Risk Assessment to determine the controls that need to be in place for any business. The use or merging of the IRGCs (International Risk Governance Council) 'unfamiliar risks' guidance with the ISACAs COBIT 2019 (overarching Governance installation), together with ITIL 4 launched in January 2019 (aligning IT services with business needs) encourages agility and an alignment to corporate objectives. A line of sight throughout the organisation is required in order to iterate Governance and resilience.

Installing COBIT 2019 alone into the Governance of the Global dependency of Satellite communications would create a huge shortfall of sight on all of the services and systems that move at pace within the Global Space sectors. Furthermore, the known critical and high risks of continuing a culture of 'legacy equipment and systems' would create yet higher stakes for Satellite communications.

Identifying the unfamiliar risks of Satellite communications, together with installing COBIT 2019 along with ITIL 4 (ITSM practices) into the Governance area of the business, from Board level, will significantly increase the chances of being able to iterate services and systems from business objectives as the need to future-proof the Governance of Satellite communications is critical. In terms of Satellites, a satellite is a resource unit within the Space economic system and in terms of Satellite communications, communications are a resource sub-unit. Adopting proven, layered and iterative methodologies together, that compliment each other, will bridge that gap and create a two-way trust ecosystem.

A simulation is a way to put forward the two frameworks, COBIT 2019 ITIL4 that have been suggested to work in unison. Layered controls equals layered resilience and whilst looking after the known challenges of cyber security, this 'combination' model offers the opportunity of iterating a combined, bespoke system within the business SOA (Service Oriented Architecture).

However, the results are clear. Not having controls in place to assist businesses in a World of 5G will have increasingly exponential consequences for enterprises of all shapes and sizes. Cyber attackers with the ever increasing sophistication of 'off the shelf' products and organised cyber crime are coming from the Dark Web and International groups.

The SOA (Service Oriented Architecture) provides the foundation for the frameworks to sit upon, whilst the CSMM (Cyber Security Maturity Model) provides a working time frame for the technology to be implemented, monitored and updated by the business. The 2 frameworks researched, COBIT 2019 and ITIL4, were both originally designed independently and have recently been updated to take into consideration that each framework can be used autonomously and by any business. Cyber attack sophistication increases the need for layering and combining the frameworks researched to address the Governance risk gaps. Additionally, identifying the unfamiliar risks, such as those presented by the recent COVID-19 Pandemic, to weave into the fabric of the organisation via Governance will stand a business in good stead.

4   Conclusion and Recommendations

The results suggest that the techniques and technologies that Satellites use are outdated when previous existing research has taken place. The advancements in techniques and technology would benefit from being aligned with an SOA (Service Oriented Architecture) model. This concludes that the provision of a solid foundation is recommended and as there is no current existing research, SOA could be explored for use in Satellite communications when considering new advances in technology.

The results also suggest that the risks of not beginning with a solid foundation (such as SOA) for frameworks to be reliably built upon would result in a high risk impact to the Governance of Satellite communications within Government departments. In addition, the conclusions highlight the SOA 'Platform' area featuring 5G wireless is also an interesting prospect and would benefit from further research as there is limited research in this area.

Whilst researching whether the COBIT 2019 framework could be used in Satellite communications it became clear that it fits into all Governance scenarios as it has a generic profile and it is therefore concluded that COBIT 2019 would fit into complex Satellite communications.

The results regarding what the current legislation is for Satellite communications highlights that the Space Treaty requires an update to consider both Governance and responsibilities in 'going to and being' in Space and 'using' Space. The conclusions are that current Government frameworks for Satellite communications are very limited and the public organisations that are taken into consideration do not seem to be joined up.

References

1. de la Plaza Ortega J (2010) Satellite communications: technological evolution. In: 2010 second region 8 IEEE conference on the history of communications
2. IRGC (International Risk Governance Council) (2015) Guidelines for emerging risk governance report—guidance for the governance of unfamiliar risks. https://epfl.ch/research/domains/irgc/concepts-and-frameworks-page-139716
3. Papadimitriou A, Adriaensen M, Antoni N, Giannopapa C (2018) Perspective on space and security policy, programmes and governance in Europe. In: 2019 Acta Astronautica, vol 161, August 2019, pp 183–191. https://doi.org/10.1016/j.actaastro.2019.05.015
4. Martinez P, Jankowitsch P, Schrogl K, Di Pippo S, Okumura Y (2018) Reflections on the 50th anniversary of the outer space treaty, UNISPACE+50, and prospects for the future of global space governance. In: 2018 ScienceDirect. https://doi.org/10.1016/J.SPACEPOL.2018.05.003
5. Tepper E (2017) Structuring the discourse on the exploitation of space resources: between economic and legal commons. In: 2018 Space policy, science direct
6. UK Government announcement (2018) Vision for growth in space. Gov.uk website. https://gov.uk/government/news/uk-space-industry-sets-out-vision-for-growth
7. UK Government announcement. (2019). Worlds first National Timing Centre. Gov.uk website. https://gov.uk/government/news/worlds-first-timing-centre-to-protect-uk-from-risk-of-satellite-failure
8. UK Government announcement (2020) UK is still a member of ESA (European Space Agency). Gov.uk website. https://gov.uk/guidance/satellites-and-space-programmes-from-1-January-2021
9. US Government guidance. GPS URE (User Range Error). GPS.Gov website. https://gps.gov/systems/gps/performance/accuracy

Enhancing Smart Home Threat Detection with Artificial Intelligence

Jaime Ibarra, Usman Javed Butt, Ahmed Bouridane, Neil Eliot, and Hamid Jahankhani

Abstract The chapter focuses on building a theoretical network, which supports the protection of home networks from critical cyberattacks. A framework is proposed which aims to augment a home router with machine learning techniques to identify threats. During the current pandemic, employees have been working from home. So it is reasonable to expect that cyberattacks on households will become more common to leverage access into corporate networks. The model described in this chapter is for a single network; however, the network would be segmented into regions to avoid a wider compromise. Since the deployment of 5G, mobile threats are rising steadily. Therefore, the UK requires a robust plan to identify and mitigate all forms of threats including nation-state, terrorism, hacktivism. Additionally, the model dynamically analyses traffic to identify trends and patterns; therefore, supporting on the building of a resilient cyber defence. The emphasis in this model is to bridge the gap of trust between the government and the public, so that trust and transparency is established by a regulatory framework with security recommendations. At present, there is no authorisation to collect this data at national level, nor is there trust between the public and government regarding data and storage. It is hoped that this model would change human perception on the collection of data and contribute to a safer UK.

Keywords Smart homes · Cyber attack · Pandemic · AI · Malicious attack · Cyber security

J. Ibarra (✉) · U. J. Butt · A. Bouridane · N. Eliot · H. Jahankhani
Northumbria University, Newcastle upon Tyne, UK
e-mail: jaime.ibarra@northumbria.ac.uk

U. J. Butt
e-mail: usman.butt@northumbria.ac.uk

A. Bouridane
e-mail: ahmed.bouridane@northumbria.ac.uk

N. Eliot
e-mail: neil.eliot@northumbria.ac.uk

H. Jahankhani
e-mail: hamid.jahankhani@northumbria.ac.uk

1 Introduction

The world has been digitalised across the last years, leading to the creation of smart homes (households with devices that have networking capabilities, known as smart objects, devices, etc.) [1]. These smart devices can provide a wide variety of household services (e.g. a smart thermostat controlling the temperature of a house).

An article written by Brandon Vigliarolo [44] shows that the International Association of IT Asset Managers (IATAM) found four main reasons for data breaches since the global pandemic which are:

- Assets are being purposely left unsecured
- Rapid addition of new hardware leaving little time for security
- Assets on home networks are fundamentally less insecure
- Unprepared users are making mistakes.

One of the issues with smart technology is that they are heterogeneous (different architectures in terms of hardware and software). There is not a standardisation for the IoT environment. Even though the most familiar communication protocol used is Bluetooth (IEEE 802.15), the messaging protocol system is different. Table 1 shows a comparison between messaging protocols used in IoT from a research presented by Naik [43], which causes concerns in terms of security [2–4] and privacy [5].

It is paramount to consider the fact that smart home devices are battery-driven and might use low-power CPUs, including lower clock rates and small throughput [45]. In [46], the authors performed security tests on a sensor with 8 MHz of CPU frequency, 10 KB of RAM memory and 48 KB of program memory, which proved that applying security mechanisms are not feasible in small IoT devices. For instance, public key algorithms such as RSA and ECC [47] are very intensive for computational processing on microcontrollers and requires many instructions to perform one security process. Therefore, tactics, techniques and procedures (TTPs) executed by malicious attackers are being developed with nefarious purposes, making more challenging the protection of home networks. A smart device can be physically accessible making them prone to tampering attacks to reduce billing costs. An example is a case published in Wired [51] where hackers can use lasers to "talk" to your amazon echo or google home, including a demonstration.

The current pandemic that lead to a global lockdown has accelerated the digital transformation of businesses. An article published by [48] shows that the strategy of digital communication transformation has been accelerated by 5.3 years in the UK based on a five-minute survey of 2,569 business decision-makers, which 300 were UK repliers. Artificial Intelligence (AI) is a cutting-edge topic for research within the cybersecurity field. Capgemini [49] published a survey interviewing 850 senior executives in a podcast discussing about AI in cybersecurity. The report [50] shows an analysis of 20 use cases of AI in cybersecurity in IT, OT and IoT and found that almost two-thirds do not think their businesses can identify threats without AI.

This chapter focuses the reviewing the landscape of smart home including taxonomies of smart devices and cyber threats. Moreover, it presents a literature

Table 1 Comparison of IoT messaging protocols

Protocol	MQTT	CoAP	AMQP	HTTP
Year	1999	2010	2003	1997
Header size	2 Bytes	4 Bytes	8 Bytes	Undefined
Message size	up to 256 MB maximum size	Normally small to fit in a single IP datagram	Negotiable and undefined	Large and undefined (depends on web server or programming technology)
Semantics/Methods	Connect, Disconnect, Publish, Subscribe, Unsubscribe, Close	Get, Post, Put, Delete	Consume, Deliver, Publish, Get, Select, Ack, Delete, Nack, Recover, Reject, Open, Close	Get, Post, Head, Put, Patch, Options, Connect, Delete
Cache/Proxy support	Partial	Yes	Yes	Yes
Quality of Service (QoS) reliability	QoS 0—delive the msg once, with no confirmation QoS 1—deliver the msg at least once, with confirmation required QoS 2—deliver the msg exactly once by using the 4-step handshake	Confirmable Message (similar to QoS 0) or Non-confirmable message (similar to QoS 1)	Settle Format (similar to QoS 0) or Unsettle format (similar to QoS 1)	Limited (via TCP)
Standards	OASIS, Eclipse Foundation	IETF, Eclipse Foundation	OASIS, ISO/IEC	IETF and W3C
Transport Protocol	TCP (MQTT-SN can use UDP)	UDP, SCTP	TCP, SCTP	TCP
Security	TLS/SSL	DTLS, IPSec	TLS/SSL, IPSec, SASL	TLS/SSL
Default port	1883/8883 (TLS, SSL)	5683 – UDP/5684 – DTLS	5671 (TLS/SSL), 5672	80/443 (TLS/SSL)
Encoding format	Binary	Binary	Binary	Text
Licensing model	Open source	Open source	Open source	Free
Organisational support	IBM, Facebook, Cisco, Red Hat, Tibco, ITSO, M2Mi, AWS, Indusoft, Fiorano	Cisco, Contiki, Erika, IoTvity	Microsoft, JP Morgan, Bank of America, Barclays, Goldman Sachs, Credit Suisse	Global Web Protocol Standard

review of existing research in artificial intelligence and machine learning techniques that could potentially be adapted to IoT infrastructures, leading to a future standardisation proposal. In addition, it shows a history of AI including Convolutional Neural Networks (CNN) and Deep Neural Networks, and an overview of some existing algorithms used for threat detection such as Multivariate Correlation Analysis (MCA). Therefore, a framework will be created to enhance smart home security divided in layers and its corresponding explanation, which includes a flow diagram for threat detection with details for each step that could be used to augment the capabilities of a home router.

The structure of this chapter is as follows. In Sect. 2, it is done a literature review of the threat landscape including taxonomies of smart devices and most relevant threats in terms of security and privacy. In Sect. 3, it is done a review of current AI and ML techniques including existing research and its impact on IoT. In Sect. 4, it is analysed some AI algorithms followed by the framework proposal. And finally, Sect. 5 concludes the chapter including some future work.

2 Smart Homes: Smart Device Taxonomies and Threat Landscape

A smart device meaning, and classification is constantly evolving as new technologies emerge. For instance, Alam et al. [6] proposed a taxonomy of smart devices, classifying a device as a sensor, a physiological or multimedia device. However, this does not include the smart devices with multiple capabilities – e.g., a bathroom scale can detect location and show the local weather forecast while measuring the user's weight. Given the diversity of devices this taxonomy offers a vague categorisation.

However, Lopez et al. [7] presented the Identity, Sensor, Actuator, Networking, Decision (ISAND) specification. This taxonomy classifies devices based on some main features, for instance, it could have an Identity, work as a Sensor, Actuator, or the device presents Networking capabilities. Moreover, the device can make Decisions. This taxonomy shows main capabilities of a device. For instance, a device can be named as a "SN-Smart object" which implies it presents Sensor and Networking capabilities. An example of a smart object is the following: A smart car can decide when the wheels pressure are not properly balanced based on its configuration/calibration, sending an alarm with a message to the display along with a tone, therefore the user can notice and make the needed arrangements for a safer driving. This taxonomy further defines the features and functionality of devices and would allow us to present a detailed approach of methods and frameworks from the security standpoint. Furthermore, recognising a legitimate device is very important, because a smart device must have a unique identifier within the network and its information can be connected to the user (e.g. web/mobile app) together with other sources (e.g. metadata).

Increased adoption of smart devices has seen an increase in cyber threats by the execution of multiple procedures in order to leak or tamper information as well as the disruption of services. Some risks are mentioned below.

- *Eavesdropping Attack*: Also called sniffing, consists about gathering real time information that smart devices, microcontrollers and smartphones transfer through a network. It can allow attackers to intercept user privacy and break data confidentiality without disturbing the transmission. In a smart home network, an eavesdrop attack can be used to steal login/password credentials when a user authenticates to the smart home app which could allow hackers to take control of the smart home environment [52].

- *Malicious Code Injection*: Malicious codes are normally scripts (software programmed), which can be inserted into the smart home app, allowing attackers to exploit vulnerabilities including authentication bypassing which allows access to unauthorised entities. In smart home environments, code injection threats present an impact on user's privacy and confidentiality including the capability of attackers to access the system, including harmful operations like stealing personal data [53].

- *Man-in-the-Middle (MitM) Attack*: An attacker can impersonate a legitimate device within the network which can steal, insert, modify or drop packets. According to [55], MitM is accomplished by the following steps. First, the attacker waits for an authentication request from a legitimate device (e.g. smart home app) in order to steal the sent request. Second, MitM creates a tunnelled protocol for authentication through a back-end server sending the stolen message to the server. Once the tunnel between the MitM and the server is set, the MitM starts forwarding legitimate client's messages of authentication across the tunnel. Finally, the MitM unwraps the messages received from the back-end server and forwards them to the legitimate client. For instance, an attacker can use this technique to affect systems such as sensor/actuator response, usage of devices that could raise bill costs [8]. As shown in Fig. 1, MitM attacks can occur between servers within the cloud, between the cloud and the internet as well as inside the home between the controller and the access point. In addition, attacks of this kind can be performed between radio base stations and the internet leveraging 4G and 5G mesh networks.

- *Man-In-The-Browser (MitB) Attack*: An attacker can leverage the usage of a malicious program to take control of data inserted by the client or recovered from the server that is normally displayed in a web browser. This can show false data regarding power consumption for instance and not legitimate readings by a smart meter [9].

- *Denial of Service (DoS) Attack*: Denial of Service (DoS) and Distributed Denial of Service (DDoS) attacks are the most common security threats. It aims to stop the availability of services as well as the communication resources [10]. It consists of flooding the home network with ECHO requests in a short period of time. Considering the hardware limits of a microcontroller (Low CPU and memory), it cannot manage high number of requests, disrupting the communication between home devices as well as the control access of the user [52] e.g. When an echo device states, "I am having trouble now".

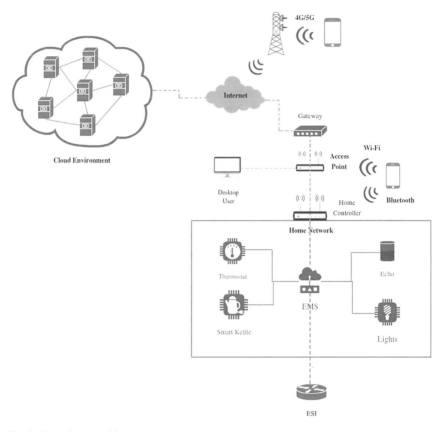

Fig. 1 Smart home architecture

• *Attacks against Home Monitoring and Control*: Fig. 1 shows a smart home architecture, where Singh et al. [11] define the Energy System Interface (ESI) and the
  Energy Management System (EMS). The ESI is connected between the home
  network and the smart grid. Some threats can include an attacker performing a
  message tampering or replay attack. For example, the attacker can imitate the
  client's cloud service sending a message to the ESI requesting to turn off all the
  devices connected to the home network because they were increasing the elec-
  tricity bill [12]. In case of message tampering or replay attack, the attacker could
  send a signal to a smart washing machine to repeatedly wash the clothes or increase
  the temperature of the oven from 120°C to 240°C. More information about this
  attack can be found on [11] which includes PLC environments and Operational
  Technology (OT).

3   Artificial Intelligence and Smart Homes: An Overview

Wilner [13] describes briefly Artificial Intelligence (AI) as scientific methods aiming to provide machines human-like capabilities such as reasoning, learning and problem-solving. This includes number of different tools, methodologies and techniques. The most noticeable named *machine learning (ML)*, which consists of teaching algorithms to recognise consistencies in rearms of data.

AI techniques have the potential to analyse large data sets faster than a human operator. The data is presented for interpretation, showing trends and possible sugges-tions in order to improve the performance of the programmed services. There is an extended literature review of the application of AI in smart homes presented in [53], categorised in five clusters which are data processing, decision-making, voice recog-nition, activity recognition and prediction-making. In addition, they divided in main functions such as healthcare, intelligent interaction, security, device management and energy management as well as the review of some products such as MATRIX, Nest Learning Thermostat. Finally, AI potentially supports some trends for smart home future applications such as automated parking for cars, "The Ubiquitous Home" [54].

Although AI provides some great future in the technology landscape it presents challenges in the existing notions of security and privacy, including human rights i.e. GDPR, and governance. The UK Information Commissioner's Office (ICO) high-lighted on its 2017 paper five different aspects of using AI in data analytics including its implications regarding data protection [56]. The five trends recognised by the ICO are:

- The use of algorithms: The ICO visualises the use of ML to analyse dataset and identify correlations to be an enhancement of traditional data analysis, including the possibility of choosing the hypothesis first and then the data extracted to validate the hypothesis.
- The opacity of processing: certain algorithms operate as a "black box" to the user. Compared with traditional data processing where there is a decision tree and a logic behind it, in ML the result of the algorithm depends on the data processing training. If trained for large datasets, it might be difficult to explain why the algorithm reached to a specific result/conclusion in a certain circumstance.
- The trend to gather "all data": ML needs large datasets to acknowledge and learn. Therefore, as much data as possible to collect and analyse is better than a random sample or statistically relevant sample.
- The repurposing of data: The data generated from one activity can be analysed for multiple purposes e.g. geo-location from a mobile phone logging into the smart home controller dashboard.
- Usage of new data types: The increase of inter-connected devices and online tracking activity leads to the automated generation of personal data, rather than provided consciously by the individual.

In this section we review some ML algorithms offering potential impact for network traffic analysis including intrusion detection and prevention. Smart home security systems must be ready to respond and mitigate cyber-attacks that could allow attackers to eavesdrop, steal personal data, or for instance, lock the person out of their own household. Cate Lawrence [57] shows an article of a device named "Dojo" by BullGuard which argues three layers of cyber defence. First is the perimeter layer that should forbid unauthorised access to a device. Second is about intrusion detection and prevention (commonly known as IDS and IPS). The third layer mentions analysis of traffic behaviour added with the detection of unusual activities or packets. There is an area or research that has been utilised to support the three layers of defence, which is called Deep Learning (DL). Figure 2 shows a categorisation of AI, which includes ML and DL.

Historically, computers struggled analysing the variety in object types, classifying as well as processing in a non-linear pattern, and accuracy [14]. Convolutional Neural Networks (CNN) were applied in late 1980s to visual tasks and were particularly inoperative until mid-2000s when the improvements in computing power and algorithm development were performed [15].

Multiple forms of ML can be separated between actions and effects. Schmidhuber reviewed unsupervised, reinforcement learning and transformative computation as well as deep supervised learning and indirect search for shortened programs encoding deep and large networks. [16]

In 2012, during the ImageNet Large Scale Visual Recognition Challenge (ILSVRC), the SuperVision team won the challenge including the classification and localisation tasks with a large-scale deep neural network, proving that CNN are more efficient [58]. However, one of the challenges of CNN is the large amount of training data that must be labelled. This requires powerful GPU to stimulate the learning process in shorter times.

Han et al. [17] proposed a two-phase technique combining CNN transfer learning along with web data augmentation. They applied an algorithm based on Bayesian

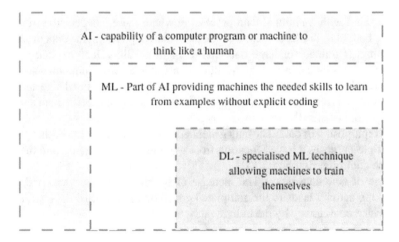

Fig. 2 Categorisation of artificial intelligence, including machine learning and deep learning

optimisation to hyper-parameters tuning which was applied to six short open datasets. Their conclusion results were the following: Traditional shallow classification methods are better than deep neural networks in small datasets; and web data augmentation can make all CNN improve their performance.

Applications of AI: An Overview and Existing Research

Liu et al. [18] argue that AI is applicable in the field of Magnetic Resonance Imaging research when applied to image detection, registration, segmentation and classification. Bychkov et al. [19] show the potential of DL in medical image classification, using CNN to predict colorectal cancer based on images of tumour samples. Rakhlin et al. [20] have found AI able to improve breast cancer diagnose accuracy, classifying breast cancer histology images.

Other area to point is security prevention and enforcement. Although we still must consider that humans are an important part of security surveillance e.g. SOCs, ML potentially provides significant assistance. Reyani and Mahdavi [21] proposed a framework for using neural networks for user identification. Sun et al. [22] set a framework aiming to check high-level facial features. Sun et al. [23] improved the previous framework including deep learning which they named it "Deepid3".

Kumar et al. [24] conducted research of the usage of neural learning for smart security in IoT environments. They used an ultrasonic accessory module in an Arduino, an Android smartphone, Feature Extraction, Neural Network training, comparison and recognition modules, obtaining a 90% of accuracy during their tests.

The investigation conducted by Gergely Acs et al. [25] show a new privacy scheme defending the smart metering system. It secures customer's privacy using holomorphic encryption while avoiding a trusted third party to exploit the perturbation algorithm. It retrieves data from the electric supplier and the smart meter, accumulating information on a certain frequency determining the total statistics instead of learning.

Yi Huang et al. [26] proposed a system that protects from injection attacks on a central controller using the cumulative sum (CUSUM) algorithm detecting attacks with a minimum number of examinations.

Eun-Kyu Lee et al. [27] presented a Frequency Quorum Rendezvous (FQR) using random wireless communication based on spectrum to protect systems against powerful attacks which include jamming attacks.

Qian Huang et al. [28] proposed a technology that improves the accuracy of indoor positioning with the assist of Li-Fi-assisted coefficient calibration [29]. This proposal leverages the existing Li-Fi lighting and Wi-Fi infrastructure providing an accurate, cost-effective and easy-to-use location framework.

Tiwari et al. [29] discusses a model for multi-device bidirectional Visible Light Communication (VLC) using the colour beams from RGB LEDs to transmit the data as well as synchronising multi-device transmission. They also proposed a modulation technique known as colour coded multiple access for multi-use VLC in smart home technologies [30]. The communication is established by using RGB Light Emission Diodes for downlink and Phosphorous-LED for uplink.

4 Secure Smart Home Network: Proposal of a Threat Detection and Secure Smart Home Framework

Figure 3 shows a proposed architecture for secure smart homes composed of the following layers:

Application Layer: A smart Home is formed by diverse applications such as monitoring, remote access, emergency, controlling. In addition, it contributes to coordinate the intelligence behind it. This model provides the services to end users in an efficient and secure manner.

Home gateway: It must provide high performance and flexibility, accepting and transferring multiple datasets from different vendors. It should have included the MCA System as well as a program to analyse potential threats within the smart home network. This device should be considered to have strong computing features

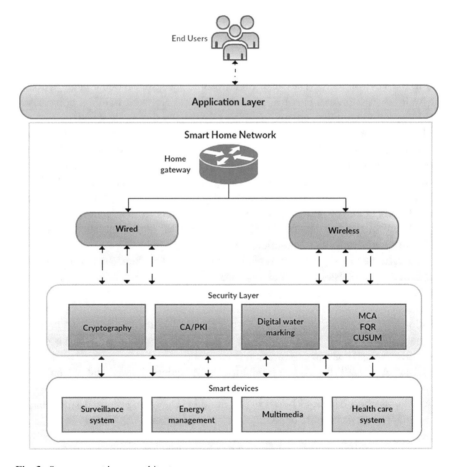

Fig. 3 Secure smart home architecture

in order to analyse and correlate all the data either inbound or outbound in order to create accurate data analytics.

Network Technology: The main role of the networking layer is to connect the smart devices allowing to share information with each other. It includes both wired and wireless devices regardless of the technology or standard that the smart device use. For wired it is considered electric wires, coaxial, optical fibres and landlines from telephones, however other standards to have in consideration are X10, KNX, LonWorks, MoCA and Insteon [34, 35]. In wireless technologies there are two kinds. The first one based on IEEE standards such as Wi-Fi, Bluetooth, ZigBee. The others are not based on standardised methods like PHY or MAC layering such as Z-Wave, SimpliciTI, EnOcean and Wavenis [34, 36]. To ensure proper authentication of legitimate devices and reliable connectivity from cloud to device the usage of message brokers is paramount for this framework e.g. MQTT [42]. Table 2 presents details of wireless communication protocols including cellular, Z-Ware and SigFox.Table 2

Security Layer: To ensure the confidentiality of data and avoid potential data leaks from outsiders it is recommended to include encryption algorithms such as holomorphic, ECDSA [37], HMAC [38]. Watermarking is used for integrity [39, 40], and a public key infrastructure [41] for inter-home authentication. In this section is where we introduce the threat detection algorithm including MCA, FQR and CUSUM.

Threat Detection Algorithm.

As shown in Fig. 5, once the new packet arrives to the gateway it receives notification checks. The OS from the network controller defines if the packet is part of the flow or from a non-existent activity (new flow). Packets from a new flow are sent to the anomaly detection system, which go through MCA, CUSUM or FQR algorithms.

MCA is an AI-based technology focused on feature extraction for original and legitimate data, which causes a significant role for data analysis as well as the framework we propose later. MCA is used to define accurate network traffic by taking the geometric correlation between the component of network traffic [31–33]. MCA detects DoS-based attacks which the algorithm presents an accuracy of 95.20% [31]. Figure 4 shows three major phases of MCA:

- Phase 1: The function is generated from the input network traffic to the internal network. The protected server resides within and is utilised to create traffic records with defined time intervals. The analysis of destination reduces the over effort of detecting malicious activity by focusing on relevant inbound traffic.
- Phase 2: MCA applies the "Generate Triangle Region Map" feature to obtain the correlation between two different components in each traffic record that came from first step. It could also extract the correlation between normalised profiles from the "Feature normalisation" module to gather traffic records.
- Phase 3: An attack detection mechanism is placed for decision-making, facilitating detection of DoS attacks for instance. Moreover, it updates constantly the signature database for the case of labour-intensive attack analysis.

Table 2 Wireless communication protocols in IoT

Name	6LowPAN	ZigBee	Bluetooth	RFID	NFC	SigFox	Cellular	Z-Wave
Standard	IEEE 802.15.4	IEEE 802.15.4	IEEE 802.15.1	RFID	ISO/IEC 14443 A&B, JIS X-6319-4	SigFox	3GPP GSMA GSM, GPRS, UMTS HSPA LTE	Z-Wave
Bands frequency	868 MHz (EU) 915 MHz (USA) 2.4 GHz (Global)	2.4 GHz	2.4 GHz	125 KHz, 13.56 MHz, 902-928 MHz	125 KHz 13.56 MHz 860 MHz	868 MHz (EU) 902 MHz (USA)	Common cellular bands	868–908 MHz
Network	WPAN	WPAN	WPAN	Proximity	P2P Network	LPWAN	WNAN	WPAN
Topology	Star, Mesh	Star, Mesh, Cluster Network	Star—Bus Network	P2P Network	P2P Network	Start Network	NA	Mesh Network
Power	Low power Consumption	30 mA Low power	30 mA Low power	Ultra-low power	50 mA low power Very Low	10–100 mW	High Power Consumption	2.5 mA Low power
Data rate	250 kbps	250 kbps	1 Mbps	4 Mbps	106, 212 or 424 kbps	100 bps (upload) 600 bps (download)	NA	40 kbps
Range	10–100 m	10–100 m	15–30 m	Up to 200 m	0-10 cm, 0-1 m, 10 cm–1 m	10 km (URBAN) 50 km (RURAL)	Several km	30 m (indoors) 100 m (outdoors)
Security	AES	AES	E0 steam, AES-128	RC4	RSA, AES	Partially addressed	RC4	AES-128

(continued)

Table 2 (continued)

Name	6LowPAN	ZigBee	Bluetooth	RFID	NFC	SigFox	Cellular	Z-Wave
Spreading	DSSS	DSSS	FHSS	GSMA	DSSS	DSSS	DSSS	No
Modulation	BPSK, O-QPSK	BPSK, O-QPSK	TDMA	FSK, PSK	ASK	UNB, DBPSK (upload) GFSK (download)	BPSK, OFDM	BFSK

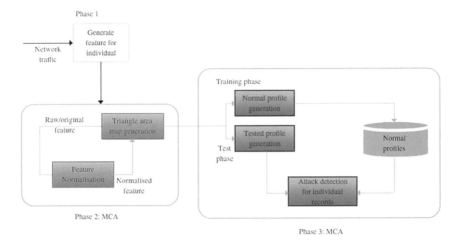

Fig. 4 MCA system

FQR is used to detect powerful jamming attacks, which the main two metrics are Time-To-Rendezvous (TTR) and Rendezvous Frequency Concentration (RFC). TTR shows the average time from a pair of nodes to rendezvous, while RFC shows the degree of the rendezvous scheme focusing on a subset of available frequencies. Jamming attacks can be leveraged if the attacker has some knowledge that frequency hopping presents a pattern on specific frequencies [27].

The Cumulative Sum (CUSUM) algorithm detects injection attacks specially addressed to smart grids. However, we consider an important element of the proposed detection algorithm because the controller and mobile devices can get access and notifications from the smart grid regarding changes on the system (i.e. EMS, ESI). On 2011, the CUSUM was proposed by [26] with two main stages: Stage 1 presents the linear unknown parameter solving technique, and Stage 2 applies a multi-thread CUSUM to deter a possible appearance of adversary at the controller. Three years later, [59] proposed a real-time detection for false data injection using CUSUM including a Markov-chain-based analytical model to determine the behaviour of the scheme. In addition, the CUSUM algorithm has been enhanced in terms of quantity evaluation based on the following metrics: False Alarm Rate (FAR), Missed Detection Ratio (MDR) and Detection Delay (T_D). This makes the algorithm more efficient and reliable in terms of accuracy.

Once the packet has been analysed by the anomaly detection mechanism, if it is a known attack the packet is discarded immediately otherwise if goes to the intelligence security analysis which its main goal is to update the database of known attacks about new attacking patterns for future detection. The Intelligence system analysis consist of the following steps: First, it creates a Data Flow Diagram (DFD) to analyse the vulnerability detected based on a template. If the vulnerability results to be true, it represents a potential threat for the home network therefore, the packet is dropped, followed by updating the attack pattern information at the known attack

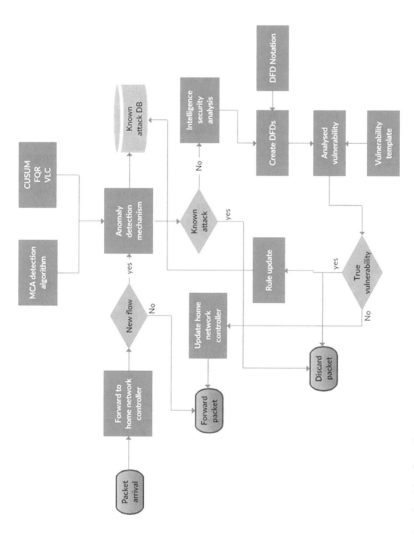

Fig. 5 Threat detection algorithm flow diagram

DB. Otherwise, the home network controller gets updated in order to forward the inbound packet to the network.

5 Conclusion

In this chapter, we have reviewed the impact of Artificial Intelligence in the protection of Internet of Things Networks with high focus in Smart Home Technologies. This allowed us to propose a framework for smart home integrating security mechanisms and introducing some AI-based algorithms to detect common cyber threats such as DoS, injection and jamming attacks. In addition, it has been proposed a threat detection algorithm flow diagram for the analysis of inbound packets to the smart home network. Even though the research is at its early stage, it is planned for future research to evaluate the efficiency of the reviewed AI algorithms in terms of efficiency and accuracy in order to investigate the probability of parsing this information in forensic data which could allow investigators to extract digital evidence without the requirement of invading personal devices. Therefore, it could support on the elaboration of reliable chains of custody as well as protecting personal data.

References

1. Ricquebourg V, Menga D, Durand D, Marhic B, Delahoche L, Loge C (2006) The smart home concept: our immediate future. In: 2006 1st IEEE international conference on e-learning in industrial electronics. IEEE, pp 23–28
2. Fernandes E, Jung J, Prakash A (2016) Security analysis of emerging smart home applications. In: 2016 IEEE symposium on security and privacy (SP). IEEE, pp 636–654
3. Simpson T (2017) Securing a heterogeneous internet-of-things. https://now.avg.com/securing-a-heterogeneous-internet-of-things
4. Violino B (2020) IoT pushes IT security to the brink. https://www.csoonline.com/article/308 1228/internet-of-things/iotpushes-it-security-to-the-brink.html
5. Energy OVO (2020) How do my smart meters communicate?. https://www.ovoenergy.com/ovo-answers/topics/smarttechnology/smart-meters/how-do-my-smart-meterscommunicate.html
6. Alam MR, Reaz MBI, Ali MAM (2012) A review of smart homes—past, present, and future. IEEE Trans Syst, Man, Cybern, Part C (Appl Rev), 42(6):1190–1203
7. López TS, Ranasinghe DC, Patkai B, McFarlane D (2011) Taxonomy, technology and applications of smart objects. Inf Syst Front 13(2):281–300
8. Kamilaris A, Tofis Y, Bekara C, Pitsillides A, Kyriakides E (2012) Integrating web-enabled energy-aware smart homes to the smart grid. Int J Adv Intell Syst 5(1):15–31
9. Dougan T, Curran K (2012) Man in the browser attacks. Int J Ambient Comput Intell (IJACI) 4(1):29–39
10. Jacobsson A, Boldt M, Carlsson B (2016) A risk analysis of a smart home automation system. Futur Gener Comput Syst 56:719–733
11. Singh S, Sharma PK, Park JH (2017) SH-SecNet: an enhanced secure network architecture for the diagnosis of security threats in a smart home. Sustainability 9(4):513
12. Singh S, Jeong YS, Park JH (2016) A survey on cloud computing security: Issues, threats, and solutions. J Netw Comput Appl 75:200–222

13. Wilner AS (2018) Cybersecurity and its discontents: artificial intelligence, the internet of things, and digital misinformation. Int J 73(2):308–316

14. Egbedion B, Wimmer H, Rebman Jr, CM, Powell LM (2019) Examining a deep learning network system for image identification and classification for preventing unauthorized access for a smart home security system. Issues Inf Syst 20(3)

15. Rawat W, Wang Z (2017) Deep convolutional neural networks for image classification: a comprehensive review. Neural Comput 29(9):2352–2449

16. Schmidhuber J (2015) Deep learning in neural networks: an overview. Neural Netw 61:85–117

17. Han D, Liu Q, Fan W (2018) A new image classification method using CNN transfer learning and web data augmentation. Expert Syst Appl 95:43–56

18. Liu J, Pan Y, Li M, Chen Z, Tang L, Lu C, Wang J (2018) Applications of deep learning to MRI images: a survey. Big Data Min Anal 1(1):1–18

19. Bychkov D, Linder N, Turkki R, Nordling S, Kovanen PE, Verrill C, Walliander M, Lundin M, Haglund C, Lundin J (2018) Deep learning based tissue analysis predicts outcome in colorectal cancer. Sci Rep 8(1):1–11

20. Rakhlin A, Shvets A, Iglovikov V, Kalinin AA (2018) Deep convolutional neural networks for breast cancer histology image analysis. In: International conference image analysis and recognition. Springer, Cham, pp 737–744

21. Reyhani SZ, Mahdavi M (2007) User authentication using neural network in smart home networks. Int J Smart Home 1(2):147–154

22. Sun Y, Wang X, Tang X (2014) Deep learning face representation from predicting 10,000 classes. In: Proceedings of the IEEE conference on computer vision and pattern recognition, pp 1891–1898

23. Sun Y, Liang D, Wang X, Tang X (2015) Deepid3: face recognition with very deep neural networks. arXiv:1502.00873

24. Kumar PM, Gandhi U, Varatharajan R, Manogaran G, Jidhesh R, Vadivel T (2019) Intelligent face recognition and navigation system using neural learning for smart security in internet of things. Clust Comput 22(4):7733–7744

25. Acs G, Castelluccia C (2012) Dream: differentially private smart metering. arXiv:1201.2531

26. Huang Y, Li H, Campbell KA, Han Z (2011) Defending false data injection attack on smart grid network using adaptive CUSUM test. In: 2011 45th Annual conference on information sciences and systems. IEEE, pp 1–6

27. Lee EK, Oh SY, Gerla M (2010) Frequency quorum rendezvous for fast and resilient key establishment under jamming attack. ACM SIGMOBILE Mob Comput Commun Rev 14(4):1–3

28. Huang Q, Li X, Shaurette M (2014) Integrating Li-Fi wireless communication and energy harvesting wireless sensor for next generation building management

29. Tiwari SV, Sewaiwar A, Chung YH (2017) Smart home multi-device bidirectional visible light communication. Photon Netw Commun 33(1):52–59

30. Tiwari SV, Sewaiwar A, Chung YH (2015) Color coded multiple access scheme for bidirectional multiuser visible light communications in smart home technologies. Opt Commun 353:1–5

31. Tan Z, Jamdagni A, He X, Nanda P, Liu RP (2013) A system for denial-of-service attack detection based on multivariate correlation analysis. IEEE Trans Parallel Distrib Syst 25(2):447–456

32. Tevari GM, Goudar RH Multivariate correlation analysis: an approach to detect DDoS attacks in FTP services

33. More KK, Gosavi PB (2016) A real time system for denial of service attack detection based on multivariate correlation analysis approach. In: 2016 International conference on electrical, electronics, and optimization techniques (ICEEOT). IEEE, pp 1125–1131

34. Mendes TD, Godina R, Rodrigues EM, Matias JC, Catalão JP (2015) Smart home communication technologies and applications: Wireless protocol assessment for home area network resources. Energies 8(7):7279–7311

35. Vanus J, Smolon M, Martinek R, Koziorek J, Zidek J, Bilik P (2015) Testing of the voice communication in smart home care. Hum-Centric Comput Inf Sci 5(1):15

36. Spadacini M, Savazzi S, Nicoli M (2014) Wireless home automation networks for indoor surveillance: technologies and experiments. EURASIP J Wirel Commun Netw 2014(1):6
37. Barenghi A, Bertoni GM, Breveglieri L, Pelosi G, Sanfilippo S, Susella R (2016) A fault-based secret key retrieval method for ECDSA: Analysis and countermeasure. ACM J Emerg Technol Comput Syst (JETC) 13(1):1–26
38. Dewan P, Dasgupta P (2009) P2P reputation management using distributed identities and decentralized recommendation chains. IEEE Trans Knowl Data Eng 22(7):1000–1013
39. Yan X, Zhang L, Wu Y, Luo Y, Zhang X (2017) Secure smart grid communications and information integration based on digital watermarking in wireless sensor networks. Enterp Inf Syst 11(2):223–249
40. Lalani S, Doye DD (2016) A novel DWT-SVD canny-based watermarking using a modified torus technique. J Inf Process Syst 12(4)
41. Im H, Kang J, Park JH (2015) Certificateless based public key infrastructure using a DNSSEC. JoC 6(3):26–33
42. EMQ Technologies Co L (2020). MQTT messaging broker solution. emqx.io. https://www.emqx.io/solutions/broker. 24 Sept 2020
43. Naik N (2017) Choice of effective messaging protocols for IoT systems: MQTT, CoAP, AMQP and HTTP. In: 2017 IEEE international systems engineering symposium (ISSE). IEEE, pp 1–7
44. Vigliarolo B (2020) COVID-19 lockdowns are causing a huge spike in data breaches. TechRepublic. https://www.techrepublic.com/article/covid-19-lockdowns-are-causing-a-huge-spike-in-data-breaches/. 5 Oct 2020
45. Bugeja J, Jacobsson A, Davidsson P (2016). On privacy and security challenges in smart connected homes. In: 2016 European intelligence and security informatics conference (EISIC). IEEE, pp 172–175
46. Gura N, Patel A, Wander A, Eberle H, Shantz SC (2004) Comparing elliptic curve cryptography and RSA on 8-bit CPUs. In: International workshop on cryptographic hardware and embedded systems. Springer, Berlin, Heidelberg, pp 119–132
47. Bafandehkar M, Yasin SM, Mahmod R, Hanapi ZM (2013) Comparison of ECC and RSA algorithm in resource constrained devices. In: 2013 International conference on IT convergence and security (ICITCS). IEEE, pp 1–3
48. ComputerWeekly.com. (2020) Covid-19 accelerates UK digital transformation efforts by over five years. https://www.computerweekly.com/news/252486191/Covid-19-accelerates-UK-digital-transformation-efforts-by-over-five-years. 5 Oct 2020
49. Capgemini Worldwide (2020) Reinventing cybersecurity with artificial intelligence: the new frontier in digital security. https://www.capgemini.com/resources/reinventing-cybersecurity-with-artificial-intelligence-the-new-frontier-in-digital-security/. 5 Oct 2020
50. Capgemini Worldwide (2020). AI in cybersecurity is the way ahead. https://www.capgemini.com/research/reinventing-cybersecurity-with-artificial-intelligence/. 5 Oct 2020
51. Greenberg A (2020) Hackers can use lasers to 'speak' to your amazon echo. Wired. https://www.wired.com/story/lasers-hack-amazon-echo-google-home/. 5 Oct 2020
52. Karimi K, Krit S (2019) Smart home-smartphone systems: threats, security requirements and open research challenges. In: 2019 International conference of computer science and renewable energies (ICCSRE). IEEE, pp 1–5
53. Guo X, Shen Z, Zhang Y, Wu T (2019) Review on the application of artificial intelligence in smart homes. Smart Cities 2(3):402–420
54. Robles RJ, Kim TH (2010) Applications, systems and methods in smart home technology: a. Int J Adv Sci Technol 15:37–48
55. Asokan N, Niemi V, Nyberg K (2003) Man-in-the-middle in tunnelled authentication protocols. In: International workshop on security protocols. Springer, Berlin, Heidelberg, pp 28–41
56. Ico.org.uk. (2020) https://ico.org.uk/media/for-organisations/documents/2013559/big-data-ai-ml-and-data-protection.pdf. 7 Oct 2020
57. Lawrence C (2020) Dojo brings critical security to smart home automation. ReadWrite. https://readwrite.com/2017/03/27/dojo-brings-critical-security-to-smart-home-automation-dl4/. 7 Oct 2020

58. Russakovsky O, Deng J, Su H et al (2015) ImageNet large scale visual recognition challenge. Int J Comput Vis 115:211–252. https://doi.org/10.1007/s11263-015-0816-y
59. Huang Y, Tang J, Cheng Y, Li H, Campbell KA, Han Z (2014) Real-time detection of false data injection in smart grid networks: an adaptive CUSUM method and analysis. IEEE Syst J 10(2):532–543

Cybercrime Predicting in the Light of Police Statistics

Jerzy Kosiński and **Grzegorz Krasnodębski**

Abstract Cybercrime is widely felt to be a huge and growing threat to individuals, companies and organizations and even entire countries. In view of the number of such incidents and their potential danger and the need to counteract them adequately, the development of cybercrime should be predict. The paper presents attempts to predict of different categories of cybercrimes in Poland in 2020 and checks whether the predicting of cybercrime gives satisfactory results, as well as to what extent the results of the predicting will indicate a trend and cyclicality in the light of random factors. For this purpose, police statistical data on cybercrime in Poland in the years 2000–2019 were used and the concept of cybercrime and its interpretation in the Polish Penal Code have been described. It also seems to be an important practical issue to determine the time horizon of predicting, based on statistical data obtained from police data, by various exploration methods, as well as the magnitude of predicting error resulting from accidental and cyclical factors. Linear trend and the Holt's method were used for predicting.

Keywords Cybercrime · Predicting · Linear trend · Holt's method

1 Cybercrime

One of the major security challenges, not only internal, is the threat of cybercrime. A further escalation of cybercrime and the use of cyberspace by criminals can be anticipated in the near future.

The concept of cybercrime can be interpreted in many ways [1]. Cybercrime can be understood in a narrow sense (computer crime) covering any illegal behaviour carried out by means of electronic activities aimed at the security of computer systems and the data processed therein. Cybercrime can also be understood in a broad sense (computer-related crime)—as any illegal behaviour committed through or in relation to a computer system or network, including offences such as the illegal possession,

J. Kosiński (✉) · G. Krasnodębski
Polish Naval Academy in Gdynia, Śmidowicza Str. 69, 81-127 Gdynia, Poland
e-mail: j.kosinski@amw.gdynia.pl

© The Author(s), under exclusive license to Springer Nature Switzerland AG 2021 55
H. Jahankhani et al. (eds.), *Cybersecurity, Privacy and Freedom Protection in the Connected World*, Advanced Sciences and Technologies for Security Applications, https://doi.org/10.1007/978-3-030-68534-8_5

offering or dissemination of information through a computer system or network. Cybercrime in this sense ranges from economic crimes such as fraud, counterfeiting, industrial espionage, sabotage and extortion, to computer piracy and other crimes against intellectual property and breaches of privacy, promotion of illegal and harmful content, facilitation of prostitution and other crimes against morals, to organized crime. The latter border is also marked by cyber-terrorism, involving attacks on public security, life and electronic warfare against critical infrastructure. In the concept of cyber-terrorism, as in cybercrime, the prefix "cyber" refers to committing a crime involving new information technologies or using cyberspace for traditional activities (e.g. planning, communication, intelligence, logistical and financial activities).

It is generally accepted that cybercrime is one of the threats that includes unauthorized conduct aimed at accessing, acquiring, manipulating or losing the integrity, confidentiality and availability of data, applications or computer systems. Cyber-threats also include cyber-terrorism, cyber-spy and cyber warfare. Cyber-threats are considered in the context of cyber-security understood as the security of globally connected information systems (e.g. internet infrastructure), telecommunication networks, computer systems and industrial control systems. Cyber-security breaches are used for a wide range of criminal activities that cause significant material and non-material damage to organisations, companies and individuals. It cannot be overlooked that this is an important problem also in the context of internal security and thus also state security.

Another source of interpretation of the notion of cybercrime could be the Communication from the Commission to the European Parliament, the Council and the Committee of the Regions of 22 May 2007 entitled "Towards a general policy on the fight against cyber crime" [2], in which the Commission drew attention to the lack of an agreed definition of cybercrime and proposed that "criminal acts committed using or directed against electronic communications networks and information systems" should be considered as cybercrimes.

The 2007 Interpol Handbook highlights four areas where cybercrime currently focuses: hacking, malware (including botnets), intellectual piracy, illegal content on digital media.

The specificity of cybercrime has been very well reflected in the Europol report for 2007 [3]. The report shows that cybercrime is presented in two shots—vertical and horizontal. The vertical approach concerns crimes that are specific to cyberspace and cannot be committed outside it. These include: hacking (DDoS attacks, botnets, zombies, etc.), crimeware (viruses, worms, Trojan horses, etc.), spamming. In the horizontal perspective, there were crimes whose accomplishment was greatly facilitated by the use of computer and IT techniques. The greatest threats were identified as: child pornography, unauthorised use of payment cards, identity theft (phishing), intellectual piracy, money laundering via the Internet (cyberlaundering), cyberterrorism.

However, the best interpretation of the cybercrime notion is based on the Council of Europe Convention of 23 November 2001 on Cybercrime [4]. The Convention contains definitions of four types of computer crime:

(1)   offences against the confidentiality, integrity and availability of computer data and systems,
(2)   crimes of counterfeiting and computer fraud,
(3)   offences related to child pornography,
(4)   offences related to the infringement of copyright and related rights.

The Convention is supplemented by the Additional Protocol to the Convention concerning criminal acts of a racist or xenophobic nature committed by means of computer systems [5].

The scope indicated in the Convention may seem rather narrow, but nevertheless, specific articles of the Polish Criminal Code (CC) can be adapted to these four categories. The categories of offences against confidentiality, integrity and availability of IT data and systems will include offences with:

- Article 165—bringing about conditions commonly dangerous to life or health,
- Article 267—unlawful obtaining of information (hacking, computer tapping, …),
- Article 268a—obstruction of information, destruction of data,
- Article 269—destruction, damaging, deletion, modification, alteration, disruption of the collection, processing of sensitive information technology data or media,
- Article 269a—interference with a computer system or an information and communication network,
- Article 269b—manufacture, sale, offering of "hacking tools".

The category of computer crimes includes acts penalized with:

- Article 270—computer counterfeiting,
- Article 285 §1—telecom counterfeiting, phreaking,
- Article 287—computer fraud,
- Article 310—counterfeiting or alteration of money, other means of payment or a document giving entitlement to receive a sum of money; placing on the market, holding such funds.

Offences relating to child pornography include acts prosecuted with:

- Article 191a—violation of sexual intimacy; perpetration of a naked person,
- Articles 198–204—sexual exploitation,
- Article 200a—grooming a minor.

The category of offences related to infringement of copyright and related rights includes acts penalized in two articles of the Penal Code:

- Article 278 §2—illegal acquisition of a computer program,
- Article 293—fencing of a computer program.

It is worth realizing that in the Polish police statistics of the above mentioned articles it is not always possible to distinguish only those acts which are connected with a computer system or ICT network.

2   Attempts to Cybercrime Prediction

2.1   Models

Commonly used models: linear trend and linear Holt model were used for prediction in this article.

The linear trend model is expressed by the equation:

$$\hat{y}_t = a \cdot t + b \tag{1}$$

where:

$$a = \frac{\sum_{i=1}^{n} (t_i - \bar{t}) \cdot y_{t_i}}{\sum_{i=1}^{n} (t_i - \bar{t})^2}, \quad b = \bar{y} - a \cdot \bar{t} \tag{2}$$

Holt's linear exponential smoothing model is used to smooth out the time series where there is a developmental trend and random fluctuations.

The prediction for moment t is as follows:

$$y^*_{t(\alpha,\beta)} = F_T(\alpha,\beta) + (t-T) \cdot S_T(\alpha,\beta) \tag{3}$$

where: T is the number of moments in time series taken into account to build a forecast, usually $T = t - 1$,

α and β—two smoothing parameters,

The Holt model is described by the following equations:

$$Ft - 1(\alpha, \beta) = \alpha \cdot yt - 1 + (1 - \alpha) \cdot (Ft - 2(\alpha, \beta) + St - 2(\alpha, \beta)) \tag{4}$$

$$St - 1(\alpha, \beta) = \beta \cdot (Ft - 1(\alpha, \beta) - Ft - 2(\alpha, \beta)) + (1 - \beta) \cdot St - 2(\alpha, \beta) \tag{5}$$

in which: $Ft - 1$ corresponds to a smoothed value from a sple exponential smoothing model (average value assessment for $t - 1$ eriod) and $St - 1$ is a smoothed trend increase for $t - 1$ period.

In order to build the Holt model, the initial values F1 and S1 are needed. Two of the standard proposals were adopted in this study: (1) based on actual values $F1 = y1$, $S1 = y2 - y1$ and (2) based on regression factors $F1 = a$, $S1 = b$ (this was not relevant to the forecast values, but (1) led to smaller errors). The problem of finding such a pair $(\alpha*, \beta*)$ for which:

$$s(\alpha*, \beta*), \quad \min_{\alpha \in [0,1], \beta \in [0,1]} s(\alpha, \beta) \tag{6}$$

where:

$$s(\alpha, \beta) = \sqrt{\frac{1}{n} \sum_{t=1}^{n} (y_t - y_t^*(\alpha, \beta))^2} \tag{7}$$

can be solved by one of the standard methods of non-linear optimization. Constant smoothening in the study was determined based on the minimal average error of expired predictions.

2.2 Cybercrimes Confirmed Prediction

The prognosis of cybercrime confirmed for 2020 based on data from 2000–2019 is dominated by very non-linear amounts of computer crime and copyright and related works (Fig. 1).

The projection of the number of identified cybercrimes confirmed for 2020 is subject to a large percentage error of 16.57% MSE (mean percentage error), which makes the projection of 45,914 identified cybercrimes unacceptable. Similarly, despite its similar value (45,031), the coefficient of determination $R^2 = 0.13$ means that the trend line matching is unsatisfactory. Visually, both prognoses are

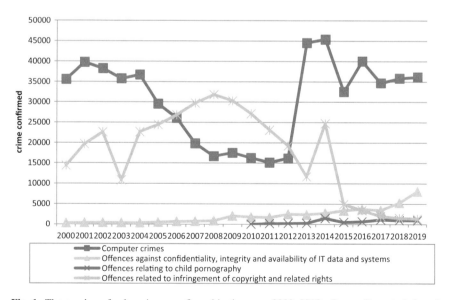

Fig. 1 The number of cybercrimes confirmed in the years 2000–2019. *Source* Own study based on Polish Police Headquarters data

shown in Fig. 2. The declining trend in the number of cybercrimes presented in Fig. 2 is in conflict with common perception and analysis [6, 7].

Predictions based on data from shorter periods of time, the last 10 or 5 years are even more mistaken, so they will not be discussed further.

As mentioned at the beginning of the article, cybercrime can be divided into four categories, so it is worth verifying the predicts for each category separately.

Computer crime
The prediction of the Holt method of computer crime (31,965 crimes) is burdened with a fairly large percentage error of 9.81% MSE, but the prediction is already admissible. The linear trend prediction (32,464 crimes) is unsatisfactory as the determination factor R^2 is 0.004. An illustration of these predictions is shown in Fig. 3.

It is worth noting that the linear trend from Fig. 3 indicates an increase in this crime.

Offences related to the infringement of copyright and related rights
Predictions of cybercrime related to copyright and related rights violations based on data from the last 19 years indicate that the trend of these events is decreasing.

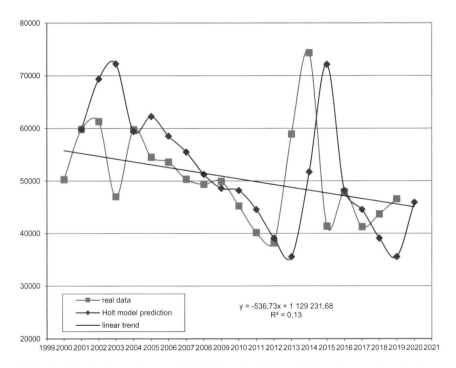

Fig. 2 Empirical values in the years 2000–2019 and predictions of cybercrimes confirmed. *Source* Own study based on Polish Police Headquarters data

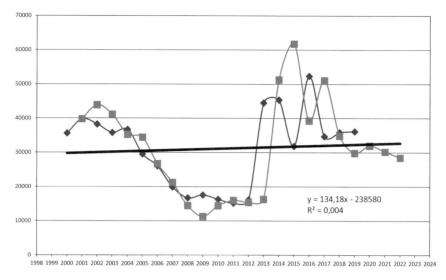

Fig. 3 Empirical values and predictions of computer crime confirmed. *Source* Own study based on Polish Police Headquarters data

Unfortunately, these predictions are statistically inadmissible (MSE—43.11%) and the determination factor $R^2 = 0.33$).

The visible trend of the decrease in the number of these crimes (Fig. 4) is understandable, because at the end of the 20th century computer and phonographic piracy in Poland was even a plague.

Offences against confidentiality, integrity and availability of IT data and systems
The number of recorded offences against confidentiality, integrity and availability of IT data and systems is significantly lower than in each of the two previous categories. However, these are mostly classic cybercrimes. Predicts indicate an increase in the number of this type of crime, with a poor match (Holt's MSE method 13.96%—an unacceptable prediction, and the determination factor $R^2 = 0.526$ of the linear match trend—weak) (Fig. 5).

Offences due to the nature of the information contained
In the case of crimes, due to the nature of the information contained, police statistics have been conducted only since 2010. Predicts have a very similar value (Holt—1145, linear trend—1226). The prediction calculated using the Holt method is acceptable (MSE—8.23%), while the second one calculated using the linear trend has a very good fit (coefficient of determination $R^2 = 0.9095$) (Fig. 6).

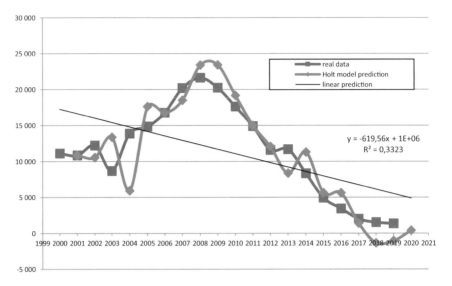

Fig. 4 Empirical values and predictions of offences related to the infringement of copyright and related rights. *Source* Own study based on Polish Police Headquarters data

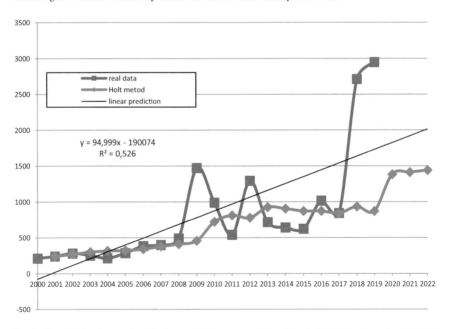

Fig. 5 Empirical values and predictions of offences against confidentiality, integrity and availability of IT data and systems. *Source* Own study based on Polish Police Headquarters data

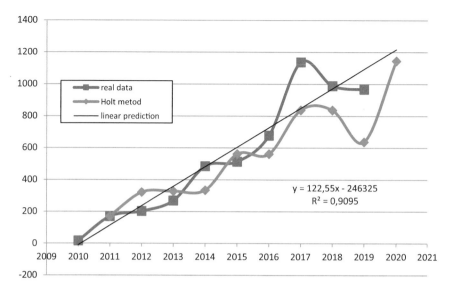

Fig. 6 Empirical values and predictions of offences due to the nature of the information contained. *Source* Own study based on Polish Police Headquarters data

2.3 Prediction of Cybercrime Detection

The detection rate is a measure of the effectiveness of police combat against crime. The detection rate is the quotient of the number of detected crimes (including those detected in the case of discontinuance) by the total number of crimes confirmed increased by the number of crimes detected after initiating proceedings discontinued in the previous year or years—expressed as a percentage. A detected crime is an crime confirmed in which at least one suspect is identified.

The prediction of computer crimes detection calculated using the Holt method indicates a significant decrease in detection (up to 41.27% in 2020 and as much as 27.15% in 2022), while the linear tremor indicates a slight increase (up to 51.13% in 2020). The Holt method prediction has a very small MSE error of 2.19% (very good prediction) and the linear trend has a very large error ($R^2 = 0.1318$). Subsequent paragraphs, however, are indented.

The detection of cybercrimes related to copyright and related rights violations has always been very high in Poland. The Holt method prognosis detection of these crimes very good (MSE 0.01%), and indicates a slight decrease to 97.01% in 2020. The linear trend, on the other hand, shows a decrease in detection rate to 96.72%, with a weak match.

The Holt method's prediction of crimes against confidentiality, integrity and availability of IT data and systems detection shows an increase in detection rates to 46.94% in 2020, but is unacceptable (MSE 21.88%). The poorly matched linear trend shows a decrease in detection rates to 23.58% in 2020.

A very good prediction of cybercrime due to the nature of the information detection contained (MSE 2.20%) shows a detection rate of 69.6% in 2020. A linear trend with a very large error indicates that this level will be higher and will amount to 75.66% in 2020.

2.4 Verification of Predictions

As part of the verification of the prognoses, the predictions presented above were calculated for 2019 and compared with the real values from 2019. The above predictions were calculated not only on the basis of the last 19 years, but also on shorter periods—the last 10 and 5 years. The linear trend prediction gave the best results when calculated for the longest period, while Holt's method proved to be the best for predicting on the basis of the last 5 years. Unfortunately, the relative error of the cybercrimes confirmed prediction was large and ranged from a fraction of a percent to, mostly, several dozen percent (max. 46%). In case of the detection rate predictions this error was much smaller (max. 19%).

3   Conclusions

The analyses presented above show that the cybercrime prediction on a global scale is very imprecise. The same statement applies to cybercrimes in the following categories: crimes of counterfeiting and computer fraud (computer crimes), crimes against confidentiality, integrity and availability of IT data and systems and crimes related to the infringement of copyright and related rights.

Cybercrime prediction may be slightly more accurate for offences due to the nature of the information contained. Despite statistically good prediction for Offences due to the nature of the information contained, one should reckon with a big mistake resulting from exogenous conditions (the numbers of crimes are largely due to random factors).

Due to the varying amounts of cybercrime committed in different categories, the predicting of cybercrime should not be conducted globally but by category or by article of the Penal Code.

Predictions about the level of the detection rate are much better, as its level is determined by the quality of police staff preparation, and this factor is not so randomly changed.

References

1. Tenth united nations congress on the prevention of crime and the treatment of offenders. Crimes related to computer networks, UN, Vienna, p 4 (2000)
2. COM (2007) 267 final. https://eur-lex.europa.eu/LexUriServ/LexUriServ.do?uri=COM:2007:0267:FIN:en:pdf
3. Europol (2007) High tech crimes within EU. Threat assessment 2007, Hague
4. https://www.coe.int/en/web/conventions/full-list/-/conventions/treaty/185
5. https://www.coe.int/en/web/conventions/full-list/-/conventions/treaty/189
6. The Global Risks Report (2018) World economic forum 13th edn. http://wef.ch/risks2018
7. Internet Organised Crime Threat Assessment (IOCTA) (2019) Europol. https://www.europol.europa.eu/sites/default/files/documents/iocta_2019.pdf

Prediction of Cyber Attacks During Coronavirus Pandemic by Classification Techniques and Open Source Intelligence

Shannon Wass, Sina Pournouri, and Gregg Ibbotson

Abstract Over the years, technology has grown rapidly and become a major part of everyday life. Due to the increased presence of technology, cybercrime is on the rise and the number of cyber-attacks has increased significantly, this has made data mining techniques an important factor in detecting security threats. This research proposes that Classification techniques can be used to reliably classify and predict cyber-attacks. This paper proposes a classification framework using data collected from Hackmagedon, a blog which contains timelines and statistics for cyber-attacks. The dataset includes cyber-attacks which occurred between 2017 and 2019 within countries in Europe. The purpose of this research is to investigate how Classification techniques can be used to better understand and predict future cyber-attacks. Different Classification techniques will be applied to the dataset to determine which technique produces the most accurate results. The model will be validated using a dataset containing COVID-19 cyber-attacks from Hackmagedon.

Keywords Cyber attack · Prediction · OSI · Pandemic · COVID-19

1 Introduction

Over the years, cyber-attacks have become an increasingly prevalent part of today's society. This is due partly to the growth in technology and it's use and dependence in everyday life. Due to the increased presence of technology, cybercrime is on the rise and the number of cyber-attacks has increased significantly.

S. Wass · S. Pournouri (✉) · G. Ibbotson
Sheffield Hallam University, Sheffield, UK
e-mail: s.pournouri@shu.ac.uk

S. Wass
e-mail: shannonwass@hotmail.co.uk

G. Ibbotson
e-mail: g.ibbotson@shu.ac.uk

In 2019, a third of businesses (32%) reported having a cyber security breach or attack. In winter 2019 and early 2020, almost half of businesses (46%) reported having a cyber breach or attack [8]. This demonstrates cyber-attacks are evolving and becoming more frequent.

Since the COVID 19 pandemic began, cyber attackers have taken advantage of the uncertainty and anxiety of the general populace as an opportunity for financial gain. In March and April alone, brute-force attacks have risen by 400% [12]. It is likely this is due to the increase in remote desktop connections, as employees have had to push technology out at an unprecedented rate, mistakes have been made, and systems are left unsecure. Under half of businesses have experienced at least one business impacting cyber attack related to COVID-19 in April 2020, such as: loss of customer, employee and confidential data, ransomware pay-outs and financial loss and theft.

In August 2020, it was discovered Blackbaud, a software provider, fell victim to a ransomware attack in May. The company is used by numerous other universities, such as Birmingham, Leeds, York, and Reading, who also suffered a data breach. The information which was breached included names and contact details for donors, alumni and stakeholders. Blackbaud paid the ransom and the data was destroyed [16].

Alongside this, in May, two companies involved in building coronavirus hospitals were hit by cyber-attacks: Bam Construct and Interserve. Bam Construct shut down its website and other systems as a precaution after being hit by a computer virus, whereas Interserve stated some operational services were affected [1].

The two attacks mentioned above were both on different work sectors: education and healthcare. This indicates cyber attackers do not discriminate on the sectors, they merely target the sector where the greatest amount of disruption and financial gain can be made.

The top sectors most likely to be affected by a cyber-attack during COVID-19 are healthcare and banking. Within the healthcare sector, malicious actors are exploiting the situation to deliver malware payloads at medical facilities to compromise the infrastructure and demand ransom to restore functionality. Due to the desperation and necessity of the compromised equipment, healthcare sectors are likely to pay the ransom as it would be more costly to not pay the ransom. In the banking sector, cyber criminals are taking advantage of the COVID-19 pandemic by creating phishing campaigns using COVID-19 misinformation to steal user data [18]. The increase in cyber-attacks and the elaborate methods used to conduct them demonstrate the need for more advanced cyber-attack detection methods. Currently, methods such as Intrusion Detection Systems are used to detect anomalies, however, false positives are frequent and IP packets can be faked. Therefore, this study aims to develop a framework using classification techniques alongside historical data to better understand and predict cyber-attacks in Covid-19 era.

1.1 Background

Many researches have been done in the field of cyber security and data mining specifically classification techniques. In this section we have reviewed few most recent studies about how classification techniques can be used for prediction of different elements in cyber security.

1.1.1 Support Machine Vector

Sarker et al. [21] proposed an Intrusion Detection model using SVM. This experiment used security datasets which represented a collection of information records consisting of security features which could be used to train the intrusion detection model. The dataset contained more than 25,000 records which were collected from a variety of intrusion detection systems simulated in a military network environment. To collect this data, an environment was created by simulating a US Air Force Local Area Network. The dataset consisted of both nominal and numerical security features:

1.   Duration logged in
2.   Number of failed logins
3.   Server error rate
4.   Protocol type.

This experiment ranked the security features and assigned them a score, this was so significant features could be chosen for further processing and irrelevant features could be removed. This made the model more effective in terms of prediction accuracy for unseen test cases. The effectiveness of the model was calculated in terms of accuracy, true positive, and false negative. To calculate the results for unseen test cases, 80% of the data was used for training and 20% for testing. The results were calculated using a matrix which reported the number of false positives, false negatives, true positives and true negatives. However, the limitations of this study were the features in the security dataset were not equally significant to build a data driven security model.

Kim Donghoon et al. [9] proposed a SVM framework to detect a class of cyber-attacks which redistribute loads, such as DDOS attacks. The dataset consisted of captured network packets, 70,000 were used for the training set and 30,000 were used for the validation set. The limitations of this experiment were the DDOS attack methods were not complicated enough compared to real world attacks, meaning the results may be inaccurate. To reliably evaluate the performance, raw data from client machines affected by a DDOS attack is needed.

Kinan Ghanem et al. [13] proposed a Network Intrusion Detection model using SVM to reduce the number of instances used during the computation of the SVM which also reduces the training time of the model. The model was evaluated using different network traffic datasets from wired and wireless networks at Loughborough University. The SVM categorised the traffic as malicious or normal, if it was malicious, it was placed into one of the four classes: Denial of Service, Remote to Local, User to Root and Probing.

A positive point to Kinan Ghanem et al. (2017)'s approach was the classifier had an accuracy of 100% for detection rate. However, the problem with this technique is the SVM required labelled training data to accurately predict the class of the data. The limitation of this study is the SVM only reached an accuracy of 81% and generated false positives of 19% for the Probing class, this could be a result of the data consisting of different factors; for example, wired and wireless connections, the different attack types and a possible imbalance in the dataset.

1.1.2   Random Forest

A Random Forest classifier was proposed by Nabila Farnaaz [11] to predict malicious intrusions using a host-based intrusion detection system. The data used for this experiment was a NSL-KDD data set, this contained records of internet traffic detected by a simple intrusion detection network, the data contained 43 features per record, 41 of these features refer to the traffic input and 2 are the labels of whether it is normal or an attack and the score given to the severity of the traffic input. Within the dataset, there were 4 different classes of attacks: Denial of Service, Probe, User to Root and Remote to Local. Their proposed approach and the method was split into 8 steps:

1. Load the dataset—the data was imported into WEKA.
2. Apply pre-processing technique—WEKA was used to replace the missing values.
3. Cluster the data into datasets—the data was clustered into groups based on their similarities.
4. Partition the data into training and test sets—the data was split into different sets.
5. Select the best set using feature selection—the best set was used based on their features e.g. type of attack.
6. Dataset was applied to the Random Forest classifier for training—the data was trained using RF.
7. Test data was given to the Random Forest classifier for classification—the data was classified using RF.
8. Calculate accuracy, detection and false alarm rate.

The proposed model yielded a high detection rate and a low false alarm rate with an overall accuracy of 99.97%

A positive point of Nabilia Farnaaz (2016)'s approach is multiple performance measures were used to evaluate the classifier: accuracy, detection rate, false positive and true negative. For example, if the classifier were only evaluated using the detection rate and everything was detected successfully, the classifier would have a success rate of 100%. However, this would not take into account the accuracy and amount of false positives vs true negatives, meaning the results of the experiment would be inaccurate.

The problem with Random Forest is many noisy trees are created which affect the accuracy and decisions for future samples. However, Nabila Farnaaz (2016) takes this into account in the feature selection process, having irrelevant features decreases the accuracy of the classifier. The features chosen were:

1. Accuracy—the ratio of correctly classified samples.
2. Detection rate—the ratio between total numbers of attacks detected by the system to the total number of attacks present in the dataset.
3. False alarm rate—this is defined as: FP/TN + FP.

An additional approach was suggested by Leyla Bilge et al. [3], a system called RiskTeller. This system used a Random Forest classifier to predict which machine was at higher risk of a cyber-attack. The dataset used for this experiment was binary file logs from 600,000 machines belonging to 18 enterprises. For each machine, 89 features were used. These features were based on the number of events, application categories, rarity of files, patching behaviour and past threat history. The features were categorised into 6 groups:

1. Volume based—percentage of events, fraction of events from top file hashes, percentage of applications.
2. Temporal—Monthly percentage of events—median/standard deviation.
3. Vulnerability/patching—percentage of patched vulnerabilities/applications.
4. Application categories—top 5 application categories with most events.
5. Infection history—fraction of events for malicious/benign/unknown files.
6. Prevalence-based—fraction of events seen in only one enterprise.

This method consisted of:

1. Data pre-processing—the file and directory names were normalized to identify those which were likely to perform the same.
2. Feature extraction—features were computed into the classifier.
3. Feature labelling—features were labelled as clean or infected machines.
4. Feature discovery—for each machine, a profile was created containing 89 different features based on events which will be used to predict the machines risk of future infection.

The problem with this method was whereas RiskTeller predicts the general risk that a machine is at risk from malware, it did not classify specific malware categories. However, a positive point is Leyla Bilge et al. (2017)'s experiment proves the concept of using machine learning to predict malware with a high accuracy rate. Risk Teller achieved a 96% true positive rate and 4% false positive rate, no previous work has been able to achieve a 96% true positive rate with a 4% false positive rate at a machine level granularity.

Rupa Ch et al. [5] proposed a computational system to detect and classify cyber-crimes using Random Forest. The data was gathered from Kaggle, a data science community and CERT-IN, an Indian Computer Emergency Response Team. The information collected from each data source consisted of 2097 records with 8 features: the type of incident, victim, access violation, harm, year, location and age of offender. After the data was pre-processed, features were extracted and used to classify the cybercrimes: Incident, Offender, Harm, Access Violation, Year and Victim.

A positive point on Rupa Ch et al. (2020)'s approach was the proposed model classified with a 99% accuracy rate. The limitations of the study was currently it only classified cybercrimes into certain groups: Identity theft, hacking, copyright attacks and other. A feature extension is required to provide countermeasures to the crime agencies, in order to reduce the frequency of cybercrimes in specific locations.

1.1.3 Naïve Bayes

Prajakta Yerpude [20] proposed a framework for predicting crime using supervised machine learning methods, such as Naïve Bayes. The communities and crime dataset from the UCI repository (a collection of databases) was used, which consisted of crime data in Chicago, containing a total of 1994 records. Included in the dataset were features such as:

1. Population—urban, rural etc.
2. Race—asian, Caucasian etc.
3. Sex—female/male
4. Police—percentage of police officers.

To select the most important features for the classifier, Feature Importance was used to assign each feature a score. The higher the score, the more significant the feature was to the classifier. The features extracted according to their feature importance scores were:

1. NumUnderPov: Number of people under the poverty line.
2. NumbUrban: Number of people in Urban Areas.
3. HousVacant: Number of vacant houses.
4. RacePctHisp: Percentage of race Hispanic.
5. LemasPctOfficDrugUn: Percentage of officers assigned to drug unit.
6. PctNotSpeakEnglWell: Percentage of people who did not speak English well.
7. acePctAsian: Percentage of race Asian.

The classifier used both clean and dirty data. The accuracy of the result for the clean data, 77.64%, was higher than the dirty data, 75.42%. This demonstrates that missing data creates inconsistencies and affects the performance of the model. A positive point of Prajakta Yerpude's approach is the feature importance score proved to be highly predictive, specifically "NumUnderPov" and "NumbUrban". The limitation of this study is the dataset consisted of every crime,this analysis can be narrowed down to categories of crime to yield more accurate results.

1.1.4 K Nearest Neighbour

Ben Abdel Ouahab Ikram et al. [2] proposed a malware classification framework using K-Nearest Neighbor alongside visualisation techniques. Using visualisation techniques, the malware binary was converted into a grayscale image, afterwards, an image descriptor was computed to classify the malware using K-Nearest Neighbour with features extracted from the data. The dataset used for this framework was Malimg, this is a malware classification dataset from the website Kaggle, containing 9339 malware samples from 25 different families in the form of grayscale images. This dataset was split into training data and test data: TestDB and TrainDB. The features exported from the dataset were: malware families names, number of images in each family.

The model was trained with different k values and a score was calculated for each case, for example: k = 2, k = 5, k = 10. Since k = 10 gave the highest training score, this was saved for future predictions. The test score yielded an accuracy of 97.92%. The model was saved and used on the TestDB with completely new, unlabelled data. The purpose of this was to see how the model would perform on unknown samples. The model yielded an accuracy of 92% with the unknown samples. Overall, the model accurately classified 46 out of 50 malware samples. The limitation of this study was the data was not cleaned before being imported into the model for training, this means predictions from the data may be inaccurate and misleading.

Masoumeh Zareapoor [25] proposed a model for predicting credit card fraud using KNN. A credit dataset was used containing e-commerce transactions with 100,000 records labelled as legitimate or fraudulent, 2293 of the records (2.8%) were fraudulent and 97,707 (97.2%) were legitimate. The model was evaluated using the following metrics: Fraud Catching Rate, False Alarm Rate and Balanced Classification Rate. Incoming transactions were classified by calculating the nearest point to the newest incoming transaction. If the nearest neighbour was fraudulent, the transaction was classified as fraudulent.

A positive point of Masoumeh Zareapoor (2015)'s approach was the possible bias in the unbalanced data was taken into account, the evaluation metrics were changed from accuracy and error rate to False Alarm Rate and Balanced Classification rate.

1.1.5 Neural Networks

Moshe Kravchik et al. [15] proposed a framework for detecting cyberattacks in industrial control systems using convolutional Neural Networks. The dataset used was a Secure Water Treatment testbed which represented a real-world industrial water plant and included 36 different cyber-attacks, consisting of 7 days of recording under normal conditions (benign) and 4 days when the 36 attacks were performed (attack) and logged on a server. The entire dataset contained 946,722 records labelled as attack or benign. The features used in this experiment were attributes from the water test sensors: flow meters, water level meters, conductivity and pH analysers. The data from the sensors was logged and used for training and testing the model. A positive point of Moshe Kravchik et al. (2018)'s approach is the model successfully detected 32 out of 36 attacks. The limitation of this study is convolutional neural networks are stateless, this means they lack the ability to learn beyond what was used as a sample.

Ihor Tereikovskyi [24] proposed a model using deep neural networks to detect cyber-attacks. The model was written using the programming language Python. The NSL-KDD dataset was used, this is a modification of KDD-99. The dataset contained a set of data to be audited, including a variety of intrusions in a military environment, the number of records was 25,192. The features used in the NSL-KDD dataset were:

1. duration—time of connection in seconds.
2. protocol_type—UDP, TCP etc.
3. service—network service.
4. flag—connection status.
5. src_bytes—Data amount transferred from source to recipient in bytes.
6. dst_bytes—Data amount transferred from recipient to source in bytes.

These features served as input variables for the neural network model, the number of input neurons was 4 which relates to the number of cyberattacks.

The features were combined into groups:

1. Basic attributes.
2. Content attributes.
3. Host traffic attributes.

The dataset contained values of each feature to detect the following types of cyber-attacks:

1. Distributed Denial of Service.
2. Probe.
3. Remote to Local.
4. User to Root.

The classifier had a 90% detection rate. A positive point of Ihor Tereikovskyi (2017)'s model is it was written in the programming language Python. Python consists of code libraries, making it easier to use and the preferred language for machine learning. The limitations of this study were the dataset only contained 25,192 records, in order for deep neural networks to perform better over other techniques, a larger dataset is required.

1.2 Summary

In this section, 2 concepts were investigated: cyber security and data mining techniques. The different components of cyber security were discussed: cyber threats, threat actors and cyber activities. Alongside this, the different data mining techniques were discussed, including their definitions and how they can be applied to cyber security.

2 Data Pre-processing

2.1 Introduction

This section aims to demonstrate the data type and pre-processing stage. Different tools which have been used for pre-processing the data will be explained.

2.2 Data Collection

This section aims to demonstrate the data collection process. The main source of data used in this research is Hackmagedon (https://www.hackmageddon.com/). Hackmagedon is a blog which contains timelines and statistics for cyber-attacks. The data collected from Hackmagedon includes cyber-attacks which occurred between 2017 and 2019 within countries in Europe. The dataset includes 1989 records of cyber-attack incidents where the type of attack is known and unknown.

2.3 Data Categorize

At this stage, the initial dataset needs to be explored in more detail. Each incident of cyber-attack comes with 11 features:

1. ID—Identifying number given to the record.
2. Date—the date the attack was recorded.
3. Author—the person behind the attack e.g. Lizard Squad.
4. Target—the name of the target.
5. Description—a brief description of the attack.
6. Attack—the type of cyber-attack e.g. Brute-Force.
7. Target Class—the business sector affected by the attack.
8. Attack Class—the category the attack falls into e.g. cyber criminals.
9. Country—the country affected by the attack e.g. UK.
10. Link—a link to a news article about the attack.
11. Tags—key words relating to the attack.

Table 1 shows an example of the structure of the dataset:

The columns that are not relevant to the research will be removed; this is because they are not useful to the predictive model. The ID, Date, Target, Description, Link and Tags columns will be removed. These columns are not an area of interest and will not improve the model. Finally, the Author column will be renamed to Attackers. The remaining columns that will be used as part of the model are: Attackers, Attack, Attack Class, Target Class and Country. After the columns are removed, the dataset is restructured, and 5 different columns remain.

Table 1 Dataset row example

Date	Author	Target	Description	Attack	Target class	Attack class	Country	Link	Tags
01/01/17	APT28 AKA fancy bear	Unnamed TV station	N/A	Targeted attack	Information and communication	CE	GB	N/A	APT2, secure works

OpenRefine will be used to cleanse the data. The data will be inputted into OpenRefine and pre-processing will begin, pre-processing includes the following steps:

1. Removing Duplicates: This will be done using facet text. In OpenRefine, Facet values provide an overview of the values in a column, making it easier to detect duplicates. For example, Account Hijacking was present twice due to a spelling mistake.
2. Capital and lower-case letters: OpenRefine can also cluster the text so lower-case and capital letters are automatically integrated. Brute-force and Brute-Force were both values in the Attack column, the difference was the capital F. The lower-case F became a capital, this merged the values into one.
3. Removing irrelevant records: This is done manually in Excel by filtering out the records which are Unidentified and contain no information about an attack.
4. Combining Values: Values need to be combined for them to be categorised for further analysis, multiple columns must be categorised:
5. Attack: The attack is categorised based on the nature of the attack. Table 2 shows the Attack categories, abbreviations and examples:

(a) Target Class: The target class is categorised based on the sector attacked. Table 3 shows the Target Class categories and examples:

(b) Attack Class: The attack class is categorised based on the class of the attack (Table 4)

Table 2 Type of threat and abbreviations

Attack	Example	Abbreviation
Account Hijacking	Account Hijacking occurs when a person's account is stolen by a hacker	AH
Brute-Force	An attacker submits multiple passwords until they gain access to a victim's account	BF
DDoS	An attacker floods a service with internet traffic to disrupt users from accessing the service as normal	DDOS
Injection	An attacker inserts malicious code into an application to manipulate the application into working a certain way	INJ
Malware	A type of software designed to cause damage to a computer: viruses, worms, trojan horses	M
MITM	An attacker listens into the data sent between two computers	MITM
Phishing	An attacker attempts to gain sensitive information from a user by pretending to be a known, trustworthy source	PH
Social Bots	Automated social media accounts	SB
Targeted Attack	An anonymous attacker actively trying to infiltrate a victim's system	TA
Unidentified	The type of attack is unknown	UID

Table 3 Type of target abbreviations

Target class	Example	Abbreviation
Administration Activities	Preparing budget, personnel management, maintenance of computer data	AA
Education and Social Work	Schools, universities and social	EASW
Financial and Communication Activities	Banking companies and communications related activities	FACA
Multiple Individuals	Several individuals	MI
Production Related Activities	Includes manufacturing companies, mining etc	PRA
Retail and Transport	Retail shops and transportation companies	RAT
Science and Technology	Companies providing technology and scientific research	SAT
Single Individual	Single individual	SI
Unidentified	The target class is unknown	UID
Utilities	Water, gas etc	UT

Table 4 Attack class abbreviations

Attack class	Example	Definition
CE	Stealing/spying on classified information from a government entity or organization	Cyber Espionage
H	The use of computer systems for politically motivated reasons	Hacktivists
CC	A criminal which uses a computer to commit crime	Cyber Criminals
CW	The use of computer systems to attack a country	Cyber Warfare

(c)   Attackers: The attack is categorised based on the group or person carrying out the attack (Table 5)

2.4   Data Statistics

2.4.1   Cyber Attacks by Attack Type

Malware attacks accounted for 43.06% of all cyber-attacks which occurred between 2017 and 2019. This is possibly due to the fact that ransomware attacks are on the rise. Whereas customer ransomware attacks are decreasing, enterprise ransomware attacks are increasing (Fig. 1).

One factor which has played a role in the increase is the number of ransomware authors, ransomware authors know there is a good chance the business will pay the ransom as it will be more costly for the business to suffer an outage than to pay the

Table 5 Attackers abbreviations

Attackers	Abbreviation
Anonymous	A
Unidentified	UID
Chinese Hackers	CH
Iranian Hackers	IH
Multiple Attackers	MA
Nigerian Hackers	NH
North Korean Hackers	NKH
Russian Hackers	RH
Turkish Hackers	TH
USA Hackers	USAH

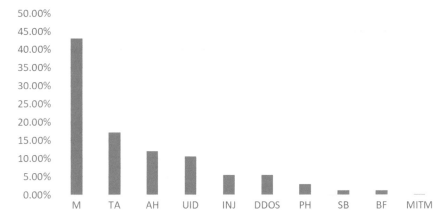

Fig. 1 Cyber attacks by type of attack

author [19]. In 2017, 39% of businesses hit by ransomware paid the author, in 2018, this increased to 45%. Finally, in 2019 this figure increased to 58% [7].

2.4.2 Cyber Attacks by Target Class

The industry most affected by cyber-attacks was Education and Social Work between 2017 and 2019, 30.98% of attacks were on this sector. In the past two years, cyber-attacks on higher education institutions exposed over 1.35 million student identities [17].

Universities are specifically targeted due to the sensitive information stored within their systems. Additionally, cyber-attacks on universities occur because the systems are large and complex, which makes implementing protections correctly difficult [4] (Fig. 2).

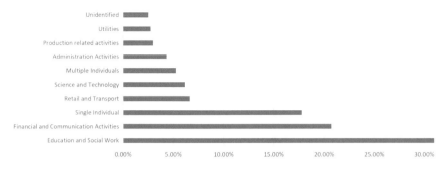

Fig. 2 Cyber attacks by industry

2.4.3 Cyber Attacks by Attack Class

Cyber Criminals were responsible for 75.70% of cyber-attacks which took place between 2017 and 2019. In 2017, the number of cyber incidents was up by 46% and in 2018, this decreased to 43% and finally, in 2019, this number decreased to 32%. One of the reasons for this decrease is between September 2017 and September 2018, the number of computer misuse incidents decreased from 1.5 million to 1 million. Another explanation for fewer businesses identifying breaches is organisations are becoming more secure and aware of cyber incidents. Since 2018, organisations have increased their defences against cyber-attacks (GOV UK Department for Digital, [14]) (Fig. 3).

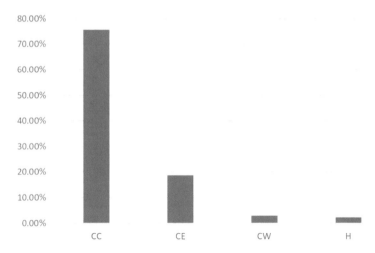

Fig. 3 Cyber attacks by attack class

2.4.4   Cyber Attacks by Country

The country most affected by cyber-attacks across all industries and attack types was the United Kingdom. In 2019, almost 50% of UK businesses suffered a security breach or cyber-attack; in the last 12 months, only half of the businesses had completed an internal and external security audit, it is highly likely they did not have the correct security protections in place to prevent a cyber-attack [23] (Fig. 4).

2.5   Summary

This section discussed the data collected and the different tools and techniques used to structure the data to prepare it for further analysis, alongside the data statistics.

3   Data Analysis

3.1   Introduction

This section will focus on investigating different classification algorithms, such as Support Vector Machine, Random Forest, K-Nearest Neighbour, Naïve Bayes and Neural Networks. Each algorithm will be applied to the dataset and the results will be analysed and compared to discover which one produces the most accurate results in predicting the type of attack used in a cyber-attack.

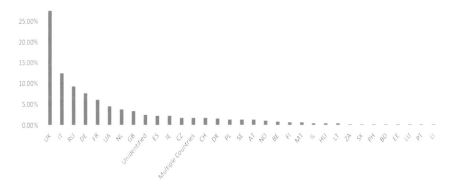

Fig. 4   Cyber attacks by country

3.1.1 Performance Evaluation Metrics

The following are the performance evaluation metrics the algorithms are going to be measured against (Table 6):

3.2 Support Vector Machine Analysis

Now data pre-processing has been completed, the analysis stage can begin. In this part of the analysis, Support Vector Machine, which was discussed in Sect. 1.1.1, will be applied to the dataset to train the data in predicting the different types of attacks used in a cyber-attack. 10-fold cross validation will be applied.

Table 6 Explanation of performance evaluation metrics

Name	Definition	
TP rate	Correctly classified values where the result is correct and is actually correct [10]	
FP rate	Incorrectly classified instances where the result is correct and is actually incorrect [10]	
Precision	The percentage of results returned which are more relevant than irrelevant [22]	Precision = TP/TP + FP
Recall	The percentage of results returned which are mostly relevant [22]	Recall = TP/TP + FN
F-measure	A weighted average of the Recall and Precision. Takes false positives/false negatives into account [10]	F-Measure = 2*(Recall*Precision) / (Recall + Precision)

```
Correctly Classified Instances      1125           63.3089 %
Incorrectly Classified Instances     652           36.6911 %
Kappa statistic                        0.3761
Mean absolute error                    0.0815
Root mean squared error                0.2855
Relative absolute error               52.1119 %
Root relative squared error          102.1648 %
Total Number of Instances           1777
```

Fig. 5 SVM summary by attack

TP Rate	FP Rate	Precision	Recall	F-Measure	MCC	ROC Area	PRC Area	Class
0.147	0.010	0.485	0.147	0.225	0.243	0.568	0.124	DDOS
0.834	0.059	0.772	0.834	0.802	0.753	0.887	0.676	TA
0.958	0.591	0.601	0.958	0.739	0.434	0.684	0.596	M
0.009	0.001	0.333	0.009	0.018	0.046	0.504	0.064	INJ
0.004	0.003	0.200	0.004	0.008	0.010	0.501	0.135	AH
0.000	0.000	?	0.000	?	?	0.500	0.015	BF
0.000	0.000	?	0.000	?	?	0.500	0.034	PH
0.000	0.000	?	0.000	?	?	0.500	0.015	SB
0.000	0.000	?	0.000	?	?	0.500	0.002	MITM
Weighted Avg. 0.633	0.297	?	0.633	?	?	0.668	0.449	

Fig. 6 SVM attack accuracy by class

3.2.1 Prediction of Attack by Support Vector Machine

Figure 5 demonstrates the Support Vector Machine classifier correctly classified 63.3% of instances and incorrectly classified 36.6% of instances based on the type of attack.

Figure 6 demonstrates the Support Vector Machine classifier yielded a TP rate of 0.633%, a FP rate of 0.297% and a Recall rate of 0.633%.

The M class has a TP rate of 0.958%, a FP rate of 0.591%, a Precision of 0.601%, a Recall of 0.958% and a F-Measure of 0.739. As the M class has a significantly higher Recall (0.958%) compared to the Precision (0.601%), this means the classifier is overclassifying instances as M. The classifier correctly classified 820 instances of M and incorrectly classified 33 as TA. Out of all the classes, M and TA were classified the most frequently.

TA has a TP rate of 0.834%, a FP rate of 0.059%, a Precision of 0.772%, a Recall of 0.834% and an F-Measure of 0.802%. The Recall and Precision are not drastically different, meaning whilst the classifier is recalling a high number of instances, the instances recalled are correct the majority of the time due to the high Precision. 287 instances of TA were classified correctly and 56 were misclassified as M. It is possible M and TA share certain patterns which makes the classifier believe TA is M and vice versa. INJ, AH and DDOS were the only other classes aside from M and TA to have correct classifications, however, the majority of the instances belonging to the classes were also misclassified as TA or M.

3.2.2 Discussion and Interpretation

Table 7 demonstrates the accuracy of the Support Vector Machine classifier:

The SVM classifier accurately classified the type of attack 63.3% of the time. The classifier performed the greatest when predicting Malware or Targeted Attack instances. It is possible this is because there are patterns within the Malware and

Table 7 SVM accuracy in predicting

Type of prediction	SVM accuracy rate
Attack	63.3%

Targeted Attack classes which the other classes do not have. The Targeted Attack class demonstrated equally high Recall and Precision, this demonstrates the classifier is selecting relevant instances which are also accurate to the class, whereas the Malware class shows a higher Recall than Precision, this means a portion of instances classified as Malware belong to another class. Aside from the classes Malware and Targeted Attack, the classifier accurately classified instances of Injection, Account Hijacking and DDOS into the correct class. However, several instances from each class were misclassified as DDOS, Targeted Attack and Malware. Out of all the classes, Malware contains the highest number of misclassified instances.

3.3 Random Forest Analysis

Random Forest, which was discussed in Sect. 1.1.2, will be applied to the dataset to train the data in predicting the different types of attacks used in cyber-attacks. tenfold cross validation will be applied.

3.3.1 Prediction of Attack by Random Forest

Figure 7 demonstrates the Random Forest classifier correctly classified 67.4% of instances and incorrectly classified 32.5% of instances based on the type of attack.

```
Correctly Classified Instances      1199            67.4733 %
Incorrectly Classified Instances     578            32.5267 %
Kappa statistic                       0.4883
Mean absolute error                   0.1041
Root mean squared error               0.2352
Relative absolute error              66.5507 %
Root relative squared error          84.1378 %
Total Number of Instances          1777
```

Fig. 7 RF summary by attack

	TP Rate	FP Rate	Precision	Recall	F-Measure	MCC	ROC Area	PRC Area	Class
	0.330	0.014	0.610	0.330	0.429	0.424	0.853	0.383	DDOS
	0.817	0.042	0.824	0.817	0.820	0.778	0.929	0.815	TA
	0.900	0.415	0.668	0.900	0.767	0.507	0.787	0.710	M
	0.264	0.013	0.569	0.264	0.360	0.361	0.759	0.260	INJ
	0.279	0.049	0.472	0.279	0.351	0.290	0.693	0.347	AH
	0.000	0.001	0.000	0.000	0.000	−0.004	0.675	0.039	BF
	0.000	0.003	0.000	0.000	0.000	−0.010	0.548	0.036	PH
	0.593	0.005	0.640	0.593	0.615	0.610	0.908	0.608	SB
	0.000	0.000	?	0.000	?	?	0.595	0.003	MITM
Weighted Avg.	0.675	0.216	?	0.675	?	?	0.796	0.598	

Fig. 8 RF attack accuracy by class

Figure 8 demonstrates the Random Forest classified yielded a TP rate of 0.675%, a FP rate of 0.216% and a Recall of 0.675%

The M class is classified more than any other class; however, it also has the highest Recall rate, 0.900%, it is likely there is a pattern within the M instances which makes the classifier believe the selected instance is M and misclassify. 770 instances of M were classified correctly, however, throughout the other classes, there are several misclassifications as M. For example, 75 instances of INJ and 45 instances of DDOS were misclassified as M. Aside from the M class, the classifier accurately classified instances of TA, DDOS, AH, INJ and SB into the correct class. However, there were also a high number of misclassifications. The classes which were misclassified the most were M, TA, DDOS and AH. The only class to have no correct classifications was MITM, all the instances of MITM were misclassified as DDOS, M and TA.

3.3.2   Discussion and Interpretation

Table 8 demonstrates the accuracy of the Random Forest classifier in terms of the different attacks within the dataset which apply to cyber-attacks:

The RF classifier accurately classified the type of attack 67.1% of the time. The classifier performed the greatest when predicting Targeted Attack, Social Bot and Malware instances, it is likely this is due to these classes sharing certain character-istics which the other classes do not have. The Targeted Attack class demonstrates equally high Recall (0.817%) and Precision (0.824%), this demonstrates the classi-fier is selecting relevant instances which are also accurate to the class, whereas the Malware class shows a higher Recall (0.900%) than Precision (0.668%), this means a portion of instances classified as Malware belong to another class. The classifier accurately classified instances of TA, DDOS, AH, INJ and SB. Out of all the classes, Man in the Middle, Brute Force and Phishing were the only classes to not have a correct classification.

3.4   Naïve Bayes Analysis

Naïve Bayes, which was discussed in Sect. 1.1.3, will be applied to the dataset to train the data in predicting the type of attack used during a cyber-attack. tenfold cross validation will be applied.

Table 8 RF accuracy in predicting

Type of prediction	RF accuracy rate
Attack	67.4%

3.4.1 Prediction of Attack by Naïve Bayes Classifier

Figure 9 demonstrates the Naïve Bayes classifier correctly classified 64.3% of instances, and incorrectly classified 35.6% of instances based on the type of attack.

Figure 10 demonstrates the Naïve Bayes classifier yielded a TP rate of 0.643%, FP rate of 0.224% and a Recall rate of 0.643%.

The TA class yields a TP rate of 0.831%, a FP rate of 0.53%, a Precision of 0.790%, a Recall of 0.877% and a F-Measure of 0.810%. The recall is slightly higher than the precision, this means the classifier is selecting relevant instances which are correct a majority of the time. For example, the classifier accurately classified 286 instances of TA. There are a few misclassifications of TA in the other classes, for example, there are 23 misclassifications of TA in the AH class. However, unlike the M class, TA has a low number of misclassifications. The M class yields a TP rate of 0.877%, a FP rate of 0.421%, a Precision of 0.659% and a Recall of 0.877%, as the recall is higher than the precision, this indicates there are patterns within the M class which causes the classifier to misclassify instances as M. The classifier believes the instances to be relevant to the M class and classifies the instances. In the INJ class, 14 instances were correctly classified and 76 instances were misclassified into the M class, it is possible this is due to M containing similar patterns which the other classes do not have which makes the classifier more likely to classify the instances as M. SB and DDOS were the only classes to have a higher Precision rate over Recall, this shows the instances recalled were relevant to the class.

```
Correctly Classified Instances        1143              64.3219 %
Incorrectly Classified Instances       634              35.6781 %
Kappa statistic                                0.4399
Mean absolute error                            0.0997
Root mean squared error                        0.2395
Relative absolute error                       63.7098 %
Root relative squared error                   85.7043 %
Total Number of Instances             1777
```

Fig. 9 NB summary by attack

	TP Rate	FP Rate	Precision	Recall	F-Measure	MCC	ROC Area	PRC Area	Class
	0.312	0.026	0.436	0.312	0.364	0.334	0.844	0.377	DDOS
	0.831	0.053	0.790	0.831	0.810	0.764	0.943	0.840	TA
	0.877	0.421	0.659	0.877	0.753	0.475	0.793	0.717	M
	0.127	0.011	0.424	0.127	0.196	0.207	0.741	0.236	INJ
	0.183	0.064	0.310	0.183	0.230	0.151	0.696	0.283	AH
	0.000	0.000	?	0.000	?	?	0.615	0.029	BF
	0.000	0.001	0.000	0.000	0.000	-0.004	0.519	0.037	PH
	0.519	0.005	0.636	0.519	0.571	0.569	0.856	0.509	SB
	0.000	0.000	?	0.000	?	?	0.410	0.002	MITM
Weighted Avg.	0.643	0.224	?	0.643	?	?	0.797	0.593	

Fig. 10 NB attack accuracy by class

Table 9 NB accuracy in predicting

Type of prediction	NB accuracy rate
Attack	64.3%

3.4.2 Discussion and Interpretation

Table 9 demonstrates the accuracy of the Naïve Bayes classifier:

The NB classifier accurately classified the type of attack 64% of the time. The classifier was more likely to recall and classify instances as Targeted Attack or Malware. Both of these classes have a higher Recall than Precision, this means the classifier is recalling instances and misclassifying. Unlike Targeted Attack or Malware, Social Bot had a higher Precision than Recall, meaning the instances recalled were accurate to the class. The classifier classified no correct instances of Man in the Middle, Phishing and Brute Force, the classifier misclassified these instances as DDOS, Targeted Attack and Malware.

3.5 K-Nearest Neighbour Analysis

K-Nearest Neighbour, which was discussed in Sect. 1.1.4, will be applied to the dataset to train the data in predicting the type of attack used during a cyber-attack. tenfold cross validation will be applied.

3.5.1 Prediction of Attack by KNN

Figure 11 demonstrates the KNN classifier correctly classified 67.2% of instances, and incorrectly classified 32.7% of instances based on the type of attack.

Figure 12 demonstrates the KNN classifier yielded a TP rate of 0.672%, a FP rate of 0.222%, and a Recall of 0.672%.

The classifier shows the TA class has a TP rate of 0.826%, a FP rate of 0.039, a Precision of 0.835% and a Recall rate of 0.826%. As the Precision is slightly higher than the Recall, this demonstrates the classifier is selecting relevant instances

```
Correctly Classified Instances        1195              67.2482 %
Incorrectly Classified Instances       582              32.7518 %
Kappa statistic                         0.482
Mean absolute error                     0.1025
Root mean squared error                 0.2391
Relative absolute error                65.5001 %
Root relative squared error            85.5589 %
Total Number of Instances             1777
```

Fig. 11 KNN summary by attack

TP Rate	FP Rate	Precision	Recall	F-Measure	MCC	ROC Area	PRC Area	Class
0.321	0.019	0.530	0.321	0.400	0.384	0.825	0.374	DDOS
0.826	0.039	0.835	0.826	0.830	0.790	0.915	0.818	TA
0.903	0.428	0.662	0.903	0.764	0.500	0.779	0.696	M
0.227	0.015	0.500	0.227	0.313	0.309	0.753	0.249	INJ
0.275	0.043	0.500	0.275	0.355	0.302	0.697	0.333	AH
0.000	0.001	0.000	0.000	0.000	-0.003	0.646	0.035	BF
0.000	0.002	0.000	0.000	0.000	-0.009	0.535	0.035	PH
0.444	0.003	0.706	0.444	0.545	0.555	0.897	0.508	SB
0.000	0.000	?	0.000	?	?	0.689	0.005	MITM
Weighted Avg. 0.672	0.222	?	0.672	?	?	0.787	0.586	

Fig. 12 KNN attack accuracy by class

and correctly classifying them as belonging to the TA class. Whereas M shows a Precision of 0.662% and a Recall of 0.903%, in this case the Recall is higher than the Precision, although the classifier is selecting relevant instances, a high proportion of the instances are being misclassified as M. For example, 50 instances of DDOS were misclassified as M.

The SB class has a TP rate of 0.444%, a FP rate of 0.003%, a Precision of 0.706% and a Recall of 0.444%. This means the classifier does not believe many instances belong in the SB class and a high proportion of the SB instances are being misclassified. For example, 12 instances of SB were correctly classified, however, 14 instances were misclassified into DDOS, TA, M and AH. This could be because the SB class shares similarities with the DDOS, TA, M and AH classes which makes the classifier believe instances of SB belong to those classes.

3.5.2 Discussion and Interpretation

Table 10 demonstrates the accuracy of the KNN classifier:

The K-Nearest Neighbour classifier using the type of attack feature accurately classified the type of attack 67% of the time. The classifier performed the greatest when predicting Targeted Attack instances. The Recall (0.826%) and Precision (0.835%) are equally as high, this demonstrates the classifier is selecting relevant instances which are also accurate to the class. The Malware class also had a high number of classifications; however, the Recall is higher than the than Precision, this a significant portion of instances classified as Malware belong to another class. Similar to Targeted Attack, Social Bot shows a higher Precision (0.706%) than Recall (0.444%), however, as the Recall is significantly lower, this demonstrates the classifier is not recalling Social Bot instances.

Table 10 KNN accuracy in predicting

Type of prediction	KNN accuracy rate
Attack	67%

3.6   Neural Networks

Neural Networks, which was discussed in Sect. 1.1.5, will be applied to the dataset to train the data in predicting the type of attack used during a cyber-attack. tenfold cross validation will be applied.

3.6.1   Prediction of Attack by NN

Figure 13 demonstrates the NN classifier correctly classified 66.8% of instances, and incorrectly classified 33.1% of instances based on the type of attack.

Figure 14 demonstrates the NN classifier yielded a TP rate of 0.669%, a FP rate of 0.212% and a Recall of 0.669%.

The M class yields a TP rate of 0.897%, a FP rate of 0.403%, a Precision of 0.674%, a Recall of 0.897% and an F-Measure of 0.770%. The Recall is higher than the Precision, this means the classifier is selecting instances it believes to be relevant to the class, however, as the Precision is also high, this is a strong indicator that the majority of the instances recalled are accurate to the M class. The classifier accurately classified 768 instances of M. Out of all of the class, the AH class had the highest number of M classifications. This indicates M and AH share similar patterns which lead the classifier to believe instances of M are AH.

```
Correctly Classified Instances      1188            66.8542 %
Incorrectly Classified Instances     589            33.1458 %
Kappa statistic                        0.481
Mean absolute error                    0.098
Root mean squared error                0.2402
Relative absolute error               62.6439 %
Root relative squared error           85.9269 %
Total Number of Instances           1777
```

Fig. 13 NN summary by attack

	TP Rate	FP Rate	Precision	Recall	F-Measure	MCC	ROC Area	PRC Area	Class
	0.330	0.023	0.486	0.330	0.393	0.369	0.823	0.401	DDOS
	0.808	0.040	0.827	0.808	0.818	0.775	0.917	0.819	TA
	0.897	0.403	0.674	0.897	0.770	0.515	0.794	0.712	M
	0.227	0.015	0.500	0.227	0.313	0.309	0.752	0.231	INJ
	0.279	0.054	0.447	0.279	0.344	0.277	0.685	0.309	AH
	0.000	0.002	0.000	0.000	0.000	-0.005	0.666	0.042	BF
	0.000	0.002	0.000	0.000	0.000	-0.008	0.568	0.042	PH
	0.519	0.005	0.636	0.519	0.571	0.569	0.885	0.507	SB
	0.000	0.000	?	0.000	?	?	0.511	0.003	MITM
Weighted Avg.	0.669	0.212	?	0.669	?	?	0.793	0.592	

Fig. 14 NN attack accuracy by class

3.6.2 Discussion and Interpretation

Table 11 demonstrates the accuracy of the Neural Network classifier:

The Neural Network classifier accurately classified the type of attack 66.8% of the time. The classifier performs the best when predicting Targeted Attacks, the Precision (0.827%) and Recall (0.808%) for this class are equally as high, this means the instances recalled are relevant and accurate. However, the classifier believes a high number of instances to be Malware, the Malware class has a Recall of 0.897% and a Precision of 0.674%. This means the classifier is recalling and misclassifying instances which do not belong to the Malware class. Unlike Targeted Attack were the Precision and Recall are equally as high, Injection shows a higher Precision (0.500%) than Recall (0.227%), as the Recall is significantly lower, this demonstrates the classifier is not recalling Injection instances.

3.7 Summary

In this section, data analysis was carried out. Support Vector Machine, Random Forest, Naïve Bayes, K-Nearest Neighbour and Neural Networks were applied to the data and various performance metrics were discussed and analysed. For each algorithm, a discussion was carried out discussing the key points made from the analysis.

4 Model Comparison

4.1 Introduction

In this section, the different algorithms are compared together in order to assess which one performs the greatest in predicting the type of attack.

Table 11 NN accuracy in predicting

Type of prediction	NN accuracy rate
Attack	66.8%

4.2 Comparison of Models

5 different models were trained for predicting the type of attack used in a cyber-attack, these models are: Support Vector Machine, Random Forest, Naïve Bayes, K-Nearest Neighbour and Neural Networks.

Out of all the models trained, the Random Forest classifier demonstrated the highest accuracy in terms of TP (0.675%) and FP (0.216%). Instances were classified correctly 67.4% of the time. The classifier successfully classified instances correctly into each class, aside from MITM. In second place, K-Nearest Neighbour demonstrated a lower TP rate (0.672%) and a higher FP rate (0.222%). However, instances were classified correctly 67.2% of the time. This is 0.2% less than the Random Forest Classifier. The RF model had less misclassifications than KNN and was more successful in classifying Social Bots, Injection and Account Hijacking instances than KNN.

Compared to Random Forest, Support Vector Machine had significantly more misclassifications, the only classes to have correct classifications were DDOS, Targeted Attack and Malware instances. The FP rate was higher, 0.297%, when compared to Random Forest's FP rate of 0.216% and the TP rate was lower, 0.633%. Instances were classified correctly 63.3% of the time, 4.1% less than Random Forest. Neural Networks and Naïve Bayes both had a lower TP rate. Compared to Random Forest, Neural Networks and Naïve Bayes demonstrated a higher number of DDOS misclassifications.

Due to the RF model demonstrating a higher TP rate, a lower FP rate and a higher accuracy in classifying instances correctly when compared against the other models, Random Forest was chosen as the most suitable algorithm for the final model (Table 12).

As shown in the table above, Random Forest demonstrated the highest accuracy in predicting the type of attack. Therefore, the model will be built using the Random Forest classification algorithm. It is possible Random Forest yielded the highest accuracy rate due to Random Forest being an ensemble model. The Random Forest model creates trees on the subset of the data and combines the output from all the trees. This reduces overfitting and reduces the variance, subsequently, this improves the accuracy of the classifier.

Table 12 Comparison of model accuracy rate

Algorithm	Accuracy rate (%)
RF	67.4
KNN	67.2
NN	66.8
NB	64.3
SVM	63.3

4.3 Summary

Based on the model comparison, Random Forest performed the best and was chosen for the final model. In the next section, in order to evaluate how successful the model is, validation data will be applied which contains unseen records of cyber-attacks which occurred during January and May 2020 during the COVID-19 pandemic.

5 Discussion

5.1 COVID 19 Data Statistics

5.1.1 Cyber Attacks by Attack Type during COVID 19

Since the COVID 19 pandemic began in January, Malware attacks have accounted for 54.54% of cyber-attacks, this is an increase of 11.48% from the end of 2019, prior to the pandemic. Similarly, Account Hijacking attacks have increased by 3%. It is likely the increase in cyber-attacks is due to the increase in remote working since the pandemic began (Fig. 15).

For example, the coronavirus pandemic has forced organisations to roll out new technologies at an unprecedented rate. Employees who have been working from home have been using virtual private networks which lack the necessary protections, this opens up a point of entry for a potential attacker.

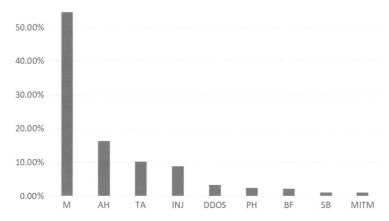

Fig. 15 Cyber attacks by type of attack during COVID19

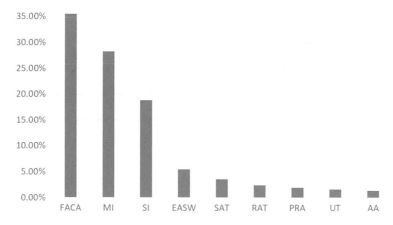

Fig. 16 Cyber attacks by target class during COVID19

5.1.2 Cyber Attacks by Target Class during COVID 19

The Financial and Communication Activities industry suffered more cyber-attacks than any other industry, accounting for 35.5% of all cyber-attacks. This is an increase of 24.4% from the end of 2019 (Fig. 16).

From February to April, banks have seen an increase in cyber-attacks by 238%. It is likely attackers are exploiting the situation of the pandemic, for example, the excessive demand for goods, the anxiety of the general population and the fact a high proportion of the population are working from home. Alongside this, attackers have been using coronavirus related information to harvest data and compromise personal data. As there is a large amount of misinformation spreading regarding the coronavirus pandemic, this means the general public are more likely to believe a phishing scam.

5.1.3 Cyber Attacks by Attack Class during COVID 19

Cyber Criminals were responsible for 89% of cyber-attacks which took place during the COVID 19 pandemic. This is an increase of 13.3% since 2019, prior to the pandemic, demonstrating that cyber-attacks undertaken by cyber criminals have significantly risen. Subsequently, cyber-attacks by the threat actor group, Cyber Espionage, have decreased by 9.5%, Cyber Warfare has also decreased by 2% and finally, Hacktivism has seen a decrease of 1.75%. However, the rate of Hacktivist attacks has fluctuated over the years. For example, in 2015 Hacktivist attacks saw a decrease of 95% and have been decreasing ever since, in 2017, IBM reported 5 incidents, two in 2018 and none in the first few month of 2019 [6] (Fig. 17).

It is possible that the decrease in Hacktivist attacks is partly due to groups such as Anonymous fading from mainstream view. Alongside this, organisations are more

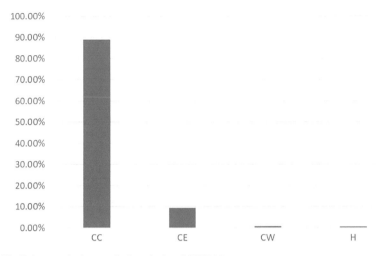

Fig. 17 Cyber attacks by attack class during COVID19

aware of Hacktivism than they were a decade ago and have the adequate protections in place to protect against attacks such as DDOS.

5.1.4 Cyber Attacks by Country During COVID 19

Based on the data, no single country has significantly been affected more than others during the COVID 19 pandemic. 73.5% of cyber-attacks affected multiple countries during one attack (Fig 18).

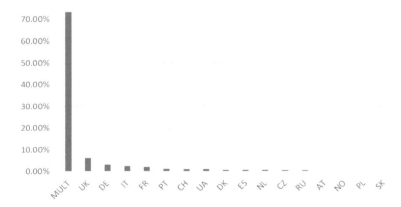

Fig. 18 Cyber attacks by country during COVID19

5.2 Evaluation of Model

To evaluate the effectiveness of the chosen model, a validation dataset is applied, the purpose of this is to see how the model performs in predicting the type of attack on unknown data. The validation dataset includes cyber-attacks which occurred during the COVID-19 pandemic between January and May 2020. One of the main differences between the validation set and the training set is the number of instances, the validation set has 35 instances and the training set has 1777. The likely reason for this difference is the time span of the data collected, for the training dataset, the data spans from 2017 to 2019, whereas the validation set spans from January to May 2020. This was taken into account when analysing the results of the validation set.

5.2.1 Validation of the Type of Attack Predictive Model

In this stage, the Type of Attack will be evaluated by applying the validation dataset to the training set. In the validation dataset, there are 35 instances where the attacks are known.

Figure 19 shows the results of the Random Forest model when the validation dataset is applied. The Random Forest correctly classified instances 62.8% of the time and incorrectly classified instances 37.1% of the time. These figures alone do not provide a clear indicator of the accuracy of the model,therefore, the results need to be analysed further.

The model demonstrated an overall TP rate of 0.629%, a FP of 0.349% and a Recall of 0.629%. The overall results of the model do not demonstrate a reliable and

```
Correctly Classified Instances          22              62.8571 %
Incorrectly Classified Instances        13              37.1429 %
Kappa statistic                          0.3309
Mean absolute error                      0.1166
Root mean squared error                  0.2499
Relative absolute error                 77.8484 %
Root relative squared error             93.5289 %
Total Number of Instances               35

=== Detailed Accuracy By Class ===
```

	TP Rate	FP Rate	Precision	Recall	F-Measure	MCC	ROC Area	PRC Area	Class
	0.000	0.029	0.000	0.000	0.000	-0.029	0.765	0.071	DDOS
	0.714	0.000	1.000	0.714	0.833	0.816	0.893	0.887	TA
	0.800	0.600	0.640	0.800	0.711	0.219	0.603	0.641	M
	0.333	0.031	0.500	0.333	0.400	0.364	0.885	0.487	INJ
	0.000	0.029	0.000	0.000	0.000	-0.029	0.706	0.063	AH
	?	0.000	?	?	?	?	?	?	BF
	0.000	0.030	0.000	0.000	0.000	-0.042	0.394	0.062	PH
	0.000	0.000	?	0.000	?	?	0.853	0.091	SB
	?	0.000	?	?	?	?	?	?	MITM
Weighted Avg.	0.629	0.349	?	0.629	?	?	0.688	0.596	

Fig. 19 Validation of type of attack predictive model

strong model, however, the individual results based on the classes provide a clearer picture.

The Targeted Attack class has the highest portion of correctly classified cyber threats, this means the model will be more accurate in predicting Targeted Attacks. Based upon the Precision, the instances recalled and classified were accurate to the class. The second highest class is Malware; however, the high FP rate suggests the model is selecting instances as Malware and classifying them as such when they belong to another class, another indicator this is the case is the higher Recall over Precision. The third highest class is Injection with a significantly lower TP rate and FP rate than Malware and Targeted Attack. However, TP/FP rate by themselves does not provide a strong indication of accuracy in terms of prediction, the Precision and Recall needs to be taken into account. For example, the Injection class has a lower TP rate, however, when the model does classify instances into the Injection class, the majority of them are correct. This is shown by the Precision being higher than the Recall.

Figure 20 shows the comparison of TP, FP and Precision for each of the different classes. The model failed to classify any instances into the Brute Force, Phishing, Social Bot and Man in the Middle classes.

Figure 21 shows the comparison of Precision and Recall for the different classes. Targeted Attack and Injection are the only classes to have a higher Precision than Recall. The Precision was not calculated for the DDOS class or the overall weight of average for the model.

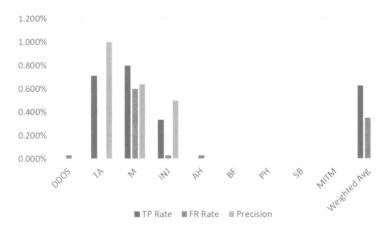

Fig. 20 TP, FP, and precision for prediction of type of attack

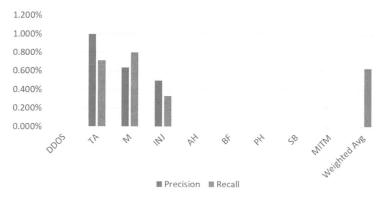

Fig. 21 Precision and recall for prediction of type of attack

5.3 Summary

In this section, the model was evaluated by applying validation data to the final model. Alongside this, variable importance was investigated to discover which variable had the largest impact on the model and the statistics for the COVID 19 validation data were discussed.

6 Conclusion

6.1 Contribution to Knowledge

Over the years, technology has been advancing at an alarming rate, alongside this, the complexity of cyber-attacks has also been progressing beyond comprehension. This means there is an increased urgency for specialists to evolve their methods in how they detect, prevent and manage cyber-attacks. Data mining is a growing technology and it is being frequently utilised in multiple sectors, such as fraud detection in the banking sector, email filtering and marketing. In recent years, data mining has been used to predict trends in cybercrimes and current or past cyber-attacks.

We have developed a predictive model based on data prior to Covid-19 pandemic, the model was applied to a COVID 19 validation dataset, this is data the model had never seen before. Based on the results of the validation data set, it was evident the model detected Malware more than any other class. However, the model demonstrated a high false positive rate when detecting Malware with the validation data. It is likely this is because there is some factor within the Malware class which makes the model inclined to overclassify instances as Malware.

Since the beginning of the COVID 19 pandemic in January, there has been a noticeable increase in cybercrime. Based on the COVID 19 data from January to

May, there is a growing trend in Malware attacks, more than half of cyber-attacks since COVID 19 began were Malware attacks, accounting for 54.5% of all cyber-attacks. Prior to the pandemic, between 2017 to 2019, Malware attacks accounted for 43.06% of cyber-attacks. As mentioned previously, enterprise ransomware is on the rise due to the increasing number of ransomware authors and the fact the attackers know there is a good chance the organisation will pay the ransom.

However, since COVID 19, attackers have been using coronavirus misinformation as a lure, such as financial scams offering financial assistance and free downloads for technology solutions which are in high demand. It is likely that due to the financial desperation COVID 19 has caused, ransomware scams have a higher likelihood of succeeding. Attackers are taking advantage of the general populations existing anxiety which was created by the pandemic and exploiting it for financial gain. Alongside this, remote working has also increased the risk of a successful ransomware attack, this is likely due to a combination of factors, such as weaker security controls and phishing scams via email.

Another observation made on the COVID 19 data is the attackers change in focus on industry. Prior the pandemic, Education and Social Work, such as universities, were more affected than any other industry, accounting for 30.98% of all attacks. However, during COVID 19, this reduced drastically to 6.3%. The highest affected industry during the first three months of the COVID 19 pandemic was Financial and Communication Activities, such as banks, accounting for 35.5% of all cyber-attacks.

Overall, using the model with the COVID 19 data, the model was successful in detecting Targeted Attacks, Malware and Injection attacks based on the high Recall and Precision rates. However, the model was unsuccessful in detecting any instances of Account Hijacking, DDOS, Phishing, Social Bot and Brute Forcing attacks. This demonstrates the model is effective in detecting certain types of attacks during the COVID 19 pandemic.

6.2 Limitations of Study

The limitations in this investigation are related to the data used. As mentioned previously, the data was collected from the Hackmageddon blog. One of the issues with this is the data came with a significant amount of irrelevant data. During the data cleansing phase of this investigation, irrelevant records were removed from the dataset and the various attack categories were reduced. The reduction of the categories was done manually and was also time consuming, this was because this was done manually. Another issue with the data is it is possible there are records stated to be a Malware attack, whereas in actuality they belong to another class. Occasionally, when the description of the attack wasn't very clear and not much information was provided, an educated guess had to be given. Due to this and the data being unbalanced as the dataset contained significantly more Malware attacks than any other class, it is likely this has added to the disproportionate amount of records identified as Malware.

6.3   Future Work

To improve and extend this research further, the following can be accomplished:

1. As mentioned in Sect. 6.2 Limitations of Study, the dataset used was unbalanced. There were a higher number of Malware instances compared to the other classes, the dataset contained 1777 records of cyber-attacks and 856 of these were Malware instances. In order to provide an accurate interpretation across the classes, the number of records in each class would need to be balanced. Alongside this, this study can be improved by using more than one attribute when predicting cyber-attacks, therefore, rather than the framework only predicting the type of attack which occurred, the framework could predict which country it's most likely to occur in.
2. An additional attribute could be used alongside the type of attack, such as the target class or country. This would allow for trends to be analysed and preventative measures put in place in specific areas, for example, which industry is more affected in France by Malware attacks compared to the United Kingdom?
3. The Target Class attribute could be used as the primary feature alongside the type of attack or country to discover which sector is affected the most by specific types of cyber attacks during the COVID-19 pandemic in the UK compared to other countries in Europe, such as Germany or France.
4. Data can be collected and used from client machines to detect whether certain attacks are more likely to occur on a wireless or wired network, or if the machine is using a remote connection.

In this investigation as per the aim and objectives, classification techniques were used. In future research, other data mining techniques can be applied alongside classification, such as clustering. K Means can be applied alongside Random Forest, K Means would classify and assign observations in the dataset into multiple groups, such as attack type or country, based on their similarities. Once the data has been separated into multiple clusters, labels will be applied based on what the clusters are, for example, a cluster representing the different types of malware. By using clustering algorithms alongside classification, clustering detects patterns within the data and assigns them into clusters based on the similarity, once multiple clusters have been created, classification is used to label and classify the data. Due to there being multiple clusters, it is likely more accurate results will be produced.

References

1. BBC (2020) Coronavirus: cyber-attacks hit hospital construction companies. Retrieved from BBC News. https://www.bbc.co.uk/news/technology-52646808
2. Ben Abdel Ouahab Ikram BM (2018) Machine learning application for malwares classification using visualization techniques
3. Bilge L (2017) RiskTeller: predicting the risk of cyber incidents

4. Carfagno D (2019) Why is higher education the target for cyber attacks? Retrieved from Cyber Shark. https://www.blackstratus.com/why-is-higher-education-the-target-for-cyber-attacks/
5. Ch R (2020) Computational system to classify cyber crime offenses using machine learning
6. Cimpanu C (2019) Hacktivist attacks dropped by 95% since 2015. Retrieved from ZDNet. https://www.zdnet.com/article/hacktivist-attacks-dropped-by-95-since-2015/
7. Coble S (2020). Ransomware payments on the rise. Retrieved from InfoSecurity. https://www.infosecurity-magazine.com/news/rise-in-ransomware-payments/#:~:text=%22This%20rise%20is%20arguably%20fueled,up%20to%2058%25%20in%202019.
8. Department for Digital Culture Media & Sport Official Statistics (2020) Cyber security breaches survey 2020
9. Donghoon K (2017) Detection of DDoS attack on the client side using support vector machine
10. Exsilio Solutions (2016) Accuracy, precision, recall & F1 score: interpretation of performance measures. Retrieved from Exsilio
11. Farnaaz N (2016) Random forest modeling for network intrusion detection system
12. Gewirtz D (2020) COVID cybercrime: 10 disturbing statistics to keep you awake tonight. Retrieved from ZDNet. https://www.zdnet.com/article/ten-disturbing-coronavirus-related-cybercrime-statistics-to-keep-you-awake-tonight/
13. Ghanem K (2017) Support vector machine for network intrusion and cyber-attack detection
14. GOV UK Department for Digital CM (2019) Cyber security breaches survey 2019
15. Kravchik M (2018) Detecting cyberattacks in industrial control systems using convolutional neural networks
16. Loader J (2020) Sheffield Hallam University confirms data breach following cyber attack. Retrieved from The Sheffield Tab. https://thetab.com/uk/sheffield/2020/07/30/sheffield-hallam-university-confirms-data-breach-following-cyber-attack-44944
17. Oskoui R (2017) The 5 industries most vulnerable to cyber-attacks. Retrieved from CDNetworks. https://www.cdnetworks.com/cloud-security-blog/the-5-industries-most-vulnerable-to-cyber-attacks/
18. Panda SK (2020) Top five sectors prone to cyber threat amid COVID-19 lockdown. Retrieved from Entrepreneur. https://www.entrepreneur.com/article/350502
19. Posey B (2019) Why enterprise ransomware attacks are on the rise. Retrieved from ITProToday. https://www.itprotoday.com/security/why-enterprise-ransomware-attacks-are-rise
20. Prajakta Yerpude VG (2017) Predictive modelling of crime dataset using data mining
21. Sarker IH (2020) IntruDTree: a machine learning based cyber
22. Saxena S (2018) Precision vs recall. Retrieved from Towards Data Science. https://towardsdatascience.com/precision-vs-recall-386cf9f89488
23. Scroxton A (2020) Almost half of UK businesses suffered a cyber attack in past year. Retrieved from ComputerWeekly. https://www.computerweekly.com/news/252480582/Almost-half-of-UK-businesses-suffered-a-cyber-attack-in-past-year
24. Tereikovskyi I (2017) Deep neural networks in cyber attack detection systems
25. Zareapoor M (2015) Application of credit card fraud detection: based on bagging ensemble classifier

Missed Opportunities in Digital Investigation

Pat Thompson and Mark Manning ⓘ

Abstract A recent strategic review of policing published by the Police Foundation (Barber in The first report of the strategic review on policing in England and Wales. Police Foundation (2020) [7]) claimed that the police service in England and Wales is not equipped to meet the scale and complexity of the various challenges it faces, one of which involves the digital elements within crime investigation. Drawing upon data gathered for an MSc dissertation evaluating practices across investigators in the South of England, monthly samples from two years of serious crime investigations established that 50% of enquiries missed all digital investigative opportunities. Where a digital opportunity was identified, potential subsequent digital enquiries were missed 47% of the time. Whilst consistent with the Her Majesty's Inspectorate of Constabulary and Fire and Rescue Services (HMICFRS) (State of Policing—The Annual Assessment of Policing in England and Wales (2018) [44]) and the Information Commissioners Office (ICO) (Mobile phone data extraction by police forces in England and Wales (2018) [55]) reports which highlight that policing capability is lagging behind modern technology and affecting public confidence; these matters will be developed and discussed leading to the conclusion that, consistent with the police foundation report, loss of public confidence will undoubtedly damage police legitimacy.

Keywords Digital · Investigative · Missed · Opportunities · Police · Legitimacy · Public · Confidence

1 Introduction

The context in which this paper will develop recognizes the rapid evolution and prevalence of complex digital elements within criminal investigations and how they

P. Thompson · M. Manning (✉)
University of Suffolk, Ipswich IP4 1 QJ, UK
e-mail: Mark.Manning.1@uos.ac.uk

P. Thompson
e-mail: P.Thompson5@uos.ac.uk

may be stretching both the legitimate boundaries of the rule of law across state boundaries and also, policing by consent [67]. It will become clear that an investigative landscape fraught with technical and procedural complexities has evolved leading to deficiencies in the police use of digital investigation opportunities within police investigations. Previous research has identified the varying guises of these complications which range from the ever-shifting developments in digital technologies affecting policing [77, 98]. Shaw [94] and Dodd [28] to the complications experienced within a Criminal Justice System anchored within historical precedence [31], and the increasingly informed expectations held by a technically educated public who simultaneously demand digital competence and digital privacy from law enforcement [30, 107].

Whilst previous research has identified these layers of complexity, a research gap exists in establishing how these complexities manifest themselves within 'practice'. The study which informed this conference paper reviewed the use of four defined digital investigative techniques and examined whether investigators were identifying digital opportunities, applying them correctly and documenting their use so as to withstand judicial review and ultimately, public scrutiny within a criminal court. The findings indicated that a knowledge gap exists across investigators in identifying suitable lines of digital investigation. Where digital lines of enquiry were identified, questions have been raised as to the effectiveness of their application. The findings also established a lack of documented rationale being applied to the investigative decisions surrounding digital investigative opportunities.

2   The Field of Digital Policing

Changes to investigation procedures [69], digital evidence handling [104], types of crime [60] and the expectations placed upon investigators [43] are just some examples of the increased investigative arena in which police investigations form part of a wider criminal justice service made up of multiple agencies and processes [78]. Complexity across these wider systems and processes compound upon a clear understanding of emerging technologies [23, 62]. These complexities also straddle a landscape of differing policing priorities across the UK set by directly elected local Policing Crime Commissioners who drive policing policy [87].

Brown [19] comments upon the complexity of issues within digital forensics citing advances in technology, disparate legal jurisdictions and a list of agencies requiring knowledge of policing processes. However, digital forensics equates to only one aspect of investigations. Further examples include issues around the under-reporting of digital criminality [108] and the implications of technology on the legal profession and its functioning within courts [110]. These examples paint a picture of technology, investigating offences and processing investigative results through the criminal justice system as being fraught with complexity. This is highlighted by Owen [82] who notes that technology outpaces policing knowledge whilst police knowledge simultaneously outpaces legislative developments. McQuade [70, p. 41] also refers to policing tactics as co-evolving with technology and criminal ingenuity, and becoming increasingly complex, via '…recurring criminal and police innovation cycles which have a ratcheting-up effect akin to a civilian arms race'. The concept of a technological arms race is not unique to law enforcement. Hofman [45] identifies the complexities of emerging technology in medicine whilst Taddeo and Floridi [103] explore the implications of artificial intelligence in warfare. Despite these "arms races" occurring across multiple sectors, perhaps unique to policing is the difficulty of matching tactics to offending and then presenting the results within a legal system rooted in historical precedent and tradition [31] and to partner agencies who may not have technological parity with the police [73].

Further technical and cultural complexities are found within society at large and the expectations and objections that may be held by the public surrounding the use of digital tactics within investigations. A contemporary example of this is offered by Spinello [101] who discusses the range of opinions surrounding the Federal Bureau of Investigations (FBI) attempts to "crack" the iPhones belonging to the 2015 San Bernadino terrorists [18] and the contrasting expectation of privacy for citizens versus the expectation to investigate terror offences. Set upon the backdrop of whistle blower Edward Snowden's revelations of widescale state sponsored surveillance style programs [91], the public have developed a complex relationship with technology and information gathering whereby large swathes of the public willingly submit huge parts of their private lives to exposure on social media [39], yet the trust held by the same public for "big tech" and government in using this data is low [46]. In the UK, organizations such as Liberty and Big Brother Watch also provide commentary against advances in law enforcement's use of technology [12, 63] whilst contrasting populist opinion suggests "If you have done nothing wrong, you have nothing to hide!" [68].

These polarizing opinions are set within a context of increasingly available sources of information, which itself causes further complexity. Baum and Potter [8] describe a disconnect between the vast availability of digital information and the levels of quality of the information available to the public. The voluminous availability of "fake news" has made the separation of fact from fiction in directing public opinion increasingly difficult [52]. When this phenomenon is applied to opinions surrounding policing and the criminal justice sector; news, fake news and opinion all add to an ever more complicated picture [84, 27]. The analysis of the expectations of digital tactic use within policing can be expanded further to explore the concept of "policing with consent" and whether this historical bedrock upon which the UK police have set their operational foundations is compatible with the pace and intrusive abilities of technology within investigations. Robertson [89] describes policing as being consented to by the public or, at least acquiesced to, due to a lack of understanding about what policing involves. The Home Office [47] attempts to codify the concept of policing by consent by providing a list of generalized aspirational police behaviors and summarizing the concept as "the power of the police coming from the common consent of the public, as opposed to the power of the state". Curiously however the Home Office [48] declare that it is not possible for a citizen to remove their consent.

These are all matters to be considered by the police service as they extend their reach into private matters through their use of technology. An example of this follows a recent trial of live facial recognition cameras by the London Metropolitan Police highlighted by Fussey and Murray [36] who discuss incorrect facial identifications and the implications of this on public confidence into the police when they comment that its usage. The ethical implications of using technology in the deployment of police resources based on algorithmic decisions is one that has implications on the model of policing by consent. Whilst the accepted model of policing by consent has held sway, Shearing and Stenning [95] previously posed long standing concerns about the emergence of private/public partnerships in law enforcement and how these mergers could affect the concept of consent.

Bayley [9] considers the tradition of policing by consent by questioning whether the concept of consent requires a review of the levels of collaboration between the police and the public. Bayley [9] posits, however, that this collaborative approach throws up its own complexities where the interests of the public, the police and public or private sector partners are all accounted for within a legal and operational model. These complexities are articulated by Sheptycki [96] who considers that the "technopoly" mergers between police and private sector partners in tackling criminality through the use of technology is drifting into a pseudo-militaristic field and that this is increasingly '…the very opposite of democratic policing when an uncomprehending public experiences a police presence that they do not endorse'. Whilst debate continues around the use of facial recognition technology by the UK police, Couchman [25] highlights concerns regarding mass surveillance aided by questions as to who owns recorded facial "data", whilst Garvie and Frankle [37] highlight issues of algorithmic racial discrimination.

It could be argued that public sector policing requires technical private sector collaborative support to provide the infrastructure and capability to keep pace with modern criminality, yet this opens a chasm between traditional notions of policing by consent and the concept of "technopoly" (an assumption that technology is always positive and of value); [92]. This shifting of the established notion of policing by consent invites polarized opinions from a more widely informed populace. In 2012, Bratton and Tumin [17] described policing in Los Angeles as having been "run into the ground" with cuts to resources and equipment. Parallels to the recent period of austerity and the effects thereof on UK policing can be drawn [67]. In a contemporary publication, the Police Foundation ([105] p. 54) highlights the current concerns around public consent and policing stating that: in recent years, a tension has emerged between the shifting focus of policing and the views of the public. With police budgets and officer numbers cut, and the balance of risk shifting from public spaces and volume crime to online threats and hidden harm, many aspects of public facing 'core' policing have effectively been de-prioritized. As a result, concerns have begun to emerge about the health of the police "covenant" with the public.

The evidence would suggest that it is difficult, however, to reconcile an absolute operational necessity to deal with increasingly inventive criminality and persistently growing personal data sets with a public who demand, from a position of education and information availability, individual attention and accountability. Having accepted that law enforcement responses to evolving criminality and the handling of digital investigative products have become more complex, it is less clear how the digitization of criminality sits within the purview of police capability.

3   The Investigation of Digital Crime

Some of the more well-known digital tactics used by investigators include, but are not limited to, the use of Automatic Number Plate Recognition [49], the use of communications data [50], the forensic analysis of digital devices [109] and the use of online research [102].

3.1   *Open Source*

Online research, often referred to as open source investigation, is a valuable tool when investigating serious crime [2, 107], though the use of open source tactics has resulted in complications in the development of policy and process at both local and national levels [3, 79].

Complications include the necessity of policing agencies to adhere to legislation that was not written with the pace of the online developments in mind [32]. Complications are also created in considering how to balance the expectations of the public who are able to complete their own open source investigations, yet, who simultaneously demand privacy from law enforcement's use of open source methodologies [30, 107]. Indeed, publicly sourced investigations via social networks have been lauded by society at large yet deemed worrisome by the police following the use of open source tools linked to vigilante behavior [6]. The ability of the public to investigate effectively via open source methodologies is both a blessing and a curse for law enforcement [53] who point to the lack of legal accountability in public investigations and a culture of skepticism within law enforcement as to the perceived benefits of public resources assisting in investigations.

The use of open source investigative methodologies has come under scrutiny by privacy campaigners. Ramakrishnan et al. [88] reported on the use of social media aggregators in predicting civil unrest across Latin America. More recently the police use of social media to monitor protestors during the Extinction Rebellion protests has been called into question by Blowe [14]. The narrative against the monitoring and recording of data by authorities across social networks is echoed by the #FREESPEACHONLINE campaign hosted by Big Brother Watch [13] who retain a stance of social media being the new "public forum" for discussion and therefore being protected from mass state snooping.

Whilst the police use of social media and open source research is limited to the prevention and detection of crime, perhaps cynicism around the leveraging of mass data sets by authorities has been fueled by recent exposés such as that around

Cambridge Analytica [21, 57]. Recent commentary around data sharing between large technology companies and police agencies has also reinforced this nervousness with one recent article by Yoannou [115] suggesting that Amazon was passing user details of "Ring" home camera systems to law enforcement along with remote access powers to allow the well intentioned switching on of cameras by police in the event of crime occurring in the locality of the device.

This cynicism about public/private partnership in harnessing and using social media in investigations can be balanced with a reported rise in complaints by victims to the police stating they have been abused in a social media space [33] and the accepted parallel of abuse and offending increasing across social media when high profile events such as terror attacks occur and are reported in the news [75]. Seemingly law enforcement is expected by victims and the public to access social media during investigations and yet, the same victims and public are simultaneously cynical and suspicious of law enforcement engagement with social media companies and private sector partner organizations.

3.2 Digital Forensics

The retrieval of evidence from digital devices, known as digital forensics, is defined by Pollitt [85] as a process to identify, preserve, analyze, and present data from digital devices. As digital devices evolve, the extraction of evidential data has become more complex as noted by Leong [61] and Agarwal et al. [1] who highlight issues between the changing technological developments in digital devices and the separation of understanding between those who have become increasingly technical in the abstraction of information and those who have the responsibility of applying legal frameworks to any recovered evidence.

This is highlighted by Mackie et al. [65] who reference the skills of digital forensics teams versus the arrival of the European Union's General Data Protection Regulations (GDPR) and the complications of bringing the two perspectives into alignment. Whilst this paper by Mackie et al [65] was largely based on the premise of data breach offences within a corporate environment, a parallel can be drawn to advances in cloud technology, the emergence of new legislation [56], and the complications of legally and technically retrieving cloud data, analyzing it correctly and presenting it in evidence. Developments in cloud technology has resulted in the holding of data on servers potentially anywhere in the world. Son et al. [100] reference cloud providers placing cloud servers across the globe to allow for balancing of resources against global demand. Pătraşcu and Patriciu [83] comment on the difficulties in retrieving data held in disparate locations and Ferguson et al. [34] discuss that this global infrastructure alone provides "severe implications for the detection, investigation and prosecution" of offences.

These "severe implications" are compounded further by considerations surrounding the proportionality of seizure and analysis of devices [10, 55, 106] (hereafter ICO), the intrusion into data held on devices about persons not believed to be linked to criminality [12] and the sheer volumes of data that needs to be stored by the police in the management of digital exhibits [86].

Certainly, issues surrounding the volumes of data recovered by police and the proportionate and relevant analysis of this data in crime investigations has led to adverse publicity, including data implications involving the collapse of the Liam Allan rape trial [99]. The subsequent review of disclosure processes and the impact digital evidence has had on investigations and the court process identified unwieldy volumes of material and a lack of management of this data as having a negative impact across the criminal justice system [5]. The proposed solutions have led to more confusion across constabularies as police officers and investigators struggle to ensure understanding across seizure, analysis and presentation of digital evidence. This confusion is equally present within the Crown Prosecution Service and the courts [15]. Criminal prosecutions within the UK rely on a burden of proof measured as "beyond reasonable doubt" [78]. Accordingly, the presentation and subsequent contesting of digital evidence in a court can be complicated without that evidence being corroborated by other means.

3.3 Communications Data

One method of achieving corroboration of some elements of digital evidence is via communications data. Communications data is "the who, where, when and how of a communication" [50]. Whilst accessing and using communications data is a valid investigative tool, its use also affords an investigator the ability to corroborate other aspects of digital evidence. Communications data is largely comprised of records of traffic across the cellular network via mobile or land-line telephony. It can also comprise mobile internet traffic records as well as more esoteric versions of communications including mobile app communications to companies such as "Uber" or "Just Eat". The Investigatory Powers Act 2016 [56] is the statutory framework that defines the circumstances and process by which investigators may apply for communications data. This process involves submitting a request which complies with the requirements of the act. This request is then triaged by a team who pass the request to an independent statutory body, "The Office for Communications Data Authorizations" (OCDA). OCDA's responsibility is to grant or deny the request having considered whether the request is lawful, necessary and proportionate [35, 40]. The proportionate lawfulness of any request for communications data is subjective and requires individual interpretation of the law against the requirements of the request. Whilst OCDA will refer to this as balancing the proportionality of the request, Gill [38] suggests this process involves the consideration of "sousveillance", the considering of public perception as to where the power balance wielded by the state is held into the intrusion of the privacy of citizens.

As the Investigatory Powers Act [56] progressed through its parliamentary approval process, Chivers [24] stated that there was a need for "a discussion about what kind of surveillance is truly necessary and proportionate in an increasingly digitalized society". Whilst legislators advised that the creation of the independent body OCDA would ensure that proportionality would remain in the forefront of communications data acquisition, others believed that the Investigatory Powers Act was a "snoopers charter" [22].

In addition to the discussions surrounding privacy and proportionality, technical complexity exists in accessing and using communications data. Lock et al. [64] reference the problems associated with the recording of communications data when routed through virtual private networks whilst Brown [19] speaks to "advancements in the functionality of information communication technologies and disparities between systems of law globally" as challenging.

Further challenges then exist in the processing of gathered communications data. Issues with the computer systems used throughout the criminal justice service [26, 111] have rendered the effective transfer of data between agencies difficult due to the incompatibility of systems used. Despite these difficulties, communications data remains a critical line of enquiry in many investigations where the linking of communication events and identification of where these events occurred is crucial to successful investigations (May, 2015).

3.4   Automatic Number Plate Recognition (ANPR)

ANPR is described by Gunawan et al. [42, p. 1973] as '…an intelligent system which has the capability to recognize the characters on vehicle number plate'.

This recognition is then overlaid onto a location, mapping where the recognition took place. Rogers and Scally [90] identify both the proactive and reactive abilities of ANPR for the investigator. Rogers and Scally [90] also identify the scale of ANPR use across the UK police establishment commenting that cameras exist on nearly all major road networks. Wright [114] corroborates that as far back as 2012, over eleven billion records of vehicle movements had been captured by the ANPR camera network.

This volume of data brings a layer of complexity to investigations surrounding the storing of images and associated vehicle, keeper and location data for an exponentially increasing data set [59]. Akhgar and Yates [4, p. 155] develop this further stating that in dealing with "big data sets" there is a requirement for '…a specific combination of tools, intel., experts and data sources along with the suitable access protocols and security solutions.'

The pattern of law enforcement difficulty with advances in technology is prevalent across all digital tactics discussed thus far. The other consistent parallel across tactics has been law enforcement's grappling with the concepts of privacy and proportionality and the varied expectations held by an increasingly vocal and digitally aware public. This theme is present across ANPR data capture and manipulation also.

Big Brother Watch [11] comment on the exponentially increasing scale of police use of ANPR data and the possible consequences of not maintaining track of police marking of the data whilst Woods [113] debates the notion that the use of ANPR data acts as an intrusion into the private lives of road users when set against both the European Convention on Human Rights and the European Union Charter.

It is acknowledged that policing has an ever-expanding remit [71, 72]. It is argued that this expansion, alongside Governmental ambitions to move policing from an "unskilled" vocation to being a formal profession [20], has led to a debate as to the overall identity and mission of the police. Amidst this identity crisis, developing measurable and successful process in the already complicated digital arena is difficult. Speaking to private sector management processes and measuring performance, Magretta and Stone [66] questioned; "given our mission, how is our performance going to be defined?". It is unclear how policing can answer this question.

4 Method

Using a County Constabulary within England as a case study, research was conducted which aimed to understand both the practical use of digital investigative tactics within 'serious crime investigations' and whether the proportionate use of such tactics was considered when they were applied.

The term serious crime was used in the sense that it is defined in statute by the Police Act 1997 [76] (Police Act 1997)

The research afforded an opportunity to explore the identified knowledge gap including topics of particular interest including: Identifying the levels of use of digital investigative tactics and establishing where opportunities are commonly missed; where digital tactics are used, establishing the levels of understanding held by investigators into the practical application of digital investigative techniques; establishing the levels of understanding of the proportionate use of digital investigative techniques; within the case study, to what extent are identified digital investigation techniques being used and understood within serious crime investigations?

This research analyzed secondary data from completed investigations as opposed to 'live' investigations. Completed investigations can be interrogated for information whereas the dynamic nature of "live" investigations risks missing the measurement of digital tactic use which may become a valid tactic later in an enquiry and therefore missed by snapshot analysis.

All of the data already existed within the policing systems of the area under research, having been collected as part of a different original purpose. One benefit of using existing data in identifying the use of digital tactics was that the investigative process was not biased by the investigator's awareness of researcher scrutiny. This knowledge may have led to the investigator including tactics that they would previously not have considered and the resulting data not being representative of the 'norm' for participants.

4.1 Data Collection Procedures

The 2019 Crime Survey of England and Wales identified 60,920 crimes in the year ending March 2019 for this particular Constabulary, not including fraud offences [81].

Utilizing the Constabulary's crime recording system, a structured query could be leveraged identifying only those reported offences which resulted in an investigation taking place.

It is acknowledged that not all reported crimes result in investigations [54]. Comparing the extracted investigations against the Home Office list of serious crimes identified which of the extracted investigations fell within the serious crime definition. Having identified a data set of crime investigations, it was necessary to identify a time period from which to collect a sample from the data.

On the understanding that 60,920 crimes were reported in year ending March 2019, it was necessary to establish how many of these resulted in investigations and subsequently, how many resulted in serious crime investigations which would produce a sample set from which to select data for this research.

The following data collection process was established:

(1) All crime investigations between January 2018 and December 2019 inclusive were extracted totaling a sample size of 13,377 investigations were exported from this two-year period. The data was cleansed and just the crime reference numbers, and offence types recorded.

(2) The Home Office list was filtered so that all offences that did not satisfy the statutory condition of potentially carrying a custodial sentence of three years or under were removed.

(3) Several offences that matched the definition of serious crime remained yet would traditionally never fall within the practicalities of serious crime investigation were removed. The rationale behind these removals were that, for example, by definition, a theft offence can carry a seven-year custodial sentence, thereby defining it as "serious". A shoplifting offence however would not be considered a "serious crime" from an investigative perspective and therefore would be outside the scope of this research. This rationale is supported by the Cambridge Harm Index [97] which suggests that not all crimes are "created equal" and that measures of seriousness can be applied to crime recording. For this research, offences have been removed which would all but exclusively be dealt with at Magistrate's court and therefore never attract the serious crime sentencing powers [93].

(4) Having filtered out offences not attracting a custodial sentence of over three years and offences not traditionally within the remit of serious crime investigations, a total of 2,275 investigations remained as in the sample providing a mean number of serious crime offences per month across the twenty-four-month sample period of 94.79.

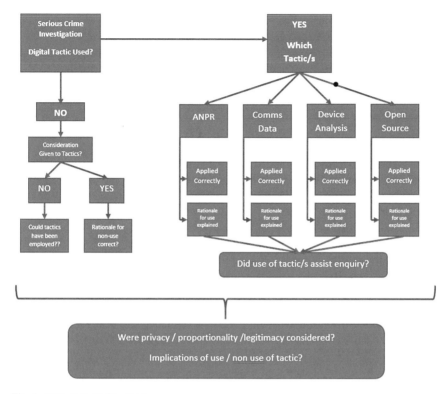

Fig. 1 Data analysis flowchart

A random selection of four investigations per month over the sample period yielded 96 investigations selected, the equivalent to an average month's serious crime investigations.

4.2 Data Analysis

Having identified 96 investigations, an analysis process was developed to capture the relevant quantitative and qualitative data required better understand the approach taken to digital elements within investigations. This analysis process is visualized in Fig. 1.

4.3 Results

Quantitative analysis of the results identifies that

- From the 96 serious crime investigations: 29 investigations did not contain circumstances where digital lines of enquiry were available or viable.
- Of the remaining investigations, 19 correctly identified one or more digital lines of enquiry. 48 investigations missed all possible digital lines of enquiry.
- 53 investigations (55%) were dealt with by officers in non-detective roles, whilst 43 investigations (45%) were dealt with by officers in detective roles.
- Within the 19 investigations where digital tactics were identified, there were instances of multiple tactics being used, with a total of 28 recorded usages of one or more of the digital tactics noted across these 19 investigations.
- Out of the 19 investigations that did apply digital tactics, it was noted that 9 of them (47.3%) missed further digital investigative opportunities having identified an initial opportunity. Of the 48 investigations which missed all digital investigative opportunities, 16 (33.4%) were missed by officers in detective roles whilst 32 (66.6%) were missed by officers in non-detective roles.
- Of the 28 recorded usages of digital tactics, 27 of them appear to have been technically applied correctly. Of the 28 recorded usages of digital tactics, 16 (57%) of them described a rationale for the use.
- Within the 48 investigations where digital enquiries were not used, the rationale for the lack of use was recorded once for each of the identified tactics.
- The legitimacy/proportionality aspects of digital tactic use were explained on 1 occasion each for the use of communications data, device analysis and open source tactics.

Qualitative analysis of the results provides the following insight into the use of (or not) of digital investigative tactics.

- Evidence existed of officers correctly identifying digital lines of enquiry, one example being an investigating officer correctly identifying that a mobile device required a more forensic download than was locally available and another officer identifying that ANPR searches could offer investigative lines of enquiry in over-laying data with a missing persons records system whereby receipts and a car parking ticket were recovered from a vehicle.
- Evidence existed of investigators incorrectly identifying digital lines of enquiry including wrong rationale for the non-utilization of ANPR searches as being given as there being insufficient cameras in an area and a time frame being too wide to provide meaningful search parameters.
- Evidence was present of the use of rationale in the consideration of use (or non-use) of digital investigative tactics including an example whereby an investigator identified issues with the accuracy of information held by the police and that this would therefore prevent the initial use of communications data applications.
- A further example demonstrates a rationale for the seizure of a victim's phone handset, the policing powers used and level of intrusiveness in searches based on quantity of data extracted. The rationale also details an explanation for the seizure and examination of the suspect's device and goes onto identify complications and missing data from these initial reviews which developed a rationale for further digital investigative work on the handsets.

- Evidence identified examples of digital investigative opportunities existing in enquiries which were missed by investigators despite the victim providing the investigator with signposting as to how to conduct relevant digital enquiries. Evidence also existed of digital enquiries not being progressed due to problems around a lack of time or an increased workload.
- These capacity issues were highlighted repeatedly and within a local area, a prioritisation process around which crimes will be dealt with and in which order was instigated, often to the detriment of progressing entirely viable digital lines of enquiry. This was adopted both by investigators and their supervisors.
- Evidence also existed of investigators not maximising the use of digital investigative methods to understand the "wider picture" including an example involving a domestic abuse offence where the investigator did not identify that more than one offence had been committed and a separate stalking report where only a single incident was investigated as opposed to the wider reported behaviours.
- Despite these problems, there also existed evidence of investigative tenacity despite digital complications including an example whereby an investigator has continued to pursue digital lines of enquiry despite initial attempts to glean evidence being frustrated by the technical architecture of the application being reviewed.

5 Discussion

The results presented from this evaluative research established a number of areas of discussion.

5.1 Missed Opportunities

The prevalence of missed digital investigative opportunities clearly identifies an area of weakness across investigations. It is, however, difficult to identify where this weakness specifically stems from.

Some of the sampled offences identified a lack of time or an increased workload as a reason as to why digital opportunities were missed.

This is consistent with recent HMICFRS reporting [44] which highlighted that a lack of capacity across investigators and police officers identified "a widening gap between the needs of the public and the police's capacity and capability to meet them".

Whilst it is accepted that lack of capacity is a partial element in the non-identification of digital lines of enquiry, it is suggested that capacity takes a less prominent role to a lack of understanding of the opportunities available in the investigation of digital elements of crime.

The same HMICFRS report (2018, p. 33), backed up by the recent ICO (2020) report identifies deficiencies in the digital elements of investigations: As long as the police persist in using 20th-century methods to try to cope with 21st-century technology and ways of life, they will continue to fall further and further behind, and the quality of justice will exponentially diminish.

This cultural difficulty in keeping pace with technology can be conceptually linked to the "Dunning-Kruger Effect" [29]; the notion that if an investigator doesn't know what opportunities are available to them in investigating crime, they would be unlikely to identify the opportunities in their investigations. This lack of knowledge, when embedded within a culture that is "using 20th-century methods to try to cope with 21st-century technology", can perhaps explain why there is such a high number of missed opportunities.

In one of the above noted examples, an exasperated victim who has been told that no further enquiries will be completed into the matter despite that victim identifying digital lines of enquiry to the investigating officer. Williams [112] discusses the public's expectation of police to tackle open source enquiries. In the example, the victim identifies communications data opportunities alongside Facebook enquiries, opportunities which have alluded the investigator. This level of education and expectation amongst the public, when faced with non-action by the police in investigating reports of crime, decreases public trust and confidence in the police.

Morrell et al. [74] suggest that one element of maintaining trust between the public and the police is to ensure that the public believe that the police are competent. In this instance, the belief of the victim is that the police are not competent. Jackson et al. [58] suggest that "when people are aligned with their society's legal structures, they are … more likely to assist the police and courts through reporting crimes, identifying culprits, and giving evidence".

The above circumstances identify how missed opportunities amongst an educated public can erode trust in the police. These missed opportunities are believed to be down to a lack of knowledge [51] and the cultural lack of modernity around policing practices [44].

Allocation Policies

Despite the filtering process applied to ensuring that only serious crime made up the research sample, a larger proportion of the sample were investigated by non-detective resources. These investigations included offences such as supplying of class A drugs, sexual assaults, fraud, and stalking offences.

The Cambridge Harm Index [97] acknowledges that not all crimes are created equally and proposes a "weighting" assessment to understand the potential harm which could be experienced from the commission of an offence.

Whilst the sample indicates that the greater number of investigations were allocated to non-detective resources, those that could be said to be of a higher harm weighting (for example rape and offences against vulnerable persons) attracted detective resources.

This can be viewed as a positive given that the sample data indicates that non-detective resources miss more digital investigative opportunities than detective resources. What the allocation process does not seem to identify however, are the potential harmful offences that start at a lower harm threshold yet repeat into a higher harm set of circumstances. The tragic case of Fiona Pilkington in 2007 highlighted that the culmination of multiple lower harm crimes can have devastating effects.

In examples given above, investigators have missed digital enquiries which would have revealed patterns of stalking behaviors or repeated domestic abuse offences, and whilst these are "serious crime" offences that would not necessarily attract detective resources, the implications of missing digital evidence which evidences ongoing domestic abuse and controlling behaviors have been repeatedly highlighted [41, 80]. It is suggested that had efforts been put into the recovery of digital evidence across these examples, the investigator would have been in a better position to risk assess and address relevant future safeguarding actions.

This lack of awareness or ability to identify digital lines of enquiry amongst non-detective resources has the potential to not only miss investigative and safeguarding opportunities, but to also erode at the public's confidence in the police's ability to investigate effectively. Whilst it is difficult to argue against prioritizing high-risk crimes against lower risk fraud offences, it is suggested that by not engaging fully in the pursuit of these crimes, officers do not learn how to develop skills in the digital investigative arena, thus reinforcing the previously referenced "Dunning-Kruger Effect" and simultaneously continuing to erode the confidence of the public into the police's abilities and legitimacy.

5.2 Digital Enquiries Applied Correctly

What is clear from the sample data is that some officers do take a positive approach to digital enquiries and will not be put off by inevitable digital complications that may arise.

The example noted above whereby an investigator continued to pursue digital lines of enquiry despite initial attempts to glean evidence being frustrated by the technical architecture of the application being reviewed demonstrates that with a willingness to pursue and learn, digital enquiries can be productive.

6 Rationale and Legitimacy

Following the collapse of the Liam Allan rape trial in 2017, scrutiny into how investigators apply rationale to their enquiries, especially around elements of digital investigation, has been the subject of much discussion. Smith [99] suggests in relation to the Liam Allen trial that: Whatever the rationale for the approach of the OIC, the material should have been scheduled and by failing to do so the CPS were denied an accurate picture of the case, leading to a flawed charging decision.

The evidence from this research suggests that it is a minority of investigators who are applying a digital rationale within investigations. Given the increasing relevance of digital data in investigations, it is suggested that documenting a rationale detailing the reasons for completing or not completing digital enquiries is as important as the investigation itself, and to miss this could lead increasingly to adverse conclusions not only within the criminal justice system but also affecting the public's wider view of the police.

The lessons from the Liam Allan case, proposed by the Attorney General's Office [5] do not seem to have filtered through to operational investigators. Whilst organizations such as Liberty and Big Brother Watch continue to publicly scrutinize law enforcement's use of technology, it is increasingly important for investigators to document a rationale, reasoning what digital enquiries are available and why they are or are not being pursued. The implications of not doing so may see repeated collapses of prosecutions and increasingly negative public scrutiny, all of which cyclically feeds back into questions around the ability of police to continue to operate with the full consent and support of the public whilst accessing the public's digital data.

This question around policing across digital data and the intrusiveness of the investigative tactics can be framed against Bayley's [9] argument that: 'policing with consent' must be rethought because the public increasingly wants direct rather than representative participation in the supervision of the police. Documenting a rationale around digital investigations requires liaison with the public to ensure that the public's consent and support is maintained. A rape victim, for example, may consent to having phone messages accessed and reviewed by investigators, but not pictures. Notwithstanding the technical complexities around this (for example, is Instagram a messaging application, an image application or a hybrid of both?), it is necessary for an investigator to balance the wishes of the victim against the evidential elements of the enquiry and the potential of scrutiny and cross examination as to investigative decisions at a later judicial hearing. This is a theme noted by the information Commissioner's Office [55] where 13 recommendations were made to constabularies around the balancing of data protections expected by the public and the competing statutory frameworks found within the criminal justice system.

Bradford [16] suggests that "the actions of police officers can have a profound effect on the legitimacy of the police". In the context of digital investigations, it could be argued that this cycle evidences a social positive feedback loop whereby publicly failing prosecutions increase public uncertainty over law enforcement's capability and competence with digital evidence. This in turn increases public scrutiny which in turn uncovers more failings. All these factors serve to augment each other.

7 Conclusions

This research has illustrated the complexity that surrounds police use of digital investigative tactics.

Complexities exist within the technical arena however perhaps more pressing are the complexities around applying the evolving pace of the modern digital world to a historically rooted criminal justice system that bases its judgements and future decisions on that of precedent. Adding to this is the public's increased knowledge of modern-day digital capabilities and the resultant expectations of law enforcement in being able to access data and process it effectively yet to simultaneously maintain privacy and to remain defensible against accusations of "big brother" interventions. These layers of complexity then feed into questions as to how UK policing can ensure that it retains legitimacy with the public and, ultimately, as to whether or not the digital elements of investigative work accord with the long-standing tradition of "policing by consent".

This research has identified that whilst a small number of investigators have an interest and an ability to positively identify and pursue the digital elements of serious crime investigations, a majority of investigators do not. This lack of ability speaks loudly to the questions posed around the legitimacy of the police and the ability of law enforcement to command the confidence of the public and ultimately to police with their consent.

This research has also identified that whilst serious crimes carrying the highest risk are largely allocated to detective resources, what could be termed as "volume-serious crime" is often allocated to non-detective resources, where the largest amount of digital opportunities are missed. As has been highlighted, it is within these cases that tragedies such as that of Fiona Pilkington are found.

This research has highlighted that despite some officers making relevant digital enquiries, the documenting of the rationale behind these enquiries is largely lacking. It is difficult to quantify the lack of a rationale behind a missed digital opportunity—has it been missed through a lack of knowledge or through choice? In either event, the lack of an accompanying rationale can have serious consequences and it exposes issues surrounding the police and their perceived legitimacy when dealing with digital enquiries.

References

1. Agarwal A, Gupta M, Gupta S, Gupta SC (2011) Systematic digital forensic investigation model. Int J Comput Sci Security 5(1):118–131
2. Akhgar B, Bayerl PS, Sampson F (2017) Open source intelligence investigation: from strategy to implementation. Springer
3. Akhgar B, Wells D (2018) Critical success factors for OSINT driven situational awareness. Eur Law Enforce Res Bull 18
4. Akhgar B, Yates S (2013) Strategic intelligence management, national security imperatives and information and communications technologies. Butterworth, Heinemann, US
5. Attorney General's Office. https://assets.publishing.service.gov.uk/government/uploads/system/uploads/attachment_data/file/756436/Attorney_General_s_Disclosure_Review.pdf. Last accessed 4 Mar 2020
6. BBC. https://www.bbc.co.uk/news/uk-england-50302912. Last accessed 28 Nov 2019

7. Barber Sir M (2020) The first report of the strategic review on policing in England and Wales. Police Foundation
8. Baum MA, Potter PB (2019) Media, public opinion, and foreign policy in the age of social media. J Polit 81(2):747–756
9. Bayley DH (2016) The complexities of 21st century policing. Polic J Policy Pract 10(3):163–170
10. Beebe N, Clark J (2005) Dealing with terabyte data sets in digital investigations. In: IFIP international conference on digital forensics. Springer, Boston, MA, pp 3–16
11. Big Brother Watch. https://bigbrotherwatch.org.uk/wp-content/uploads/2013/03/ANPR-Report.pdf. Last accessed 8 Mar 2020
12. Big Brother Watch. https://bigbrotherwatch.org.uk/wp-content/uploads/2017/11/Police-Access-to-Digital-Evidence-1.pdf. Last accessed 17 Feb 2020
13. Big Brother Watch. https://bigbrotherwatch.org.uk/campaigns/freespeechonline/#introduction. Last accessed 28 Feb 2020
14. Blowe K. https://freedomnews.org.uk/police-surveillance-a-note-for-extinction-rebellion-campaigners. Last accessed 28 Feb 2020
15. Bowcott O. https://www.theguardian.com/law/2019/may/01/explosion-in-digital-evidence-has-left-cps-struggling-says-union. Last accessed 4 Mar 2020
16. Bradford B (2014) Policing and social identity: procedural justice, inclusion and cooperation between police and public. Polic Soc 24(1):22–43
17. Bratton W, Tumin Z (2012) Collaborate or perish!: reaching across boundaries in a networked world. Crown Business, pp 16–79
18. Braziel R, Straub F, Watson G, Hoops R (2016) Bringing calm to chaos: a critical incident review of the San Bernardino public safety response to the December 2, 2015, terrorist shooting incident at the Inland Regional Center. In: United States. Department of Justice, Office of Community Oriented Policing Services. United States. Department of Justice. Office of Community Oriented Policing Services
19. Brown CSD (2015) Investigating and prosecuting cyber crime: forensic dependencies and barriers to justice. Int J Cyber Criminol 9(1):55–119
20. Brown (2018) Extending the remit of evidence-based policing. Int J Police Sci Manage 20(1):38–51
21. Cadwallader, Graham-Harrison E (2018) Revealed: 50 million Facebook profiles harvested for Cambridge Analytica in major data breach. The Guardian 17:22
22. Carlo S. https://www.independent.co.uk/voices/snoopers-charter-theresa-may-online-privacy-investigatory-powers-act-a7426461.html. Last accessed 6 Mar 2020
23. Chan JBL (2001) The technological game: How information technology is transforming police practice. Criminol Crim Just 1(2):139–159
24. Chivers W (2015) A mature debate on communications surveillance? Available: http://orca.cf.ac.uk/95069/1/index.html. Last accessed 31st July 2020
25. Couchman H. https://www.libertyhumanrights.org.uk/sites/default/files/Liberty%27s%20Briefing%20on%20Facial%20Recognition%20-%20October%202019.pdf. Last accessed 21 Feb 2020
26. Crown Prosecution Service. https://www.cps.gov.uk/legal-guidance/disclosure-guidelines-communications-evidence. Last accessed 8 Dec 2020
27. Dentith MR (2016) The problem of fake news. Public Reason 8(1–2):65–79
28. Dodd V. https://www.theguardian.com/uk-news/2020/feb/26/extra-officers-must-lead-to-less-priti-patel-tells-police-chiefs. Last accessed 10 Mar 2020
29. Dunning D (2011) The Dunning-Kruger effect: on being ignorant of one's own ignorance. Adv Exp Soc Psychol 44:247–296
30. Egawhary EM (2019) The surveillance dimensions of the use of social media by UK police forces. Surveill Soc 17(1):89–104
31. Entchev I (2011) A response-dependent theory of precedent. Law Philos 30(3):273–290
32. Eskens S, Van Daalen O, Van Eijk N (2016) 10 standards for oversight and transparency of national intelligence services. J Nat Security Law Policy 8(3):1–38

33. Evans M, Police facing rising tide of social media crimes. Telegraph, 20156/15
34. Ferguson I, Renaud K, Irons A (2018) Dark clouds on the horizon: the challenge of cloud forensics. Cloud Computing 61
35. Forensic Analytics. https://www.forensicanalytics.co.uk/the-investigatory-powers-act-and-access-to-comms-data/. Last accessed 17 Dec 2019
36. Fussey P, Murray D (2019) Independent report on the London metropolitan police service's trial of live facial recognition technology. Human Rights Centre, University of Essex
37. Garvie C, Frankle J (2016) Facial-recognition software might have a racial bias problem. The Atlantic 7
38. Gill P (2013) Should the intelligence agencies 'show more leg' or have they just been stripped naked? Inf Security 30(1):11–31
39. Goodman M (2015) Future crimes: inside the digital underground and the battle for our connected world. Random House
40. Gov. uk https://www.gov.uk/government/organisations/office-for-communications-data-authorisations/about. Last accessed 17 Dec 2019
41. Grierson J, https://www.theguardian.com/society/2020/apr/15/domestic-abuse-killings-more-than-double-amid-covid-19-lockdow. Last accessed 4 June 2020
42. Gunawan TS, Mutholib A, Kartiwi M, Performance evaluation of automatic number plate recognition on android smartphone platform. Int J Electri Comput Eng 7(4):1973
43. Heaton R (2012) Police resources, demand and the Flanagan report. Police J 1–22
44. Her Majesty's Chief Inspector of Constabulary (2018) State of Policing – The Annual Assessment of Policing in England and Wales 2018. https://www.justiceinspectorates.gov.uk/hmicfrs/wp-content/uploads/state-of-policing-2018.pdf. Last accessed 3 June 2020
45. Hofmann BM (2015) Too much technology. BMJ 16(2):705
46. Holmes PG, Burum S (2016) Apple v. FBI: Privacy vs. Security? National Soc Sci Proc 62(1):24–41
47. Home Office. https://www.gov.uk/government/publications/policing-by-consent/definition-of-policing-by-consent. Last accessed 18 Feb 2020
48. Home Office. https://www.gov.uk/government/collections/communications-data. Last accessed 10 Nov 2019
49. Home Office. https://assets.publishing.service.gov.uk/government/uploads/system/uploads/attachment_data/file/806674/NASPLE_-_January_2019_. Last accessed 10 Nov 2019
50. Home Office: Statement to Parliament: Home Secretary: Publication of draft Investigatory Powers Bill. November 4, 2015. House of Commons, London (2015)
51. Honess R (2020) Mandatory police training: the epitome of dissatisfaction and demotivation? Polic J Policy Pract
52. Hooper V (2019) Addressing the challenge of guiding our students on how to deal with fake news. In: InSITE 2019: informing science + IT education conferences: Jerusalem, 021–032
53. Huey L, Nhan J, Broll R (2013) Uppity civilians' and 'cyber-vigilantes': The role of the general public in policing cyber-crime. Criminol Criminal Justice 13(1):81–97
54. Hymas C. https://www.telegraph.co.uk/politics/2019/04/23/police-chief-admits-60-per-cent-crime-not-fully-investigated/. Last accessed 21 Apr 2020
55. Information Commissioner's Office (2020). Mobile phone data extraction by police forces in England and Wales. https://ico.org.uk/media/about-the-ico/documents/2617838/ico-report-on-mpe-in-england-and-wales-v1_1.pdf. Last accessed 29 July 2020
56. Investigatory Powers Act (2016)
57. Isaak J, Hanna MJ (2018) User data privacy: Facebook, Cambridge Analytica, and privacy protection. Computer 51(8):56–59
58. Jackson J, Bradford B, Hough M, Kuha J, Stares S, Widdop S, Fitzgerald R, Yordanova M, Galev T (2011) Developing European indicators of trust in justice. Eur J Criminol 8(4):267–285
59. Jaques P. https://www.policeprofessional.com/news/huge-volume-of-visual-evidence-puts-investigators-under-extra-pressure/. Last accessed 20 Dec 2019

60. Kirby S, Penna S (2011) Policing mobile criminality: implications for police forces in the UK, Policing. Int J Police Strat Manage 34(2):182–197
61. Leong RS (2006) FORZA–Digital forensics investigation framework that incorporate legal issues. Digital Invest 3:29–36
62. Levi M, Leighton Williams M (2013) Multi-agency partnerships in cybercrime reduction: mapping the UK information assurance network cooperation space. Inf Manage Comput Security 21(5):420–443
63. Liberty Human Rights. https://www.libertyhumanrights.org.uk/news/press-releases-and-sta tements/liberty-client-takes-police-ground-breaking-facial-recognition. Last accessed 17 Feb 2020
64. Lock R, Cooke L, Jackson T (2013) Online social networking, order and disorder. Elect J E-Govern 11(2):229–240
65. Mackie J, Taramonli, Bird R (2017) Digital forensics and the GDPR: examining corporate readiness. In: European conference on cyber warfare and security, pp 683–691
66. Magretta J, Stone N (2002) What management is: how it works and why it's everyone's business. Free Press, New York, NY
67. Manning M, Agnew S (2020) Policing in the era of AI and smart societies: austerity; legitimacy and blurring the line of consent. In: Jahankhani H, Akhgar B, Cochrane P, Dastbaz M (eds) Policing in the era of AI and smart societies. Springer
68. Marwick A, Hargittai E (2019) Nothing to hide, nothing to lose? Incentives and disincentives to sharing information with institutions online. Inf Comm Soc 22(12):1697–1713
69. McCartney C, Shorter L (2019) Police retention and storage of evidence in England and Wales. Int J Police Sci Manage
70. McQuade S (2006) Technology-enabled crime, policing and security. J Technol Stud 32(1):32–42
71. Millie A (2013) The policing task and the expansion (and contraction) of British policing. Criminol Criminal Justice 13(2):143–160
72. Millie A, Bullock K (2012) Re-imagining policing post-austerity. British Acad Rev 19:16–18
73. Mishra JL, Allen DK, Pearman AD (2011) Information sharing during multi-agency major incidents. Proc Am Soc Inf Sci Technol 48(1):1–10
74. Morrell K, Bradford B, Javid B (2020) What does it mean when we ask the public if they are 'confident' in policing? The trust, fairness, presence model of 'public confidence. Int J Police Sci Manage 22(2):111–122
75. Müller K, Schwarz C (2019) Fanning the flames of hate: Social media and hate crime. SSRN Electronic J 3082972
76. National Archives. http://www.legislation.gov.uk/ukpga/1997/50/section/93. Last accessed 6 Apr 2020
77. National Police Chiefs Council. https://www.npcc.police.uk/documents/Policing%20Vision. pdf. Last accessed 29 Sept 2019
78. Newburn T (2017) Criminology, 3rd edn. Routledge. ps, London, pp 579–580
79. Nhan J, Huey L (2008) Policing through nodes, clusters and bandwidth. Technology, crime and social control, Technocrime, pp 1–13
80. Office for National Statistics. https://www.ons.gov.uk/peoplepopulationandcommunity/cri meandjustice/datasets/policeforceareadatatables. Last accessed 21 Apr 2020
81. Office for National Statistics. https://www.ons.gov.uk/peoplepopulationandcommunity/cri meandjustice/articles/homicideinenglandandwales/yearendingmarch2018#how-are-victims-and-suspects-related. Last accessed 4 June 2020
82. Owen R (2018) Law enforcement's dilemma: fighting 21st century encrypted communications with 20th century legislation. Homeland Security Affairs
83. Pătraşcu A, Patriciu VV (2013) Beyond digital forensics—a cloud computing perspective over incident response and reporting. In: IEEE 8th international symposium on applied computational intelligence and informatics, pp 455–460
84. Peters MA (2018) The information wars, fake news and the end of globalization. Edu Phil Theory 50(13):1161–1164

85. Pollitt M (2004) Six blind men from Indostan. In: Digital forensics research workshop (DFRWS)
86. Quick D, Choo KKR (2014) Impacts of increasing volume of digital forensic data: a survey and future research challenges. Digital Invest 11(4):273–294
87. Raine JW, Keasey P (2012) From police authorities to police and crime commissioners. Int J Emer Ser
88. Ramakrishnan N, Butler P, Muthiah S, Self N, Khandpur R, Saraf P, Wang W, Cadena J, Vullikanti A, Korkmaz G, Kuhlman C (2014) 'Beating the news' with EMBERS: forecasting civil unrest using open source indicators. In: Proceedings of the 20th ACM SIGKDD international conference on knowledge discovery and data mining, pp 1799–1808
89. Robertson A (2016) Policing by consent: some practitioner perceptions (Doctoral dissertation). University of Sunderland
90. Rogers C, Scally EJ (2018) Police use of technology: insights from the literature. Int J Emer Ser 7(2):100–110
91. Scheuerman WE (2014) Whistleblowing as civil disobedience: the case of Edward Snowden. Phil Soc Critic 40(7):609–628
92. Segal HP (1993) Technopoly: the surrender of culture to technology. The Organization of American Historians
93. Sentencing Council. https://www.sentencingcouncil.org.uk/the-magistrates-court-sentencing-guidelines/. Last accessed 22 Apr 2020
94. Shaw D. https://www.bbc.co.uk/news/uk-44884113. Last accessed 21 Dec 2019
95. Shearing CD, Stenning PC (1983) Private security: implications for social control. Soc Probl 30(5):493–506
96. Sheptycki J (2019) Technopoly and policing practice. Eur Law Enforce Res Bull (4 SCE):133–139
97. Sherman L, Neyroud PW, Neyroud E (2016) The Cambridge crime harm index: measuring total harm from crime based on sentencing guidelines. Policing 10(3):171–183
98. Slessor J. https://www.accenture.com/gb-en/insights/public-service/reimagining-police-workforce-future-vision. Last accessed 28 Sept 2019
99. Smith T (2018) The "near miss" of Liam Allan: critical problems in police disclosure, investigation culture, and the resourcing of criminal justice. Criminal Law Rev (9)
100. Son S, Jung G, Jun SC (2013) An SLA-based cloud computing that facilitates resource allocation in the distributed data centers of a cloud provider. J Supercomput 64(2):606–637
101. Spinello RA (2019) Ethics in cyberspace: freedom, rights, and cybersecurity. Next-Generat Ethics Eng Better Soc 454
102. Staniforth A (2016) Police use of open source intelligence: the longer arm of law. In: Open Source Intelligence Investigation. Springer, Cham, pp 21–31
103. Taddeo M, Floridi L (2018) Regulate artificial intelligence to avert cyber arms race. Nature 556(7701):296–298
104. Taylor M, Haggerty J, Gresty D, Hegarty R (2010) Digital evidence in cloud computing systems. Comput Law Security Rev 26(3):304–308
105. The Police Foundation (2020) Public Safety and Security in the 21st Century. Available: https://policingreview.org.uk/wp-content/uploads/phase_1_report_final-1.pdf. Last accessed 31st July 2020
106. Trenwith PM, Venter HS (2013) Digital forensic readiness in the cloud. In: Information security for South Africa. IEEE, pp 1–5
107. Trottier D (2015) Open source intelligence, social media and law enforcement: visions, constraints and critiques. Eur J Cult Stud 18(4–5):530–547
108. United Nations Office on Drugs and Crime. https://www.unodc.org/documents/organized crime/UNODC_CCPCJ_EG.4_2013/CYBERCRIME_STUDY_210213.pdf. Last accessed 1 Nov 2019
109. Van Baar RB, Van Beek HMA, Van Eijk EJ (2014) Digital forensics as a service: a game changer. Digital Invest 11:54–62

110. Wall DS, Johnstone J (1997) The industrialization of legal practice and the rise of the new electric lawyer: the impact of information technology upon legal practice in the UK. Int J Sociol Law 25(2):95–116
111. Waterhouse J. https://www.bbc.co.uk/news/uk-46964659. Last accessed 20 Dec 2019
112. Williams J (2017) Legal and ethical issues surrounding open source research for law enforcement purposes. In. ECSM 4th European conference on social media. Academic Conferences and Publishing Limited
113. Woods L (2017) Automated number plate recognition: data retention and the protection of privacy in public places. J Inf Rights Policy Pract 2(1):1–21
114. Wright S. http://eprints.leedsbeckett.ac.uk/2096/3/Watching%20Them%20Watching%20Us.pdf. Last accessed 20 Dec 2019
115. Yoannou CJ. https://www.prindlepost.org/2020/02/sensorvault-and-ring-private-sector-data-collection-meets-law-enforcement/. Last accessed 28 Feb 2020

Cyber-Disability Hate Cases in the UK: The Documentation by the Police and Potential Barriers to Reporting

Zhraa A. Alhaboby [ID]**, Haider M. Al-Khateeb** [ID]**, James Barnes** [ID]**,
Hamid Jahankhani** [ID]**, Melanie Pitchford** [ID]**, Liesl Conradie** [ID]**,
and Emma Short** [ID]

Abstract Disability hate crime is under-reported in the UK with perceived limited support given to the victims. The use of online communication resulted in cyber-disability hate cases, recognised by the Police with the addition of an 'online-flag' in the documentation. However, the cases remain under-reported, with potential individual, societal and organisational barriers to reporting especially during a pandemic. This paper aims to contextualise the reporting of cyber-disability hate cases, identify potential barriers, and provide recommendations to improve support to victims by the Police. The retrospective examination was carried out on disability-related cyber incidents documented by a police force in the UK for 19 months. Among 3,349 cyber-crimes, 23 cases were included. The analysis covered descriptive statistics and qualitative document analysis (QDA). Only 0.7% of cyber incidents or 6.7% of cyber-hate incidents were disability related. The age of victims ranged between 15 and 61 years, with a mean of 25.8 years. Most of the victims (78%) were from White ethnic background, and the majority were females (61.5%). Three overarching themes emerged from the qualitative data as influencers of reporting or documentation, these were: psychological impact, fear for safety, and the type of disability. Cyber-offences resulted in a serious impact on wellbeing, however, cases that included people with visible disabilities were more documented. Further awareness-raising targeting the

Z. A. Alhaboby (✉) · H. M. Al-Khateeb
University of Wolverhampton, West Midlands, Wolverhampton WV1 1LY, UK
e-mail: Z.Alhaboby@wlv.ac.uk

J. Barnes
Fatima College of Health Sciences, Abu Dhabi, UAE

H. Jahankhani
Northumbria University London Campus, London 1 7HT, UK

M. Pitchford
University of Bedfordshire, Bedfordshire 1 3JU, UK

L. Conradie
Open University, Milton Keynes MK7 6AA, UK

E. Short
De Montfort University, Leicester L1 9BH, UK

© The Author(s), under exclusive license to Springer Nature Switzerland AG 2021
H. Jahankhani et al. (eds.), *Cybersecurity, Privacy and Freedom Protection in the Connected World*, Advanced Sciences and Technologies for Security Applications, https://doi.org/10.1007/978-3-030-68534-8_8

police and public is needed to understand the impact of cyber-offences and recognise the different types of disabilities, which might encourage both reporting and documentation.

Keywords Incident response · Law enforcement · Online hate crime · Disabled people · Justice · Law

1   Introduction

Disability is one of the protected characteristics in the UK. It is defined under the Equality Act 2010 as a "physical or mental impairment and the impairment has a substantial and long-term adverse effect on his or her ability to carry out normal day-to-day activities" [1, p. 7]. More than 11 million individuals in the UK live with impairment, disability and/or a long-term condition [2]. A substantial proportion of them face challenging circumstances that are considered social determinants of health such as living standards, employment issues, and education [2]. Disabled people are also more likely to experience unfair treatment, discrimination, and crime [2]. These issues necessitate collaborative work to facilitate health-management, support, as well as overcoming the disabling barriers in society.

The victimisation of disabled people is well documented in the literature [3, 4]. Victimisation is any repeated negative behaviour or attention over time by an individual or a group towards the "victim" [5]. It can range from harassment incidents [6–8] to disability hate crimes [9].

Hate crimes include a range of criminal behaviours motivated by hostility towards protected characteristics such as disability, race, religion, sexual orientation or transgender identity [10]. Accordingly, disability hate crime is defined by the Association of Chief Police Officers (ACPO) and the Crown Prosecution Service (CPS) as "Any criminal offence which is perceived, by the victim or any other person, to be motivated by a hostility or prejudice based on a person's disability or perceived disability" [11]. Such experiences include harassment, intimidation, damaging property or violence. The reporting of disability hate crimes increased over time from 1,748 cases in 2011/2012, to 2,020 in 2013/2014 and reached 7,226 cases in 2017/2018 [10]. In the three years ending March 2018, a total of 52,000 disability hate incidents and crimes were reported in England and Wales [10].

Disability hate crime remains an ongoing issue in the UK, despite the continuous efforts to identify the underlying factors and the systematic consequences [12]. One of the acknowledged issues is the under-reporting and the barriers to the criminal justice system [13]. In a quantitative study, hate crime data over 10 years from 2005 to 2015 were examined. It was observed that disability hate crime is under-reported compared to other categories of hate crime. It was estimated that 56% of disability hate incidents were reported to the police, compared to 42% of race incidents, however, the police were less likely to investigate disability incidents (10%) compared to other crimes such as race incidents (16%) [14]. A potential issue that undermined reporting is the normalisation of hate speech over time, probably due to external factors such as tax-paying [15] and the stereotyping in media representations [16].

Discrimination is an ancient but not a static phenomenon; the increasing use of technology has resulted in 'cyber-victimisation' cases. These are either pure online offences (cyber dependent) or as a continuation of traditional crimes using electronic communication (cyber-enabled) [17]. Online offences were found to have no less devastating multi-faceted impact on the victims compared to their offline counterparts [18, 19]. Cyber experiences are complicated by the anonymity of offenders, the availability of a broad range of means to employ, longevity of exposure due to permanent comments online and unavoidability by physical absence from a specific context [20]. In the UK, it was found that one in every four adults with disabilities (23.1%) experienced crime including electronically facilitated crimes [21]. Due to the variations in definitions among researchers, disciplines and stakeholders [18], hate incidents that include online communication will be addressed under the umbrella term 'cyber-victimisation' in this paper.

Cyber-victimisation imposes a huge impact upon victims, such as an individual's wellbeing, social relations and can result in long term consequences such as mental health illnesses or mortality [18]. Such impact requires collaborative work to identify risks and provide proper support [22]. A recent investigation by the House of Commons examined the online experiences of disabled people in the UK [23]. The report acknowledged the importance of online communication for people with disabilities, the impact of hostility in online communication, and recognised that current laws were unfit for purpose to protect people with disabilities. Moreover, there is a risk of escalation of disability discrimination during the current COVID-19 pandemic. There are continuous calls to ensure the response to COVID-19 is inclusive and fair to disabled people [24, 25]. However, many UK-based organisations and activists raised concerns over the rights of disabled people during the pandemic in response to perceived discrimination in regulations and practice [26]. Hence, the current situation raises further concerns over cyber-victimisation and requires support channels to provide appropriate response and support to the victims.

The Police are one of the major instrumental support channels approached by victims of cyber-offences [27]. The "online flag" in police records became mandatory in April 2015 [10]. It is used to help identify the extent of using electronic communications to facilitate crimes nationally. The online flag is used with offences that were committed or facilitated through computers, computer networks or computer enabled devices. In an analysis of using the online flag by 30 out of 40 police forces in the UK, it was found that racially motivated crimes were highest in numbers (928 offences), followed by sexual orientation (352 cases), disability (225 cases), religion (2010 cases) and transgender (69 cases). Race was also identified as the most common motivating factor reported for hate crimes, however, as a proportion, only 2% of racially motivated hate crimes were online. Thus, after putting these numbers as proportions, the use of electronic communication was commoner in transgender, disability and sexual orientation hate crimes, with a frequency of 6%, 4%, and 4% respectively [10].

Victims of cyber-offences perceived that the support channels did not take them seriously [19, 27], and this is also demonstrated in cyber-disability hate cases [23]. There are ongoing efforts in the UK for police cyber-crime training, and also encouraging researchers to narrow the gap between theory and practice [28]. This paper examines the documented cases of cyber-disability hate by a police force in the UK, to situate cyber-disability hate among other offences, identify the impact upon victims, patterns of reporting/documentation, and guide future work.

2   Methods

An opportunistic retrospective examination of cyber incidents was carried out on police records provided by a police force in the UK, documented over 19 months (between July 1st, 2014 and January 31st, 2016). Qualitative Document Analysis (QDA) was conducted, it is a systematic approach for evaluating and interpreting written documents to elicit meaning [29, 30]. The analysis steps included: (1) setting document selection criteria; (2) identifying key areas of analysis; (3) coding, and analysis [31]. The analysis stage generally includes content analysis, thematic analysis or both [32]. This approach is suitable as a stand-alone method or triangulated with other qualitative methods to increase confidence in recommendations [29].

The analysis in this paper was underlined by phenomenological philosophy, which looks at official reports as social constructs [33] and it is in-line with previous work carried by the authors in this area [19, 27, 34]. Hence, thematic analysis was employed in the last stage. The advantages of using QDA with the Police records include: (1) its efficiency as a research method; (2) lack of reactivity, i.e. unaffected by research process and the stability of data without alteration by researchers; and (3) exactness of police records. This approach helped to address patterns that construct reporting of cyber-victimisation cases targeting disabled people. However, the QDA here is limited by partial reporting of some cases.

Ethical approval was granted by the University Research Ethics Committee (UREC) at a UK University. Cases eligibility criteria included the following: (1) the case is identified via the online flag in the police records; or (2) using a cyber-related keyword search in the documentation; and (3) the victim has a disability and/or a long-term health condition that is documented within the case.

2.1   Case Selection Process

A total of 3,349 cyber-crime cases were identified from the police records over the specified timescale. This data set included 1,493 cases with "online flag", and 1,856 cases identified using cyber-related keyword searches. From the overall dataset, two subsets were extracted: (1) cyber-hate cases, and (2) disability-related cases.

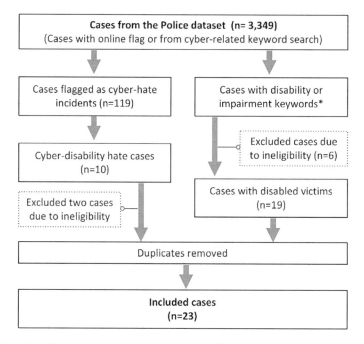

Fig. 1 Flowchart illustrating the case selection process. *These cases were extracted via systematic search using keyword-based queries

The cyber-hate dataset included all cyber-related cases categorised as hate incidents following the Crown Procession Service (CPS) guidelines, the total was 119 cases. These were scanned for eligibility criteria, 10 cases were shortlisted, and two were excluded due to not looking at a victim with a disability; one case included a comment that implies the offender is speaking on behalf of people with disabilities. The disability data set was extracted following a search using relevant keywords, including "disability", "impairment", "chronic illness", "chronic condition", "long term condition" or "long term illness". The disability dataset initially included 25 cases, of which 6 were excluded mainly due to the victim not having a disability/long term condition, for example, the offender has a mental health condition. From both, cyber-hate and disability-related datasets, a total of 23 cases were eventually included in this analysis. The case selection process is illustrated in Fig. 1.

2.2 Analysis

This paper presents the results collectively without references to individual cases. Free text written in each included case was extracted into a separate document. The guidelines for thematic analysis were followed [35]. The text was read, re-read and then open codes were applied. Codes were examined and further grouped into categories, then themes were arranged around a central concept, focusing on the reporting and impact of cyber-victimisation of disabled people. Demographics of victims and alleged offenders were extracted from included cases to situate the sample; however, the data was incomplete in some instances.

3 Results

Only 0.7% of the overall cybercrime incidents, or 6.7% out of cyber-hate incidents, were disability related. Incident were reported from 6 different cities and towns, and the included cases were recorded by the police under different crime groups as visualised in Fig. 2. Harassment was the highest group 10 (44%), one of which was a single incident, followed by sending grossly offensive materials via electronic communication (18%), disability hate 3 (13%), domestic incident 2 (9%), racial hate incident 1 (4%), and one racial hate incident with an injury, followed by 1 (4%) sending letters to cause distress, and one (4%) public fear/distress case.

3.1 Demographics

Most of the victims were females 16 (61.5%) compared to 11 (42.3%) male victims. These numbers are higher than case numbers due to the involvement of more than one victim in one case. The age of victims ranged between 15 and 61 years, with a mean of 25.8 years. In one case there were 3 victims aged 13, 13 and 12 years. The

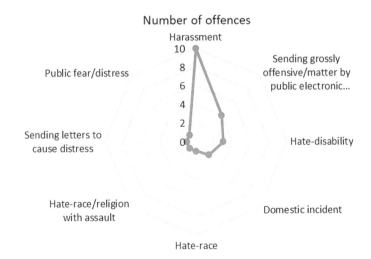

Fig. 2 Radar chart illustrating how disability-related cases were recorded by the police

majority of victims 18 (78%) were from White ethnic backgrounds, 2 (8.7%) from Black ethnic backgrounds, one (4.3%) victim was from Asian ethnic background, and 2 (8.7%) unreported cases.

The alleged perpetrators were equally males (n = 10), and females (n = 10), and 3 cases with unreported gender. The age range of offenders was 14 to 51 years with a mean of 31.8 years in 17 reported cases. Ethnicity wise, 15(65.2%) offenders were White and one offender (4.3%) was from Black ethnic background, and 7(30.4%) unreported cases.

Offender's relationship to the victim was mainly acquaintance 7 (31%) and ex-partner 6 (26%), followed by stranger 4 (17%), neighbour 2 (9%), and 4 (17%) unreported cases.

3.2 Qualitative Findings

Upon examining the written text in the documentation, three overarching themes emerged from included cases. The themes below do not include direct quotations to ensure the anonymity of reporting such sensitive issues.

Theme 1: Psychological impact. The most common impact shared in documented cases was psychological. Reports frequently mentioned the victim being distressed, scared, worried, anxious, alarmed, stressed, upset, crying or being afraid. One case explicitly included that the victim's illness has deteriorated following the incident. One case mentioned sleep disturbances. Other psychosomatic or behavioural effects reported were also influenced by psychological impact and interacted with the second theme.

Theme 2: Fear for Safety. This theme emerged from reports where victims shared threats received or abusive communication that lead them to fear for their safety or the safety of their family members. These threats or abusive comments were mainly through electronic communications such as phone messages or photos. However, in a few cases, this was associated with offline actions such as knocking the door or breaking windows, in these cases the threats were taken more seriously by the victims. Such fears resulted in avoiding being outdoors, taking leave from work or not sending children to school.

Theme 3: The Type of Disability. Most of the reports in this recurrent theme implied in the incident records that the victim had a visible impairment. This was apparent through comments of harassment or disability hate incidents where the offender used words such as "crippled", "spastic" or referring to how the victims look physically. Some cases involved referring to the victim's use of disability aids. This was followed by a few cases in which the victim had a learning disability or was attending a special education school. In some of the cases of learning disability, the offenders also has other disabilities. While few cases involved victims having mental health illness. No other types of disabilities or impairments were mentioned in the records by the police.

4 Discussion

The documented patterns of reporting showed a triad of three factors that could have influenced the victims' decision into taking the case to the police. These themes were mainly developing psychological consequences, receiving threats, and living with certain types of impairments. These factors could be summarised in Fig. 3.

The psychological impact is a consistent finding in the literature that is associated with cyber-victimisation experiences [18]. Hence, triggering stress seems to be an important factor contributing to the decision of contacting the police. The second factor is receiving threats or extremely abusive comments, which is documented with online and offline victimisation [36, 37].

The third factor, having certain types of impairments could be a major contributor to under-reporting. The findings here suggest that having a visible disability or to a less extent learning disability are potentially related to reporting being a victim of cyber incidents. This is relevant to a UK-based survey that examined the perceived motivation of offenders in disability hate incidents [9]. The motivation of offenders ranged from hate and jealousy to accusations of fraud because of the relative invisibility of some impairments. Other work suggests that people with invisible disabilities such as Myalgic encephalomyelitis or epilepsy are also targeted, and when encouraged they are keen to share their experiences of online abuse [34]. This indicates under-reporting of incidents involving invisible impairments.

The under-reporting of disability hate crimes is a multifaceted issue. One aspect lies in public awareness and the stereotyping around disability. This could be linked to the ongoing debate about adopting the medical model and the social model of disability. The medical model focuses on impairments, and the medical diagnoses lead to a legitimised 'sick role' in society [38]. However, this model has the potential

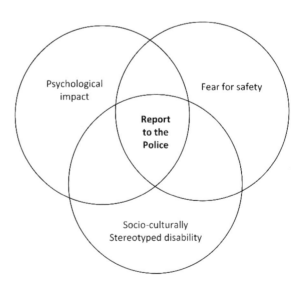

Fig. 3 The interplay of identified factor in reporting cyber-victimisation by disabled people to the police

of exacerbating discrimination due to focusing on deficits. Hence, the social model was introduced with the argument of separating the impairment from the disability, i.e. people with impairments are disabled by their surrounding societies. Activists in disability campaigns advocate for adopting the social model but the medical model and diagnosing medical conditions can influence how the public view and stereotype disabled people. The legislations in the UK to protect disabled individuals might have created an inherited prejudice towards people who are being put in vulnerable situations because of their immediate socio-cultural context [12]. A suggested official explanation of under-reporting focused on people's understanding of the questions and that responses were based on 'perceived vulnerability' [10]. This might indicate the police documentation is reliant on the medical model of disability and the legitimised 'sick role' in society [38]. Resulting in a complicated situation in which people with visible disabilities were discriminated against, and those with invisible disabilities to be abused for not looking physically ill. Hence, further, awareness-raising is needed to understand the different types of disabilities, and to encourage people with all types of impairments to come forward.

Under-reporting of disability hate incidents is not a new phenomenon [14], and the results in this paper indicate that the 'cyber' aspect had worsened the situation. This could be because of victims' issues in trusting the support channels [19], or due to issues in training the police to deal with cyber-crimes. In a recent study, the majority of participants from a police force in the UK (56% n = 163) did not feel confident to deal with cyber-crime cases [17]. This issue is of concern especially in pandemic circumstances where appropriate support is needed. During the pandemic, people relied more on online communication, and concerns over escalations of disability discrimination were constantly shared [25, 26].

Most of the victims in this paper were from White ethnic backgrounds. This information should be treated sensitively and requires further research because disability-hate is under-reported compared to other hate crimes [14]. However, it is also to be acknowledged that in the records a hate crime can be recorded under more than one motivating factor or flag, which potentially requires further investigations and leads to undercounting [10]. Hence, race is also a protected characteristic and its interplay with disability, if not appropriately addressed, could result in less trust in the criminal justice system and exacerbate marginalisation. Additionally, it is important to note that some of the excluded cases from the analysis here could have been involving an individual with a disability but were missed if this was not documented by the Police or brought up by the victim. This also indicates the need for a coherent multiagency system and consistency in documentation.

5   Conclusion

In conclusion, cyber-disability hate crime is an under-reported issue in the UK compared to other hate crimes and cyber offences. This could be influenced by several factors such as race, type of disability, in addition to fear and distress. Therefore, it is recommended to appreciate this gap as part of training programmes delivered to police personnel. This will help to raise awareness about the different types of disabilities and improve the training on the impact of cyber-crime and its documentation. Raising awareness among the public is also indicated to support the role of the police in tackling disability discrimination, which includes cyber offences and covers all types of disabilities.

Acknowledgements This publication was partially supported by the Police Innovation Fund 2016/17 from the Home Office, UK. It aims to advance incident response against cyberharrasment. Its contents are solely the responsibility of the authors and do not necessarily represent the official views of funder.

References

1. Equality Act (2010) Guidance on matters to be taken into account in determining questions relating to the definition of disability. https://assets.publishing.service.gov.uk/government/upl oads/system/uploads/attachment_data/file/570382/Equality_Act_2010-disability_definition. pdf. 15 Oct 2016
2. Office for Disability Issues (2014) Disability facts and figures. https://www.gov.uk/government/ publications/disability-facts-and-figures/disability-facts-and-figures#fn:11. 20 Aug 2020
3. Blake JJ et al (2012) National prevalence rates of bully victimization among students with disabilities in the United States. Sch Psychol Q 27(4):210–222
4. Fridh M, Lindström M, Rosvall M (2015) Subjective health complaints in adolescent victims of cyber harassment: moderation through support from parents/friends—a Swedish population-based study. BMC Public Health 15(1):949–949
5. Kouwenberg M et al (2012) Peer victimization experienced by children and adolescents who are deaf or hard of hearing. PLoS ONE 7(12):e52174–e52174
6. Fekkes M, Pijpers FI, Verloove-Vanhorick SP (2004) Bullying behavior and associations with psychosomatic complaints and depression in victims. J Pediatr 144(1):17–22
7. Sentenac M, et al (2011) Peer victimization among school-aged children with chronic conditions. Epidemiol Rev mxr024
8. Barakat LP, Wodka EL (2006) Posttraumatic stress symptoms in college students with a chronic illness. Soc Behav Pers: Int J 34(8):999–1006
9. Quarmby K (2015) Disability hate crime motivation survey—results. https://katharinequarmby. wordpress.com/. 28 Sept 2015
10. Home Office (2018) Hate crime, England and wales, 2017/18. https://assets.publishing.ser vice.gov.uk/government/uploads/system/uploads/attachment_data/file/748598/hate-crime-1718-hosb2018.pdf. 20 Aug 2020
11. Crown Prosecution Service (2018) Disability hate crime and other crimes against disabled people—prosecution guidance. https://www.cps.gov.uk/legal-guidance/disability-hate-crime-and-other-crimes-against-disabled-people-prosecution-guidance. 17 Oct 2018
12. Emerson E, Roulstone A (2014) Developing an evidence base for violent and disablist hate crime in britain: findings from the life opportunities survey. J Interpers Violence
13. Sin CH, et al (2009) Disabled people's experiences of targeted violence and hostility. In: Equality and human rights commission research report series
14. Macdonald SJ, Donovan C, Clayton J (2017) The disability bias: understanding the context of hate in comparison with other minority populations. Disabil Soc 32(4):483–499
15. Burch L (2018) 'You are a parasite on the productive classes': online disablist hate speech in austere times. Disabil Soc 33(3):392–415

16. Alhaboby ZA, et al (2017) Disability and cyber-victimisation in the image of disability: essays on media representations. In: Schatz AEGJL (ed). McFarland Press, North Carolina, United States, pp 167–180
17. Eze T, Hull M, Speakman L (2019) Policing the cyber threat: exploring the threat from cyber crime and the ability of local law enforcement to respond. In: European intelligence and security informatics conference. IEEE, Karlskrona, Sweden
18. Alhaboby ZA et al (2017) Cyber victimisation of people with chronic conditions and disabilities: a systematic review of scope and impact. Trauma, Violence Abus 20(3):398–415
19. Alhaboby ZA et al (2016) 'The language is disgusting and they refer to my disability': the cyberharassment of disabled people. Disabil Soc 31(8):1138–1143
20. Anderson J, Bresnahan M, Musatics C (2014) Combating weight-based cyberbullying on Facebook with the dissenter effect. Cyberpsychology, Behav Soc Netw 17(5):281–286
21. Office for National Statitics (2019) Disability and crime, UK. https://www.ons.gov.uk/peoplepopulationandcommunity/healthandsocialcare/disability/bulletins/disabilityandcrimeuk/2019. 20 Aug 2020
22. Alhaboby ZA, et al (2018) Understanding the cyber-victimisation of people with long term conditions and the need for collaborative forensics-enabled disease management programmes. In: Cyber criminology. Springer, Berlin, pp. 227–250
23. House of Commons (2019) Online abuse and the experience of disabled people. https://publications.parliament.uk/pa/cm201719/cmselect/cmpetitions/759/75902.htm.
24. World Health Organization (2020) Disability considerations during the COVID-19 outbreak. World Health Organization.
25. Armitage R, Nellums LB (2020) The COVID-19 response must be disability inclusive. Lancet Public Health 5(5):e257
26. Disability Rights UK (2020) Covid 19 and the rights of disabled people. https://www.disabilityrightsuk.org/news/2020/april/covid-19-and-rights-disabled-people. 1 July 2020
27. al-Khateeb HM, et al (2017) Cyberstalking: investigating formal intervention and the role of corporate social responsibility. Telemat Inform 34(4):339–349
28. Cockcroft T, Schreuders ZC, Trevorrow P (2018) Police cybercrime training: perceptions, pedagogy, and policy. Polic: J Policy Pract
29. Bowen GA (2009) Document analysis as a qualitative research method. Qual Res J 9(2):27–40
30. Baxter S et al (2016) Evaluating public involvement in research design and grant development: using a qualitative document analysis method to analyse an award scheme for researchers. Res Involv Engag 2(1):1
31. Wach E, Ward R (2013) Learning about qualitative document analysis
32. Seale C (2012) Researching society and culture. Sage 555
33. Bowling A (2009) Research methods in health: Investigating health and health services, 3rd edn. Open University Press, Maidenhead
34. Alhaboby ZA, et al (2017) Cyber-victimisation of people with disabilities: challenges facing online research. Cyberpsychology. J Psychosoc Res Cyberspace J Psychosoc Res Cyberspace 11(1)
35. Braun V, Clarke V (2006) Using thematic analysis in psychology. Qual Res Psychol 3(2):77–101
36. Sheridan L, Roberts K (2011) Key questions to consider in stalking cases. Behav Sci Law 29(2):255–270
37. Sheridan LP, Grant T (2007) Is cyberstalking different? Psychol, Crime Law 13(6):627–640
38. Haegele JA, Hodge S (2016) Disability discourse: Overview and critiques of the medical and social models. Quest 68(2):193–206

Smart Secure USB SSU-256

Muhammad Ehsan ul Haq, Zeeshan Ali, Muhammad Taimoor Ali, Ruqiya Fazal, Waseem Iqbal, and Mehreen Afzal

Abstract USBs are the most common devices for data sharing and transferring either for personal day to day use or at the organizational level. Its usage is increasing exponentially despite the data breaches occurring due to the noncompliance of security measurements. Consumers are at risk when sensitive data is stored on unsecured USBs. The consequences of losing drives (or when picked up by unauthorized persons) loaded with sensitive information can be significant, including the loss of customer data, financial information, business plans and other confidential/sensitive information, risk of reputation damage. Apropos, this problem of keeping the data confidential from unauthorized users need to be addressed immediately. Therefore, in this paper we present an indigenous solution for this problem which can easily be used by general users and sensitive organizations (strategic, banks, academia, law enforcement, armed forces, telco's and many others) to overcome the above stated confidentiality problem. Our proposed Smart Secure USB (SSU-256), will serve as secure channel for both data storage and transfer.

Keywords USB · Authentication · SSU · Cyber attack · Encryption

M. Ehsan ul Haq · Z. Ali · M. T. Ali · R. Fazal · W. Iqbal (✉) · M. Afzal
Department of Information Security, National University of Sciences and Technology, NUST, Islamabad 44000, Pakistan
e-mail: waseem.iqbal@mcs.edu.pk

M. Ehsan ul Haq
e-mail: 354721ehsan@gmail.com

Z. Ali
e-mail: zeeshan.ali92171@yahoo.com

M. T. Ali
e-mail: muhammadtaimoor1999@gmail.com

R. Fazal
e-mail: ruqiya830@gmail.com

M. Afzal
e-mail: mehreen.afzal@mcs.edu.pk

© The Author(s), under exclusive license to Springer Nature Switzerland AG 2021   135
H. Jahankhani et al. (eds.), *Cybersecurity, Privacy and Freedom Protection in the Connected World*, Advanced Sciences and Technologies for Security Applications, https://doi.org/10.1007/978-3-030-68534-8_9

1 Introduction

In year 2000 USB memory stick was launched which was light in weight, portable, comparatively cheap and offered high transfer rate. Since then the usage of USBs is increasing exponentially ignoring the fact that they offered no kind of security to user's data (Fig. 1).

As both consumers and businesses have increased the demand for these drives, manufacturers are producing faster devices with greater data storage capacities [2], despite the data breaches occurring due to the noncompliance of security measurements. To avoid these data losses secure USBs must be brought into use (Fig. 2).

Data breaches and cyber-attacks are anticipated to increase in the due course of time as the computer networks are expanding. As technology progresses, more and more of our information is at risk. As a result, cyberattacks have become increasingly common and costly. As per Ponemon report 70% of businesses have traced the loss of sensitive or confidential information to USB memory sticks [3]. It is crucial and mandatory to protect our data to get rid of cyberattacks (Figs. 3 and 4).

The basic idea behind the project being proposed is to design a secure USB which will serve as secure channel for both data storage and transfer. This project is going to be the productive insight in the industry due to its Multi-layer security i.e. Biometric Authentication and Encryption which will work completely independent of each other, beside that it provides User specific Compartments and an Admin for users registration and deletion as well as being Indigenous, Cost Effective, assuring integrity and an easy to use, plug n play portable device.

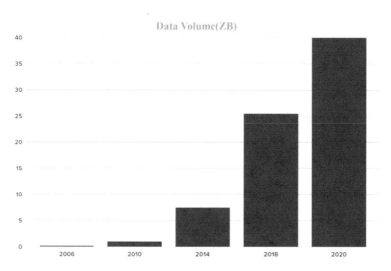

Fig. 1 Global trends of data volume [1]

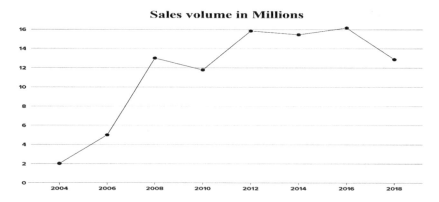

Fig. 2 USBs sales volume [2]

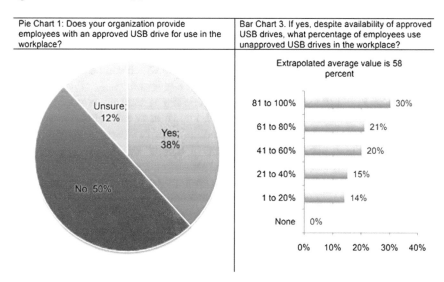

Fig. 3 USBs data loss [3]

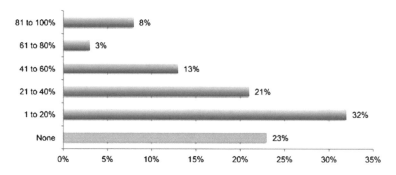

Fig. 4 What percent of USB drives used in the workplace are safe and secure? [3]

2   Literature Review

Biometric authentication is a process which uses specific modality to verify user such as fingerprint, retina etc. But we are using fingerprint for authentication purpose. Since everyone on Earth has unique and distinctive fingerprint and it is one of the most used modalities for authentication. It is also quite accurate and quick process.

Advanced Encryption Standard AES comes under the category of symmetric cryptography. In symmetric cryptography same key is used to encrypt and decrypt message. That key is called shared key/private key. AES is based upon substitution/permutation and works on block of either 128,192 or 256 bits. Main advantage of using symmetric encryption over asymmetric is that its fast and efficient for larger data. And there is need to keep the key secret in asymmetric encryption. Amongst other modes of AES, XTS mode [4] is selected.

XTS was added to the catalogue of AES block cipher modes back in 2010 by NIST of Science and Technology). It is used by Data Traveler 4000G2 and Data Traveler Vault Privacy 3.0. The intention of designing AES-XTS was to develop such a move that vanquishes the shortcomings of other modes of AES. It eradicates the possible vulnerabilities linked with some side channel attacks which can be used to exploit drawbacks of other modes.

Two AES keys are used by XTS. Where one key is used for block encryption whereas the other is used to encrypt a value known as tweak value. Galois polynomial function is used to bring further modification in tweak value and then it is XORed with both plaintext and ciphertext of each block. The purpose of GF is to provide diffusion and to make sure that identical ciphertexts are not produced by identical plaintexts. This factor allows XTS to produce unique ciphertexts from identical plaintext without making use of initialization vector and chaining. Consequently, the text is now double encrypted using two independent keys, and text is decrypted by reversing the process. Since there is no chaining involved so if stored cipher is damaged or becomes corrupted, only the data of that specific block will be affected. With chaining involved there is possibility of error propagation when decrypted (Fig. 5).

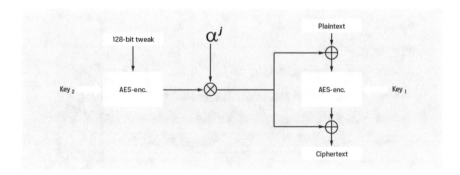

Fig. 5 XTS mode of AES [5]

For authentication purpose we will be using R305 sensor [6]. Users are authenticated through their fingerprint, R305 is one of the most widely used module for fingerprint authentication, with the assistance of DSP in its core. User communicates with this module by using hex codes in a certain specific format command. The red LED on this scanner indicates regarding the state of sensor (ON or OFF).

Each R305 module has specific identifying address and while communicating with another system each instruction is transferred in the form of packet which indicates the address of the device. This module only gives response to those data packages whose address is same as its identifying address. This address is 4 bytes long. By default, its factory value is 0xFFFFFFFF. Almost 500 ms are required for initialization of R305.

Safe Authentication Protocol for Secure USB Memories [7], User and device Authentication protocol is way more important than data encryption itself, manufacturers mainly focus on keeping data secure which leaves a flaw in the security of secured USB drives making them vulnerable, Fingerprint is the most used biometric modality because of its two traits uniqueness and permeance Fingerprint based symmetric cryptography [8] uses same fingerprint for encrypting and decrypting data, it uses a string of binary number extracted from fingerprint template act as cryptographic key. This cryptographic key is used for encrypting messages and for decryption process the key is generated from fingerprint instance and then both cryptographic keys are compared. Fingerprint is the most used biometric modality because of its two traits uniqueness and permeance Apart from being user friendly this algorithm got vulnerable due to fingerprint cloning. Other issues related with integration of biometric system with cryptographic system [9]. Techniques which are based on biometric authentication generate a biometric key which is not beneficial in cryptographic applications since they involve sharing unencrypted biometric data over an insecure channel thus, such applications require spawning of biometric keys to release the secret encrypted message that was sent.

Some other kind of secured USB drives like EAGET, having features like fingerprint encryption, user segregated compartments, and dual storage. FU5 is combination of biometric technology and storage. The data stored in USB can only be accessed by authentication through fingerprint. Extreme by SanDisk and Kingston's defender providing symmetric encryption only.

3   Proposed Solution

Smart and secure USB which will serve as authenticated storage medium for data transfer our proposed designs will fulfill following requirements.

The biometric authentication will be done using fingerprint authentication. Data is further secured through Encryption standards i.e. AES-256, which further includes file-based encryption. An administrator in the form of admin is there to register and delete users. This project will help to protect and secure user's data from unauthorized access by having user segregated compartments. Keeps record of all the recent plugins. It will need no external power source. Providing top level features in a minimal price, it's an indigenous and first of its kind smart secure USB. It uses Raspberry pi zero having R305 attached to it with an external SD card to have user segregated compartments. Architectural design of it is as follow.

- Authentication: In this part our device will authenticate its owner and other users.
- Encryption: For security purpose we will be encrypting user's data with AES-256.
- User specific Compartments: Each user will be allowed to have specific compartment and no one else would be allowed to have access to that compartment.
- Multi- Layer Security: Both encryption and authentication together will provide multi- layer security (Figs. 6 and 7).

Experimental setup:

For authentication purpose SSU is using R305 sensor. R305 is one of the most widely used module for fingerprint authentication, with the assistance of DSP in its core. User communicates with this module by using hex codes in a certain specific format.

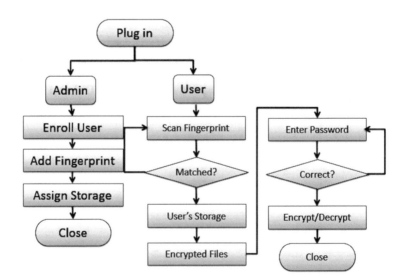

Fig. 6 Block diagram of SSU working

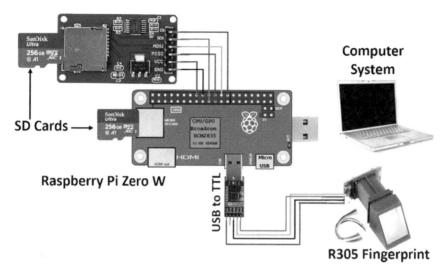

Fig. 7 SSU-256 proposed circuitry

For enrollment and deletion of fingerprints we have used library of Python which is named as "PyFingerprint".

User Enrollment:

initialize fingerprint sensor
new_image = readimage ()
if new_image in stored_images
Print (ERROR! This user already exists)
else
verify_image = readimage ()
if new_image = = verify_image
image_hash = hash(image)
save_image(new_image)
username = input(username)
assign_image_hash_to_username ()
assign_storage_to_user ()
else
print (ERROR! Image not matched)

Code:

```
print('Waiting for finger...')
## Wait for fingerprint to scan
while (f.readImage() == False):
    time.sleep(1)
    pass
f.convertImage(0x01)
result = f.searchTemplate()
positionNumber = result[0]
if (positionNumber >= 0):
    print('Template already exists at position #' + str(positionNumber))
    exit(0)
print('Remove finger...')
time.sleep(2)
print('Waiting for same finger again...')
## Wait that finger is read again
while (f.readImage() == False):
    time.sleep(1)
    pass
time.sleep(1)
## Converts read image to characteristics and stores it in charbuffer 2
f.convertImage(0x02)
## Compares the charbuffers
if (f.compareCharacteristics() == 0):
    raise Exception('Fingers do not match')
## Creates a template
f.createTemplate()
## Saves template at new position number
positionNumber = f.storeTemplate()
print('Finger enrolled successfully!')
```

Delete User:

username = input(username)# *Enter username of user to be deleted*
If username exist
image_hash = read_hash(username)
Delete(image_hash)
Delete(username)
Else
Print (ERROR! This user does not exist)

Code:

```
try:
    positionNumber = input('Please enter the template position you want
to delete: ')
    positionNumber = int(positionNumber)

    if ( f.deleteTemplate(positionNumber) == True ):
        print('Template deleted!')
except Exception as e:
    print('Operation failed!')
    print('Exception message: ' + str(e))
    exit(1)
```

SSU-256 is using XTS mode of AES 256. It is the standard mode of AES 256 used in market since, it caters for the issues in other modes. Lack of diffusion is exhibited by ECB mode since it generates same ciphertexts for identical plaintexts. Whereas in

CBC encryption of each block is sequential so each block needs to be calculated to calculate next block. So, encryption/decryption cannot be parallelized in this mode. In this XTS mode randomization and efficiency is achieved through Tweak value and Galois function.

In XTS each data block is assigned a positive integer value known as tweak value. It starts off from a random integer and then assigned one after the other. Tweak value must be converted to little endian byte array.

For the sake of key derivation SSU-256 is using a function called Password Based Key Derivation Function (PBKDF-2) [10]. It is just a simple cryptographic key derivation function. It is immune to dictionary attacks and rainbow table attacks. It takes several parameters as an input and generates a derived key as an output with following intakes:

- Password
- Salt
- Count of iterations
- Hash Function
- Derived-key-length.

For Encryption:

Select file to encrypt
plaintext = read(file)
password = input(password)
cfm_password = input(cfm_password)
if password_validity = = True
if password = = cfm_password
hash = hash(password)
# Store hash to be used for decryption
save(hash)
key = PBKDF2(password)
ciphertext = encrypt (AES, key, plaintext)
else
Print (ERROR! Password Mismatched)
else
Print (ERROR! Invalid_Password)

Code:

```
if mode == 'encryption':
    print("Plaintext= ")
    print(text)
    encryptor = xts_aes.encrypt
    ciphertext = encryptor(text)

print('{ciphertext_type}:{ciphertext}'.format(ciphertext_type=TEXT_TYPES[
inverse_mode],ciphertext=binascii.hexlify(ciphertext).decode()))
```

For Decryption:

file = input (filename of encrypted file) *# Enter same password used at the time of encryption*
password = input(password)
hash = hash(password)
verify_hash = read(stored_hash)
If hash = = verify_hash
key = PBKDF2(password)
ciphertext = read(filename)
plaintext = decrypt (AES, key, ciphertext)
Else
Print (ERROR! Wrong Password)

Code:

```
if mode == 'decryption':
    print("Plaintext= ")
    print(text)
    encryptor = xts_aes.decrypt
    plaintext = encryptor(text)

print('{plaintext_type}:{plaintext}'.format(ciphertext_type=TEXT_TYPES[in
verse_mode],plaintext=binascii.hexlify(plaintext).decode()))
```

We have segregated the memory into user specific compartments. Each user will be allotted a specific compartment and no user will be allowed to look into anyone else's compartment. We are using FAT32 file system. File system basically controls how the data is stored or retrieved. If there was no file system, then the data would be stored as a single piece of data like user cannot tell when one data has stopped and the next has begun. Data is segregated into pieces where each piece is properly identified by giving it a specific name.

Encryption along with authentication provides multi -layer security to the data of user. Plus, our idea also facilitates user with file-based encryption, Additional feature of PBKDF2 makes it harder for attacker to brute force the password.

4 Analysis of the Proposed Solution

SSU-256 stands out tall amongst its market competitors.

- Providing user, a two-layer security i-e Authentication and encryption.
- File base encryption makes separate passwords for all files hence adding more protection to the secured data.
- Keeps record of all the recent plugins.
- Pakistan based Indigenous solution to unsecured USBs data breaches.
- No external power source is needed.
- Providing top level features in a minimal price (Fig. 8).

Features	SanDisk	Eaget	Kingston	SSU-256
Authentication	✗	✓	✗	✓
Encryption	✓	✗	✓	✓
Multi-users	✗	✓	✗	✓
Multi-layer security	✗	✓	✗	✓
Admin rights	✗	✓	✗	✓
Recent login history	✗	✗	✗	✓

It can serve many different fields of life such as law Enforcements, strategic organizations, Telco's, IT sector, Banks, Corporate Sector and General digital Users (Fig. 9).

Fig. 8 SSU-256

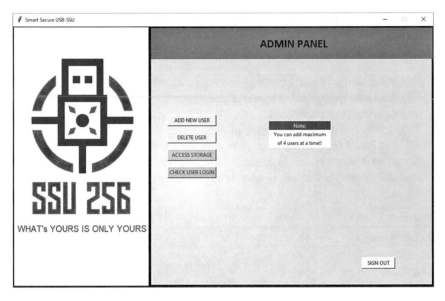

Fig. 9 GUI of SSU-256

5   Conclusion

SSU-256 is going to stand out as a best external hard drive that gives precedence to the security of user's data which they carry and this property will make it one of the indispensable tool for those who carry sensitive data with them. In this era where everyone needs to move data from one system to other, most of the companies have banned the usage of these devices in their office premises which makes life of employers quite tough. So, SSU-256 will serve as a sound solution to these problems of data leakage, privacy issues and other attacks possible through USBs. In this modern era of data where it is the most invaluable thing and every other person wants to get hold of other's data. It is the hour of need to have such a project that is capable enough to combat these problems linked with data security.

SSU-256 with other advancements on encryption techniques, modern authentication techniques can easily enhance scope of SSU and will be even more beneficial for companies and general users.

References

1. https://www.researchgate.net/Global-growth-trend-of-data-volume-2006–2020-based-on-The-digital-universe-in-2020-researchgate_net
2. https://www.statista.com/statistics/485531/sales-volume-usb-flash-drives-germany/
3. U.S. survey of IT and IT security practitioners

4. HARDWARE ENCRYPTED USB USING AES-256 by Amir Abbas, Marvi Waheed, Kamran Anwar
5. https://medium.com/asecuritysite-when-bob-met-alice/who-needs-a-tweak-meet-full-disk-encryption-437e720879ac
6. Python-R305 Documentation, Release 1.0.0
7. Lee K, Yeuk H, Choi Y, Pho S, You I, Yim K Safe authentication protocol for secure USB memories. https://isyou.info/jowua/papers/jowua-v1n1-4.pdf
8. Barman S, Chattopadhyay S, Samanta D Fingerprint based symmetric cryptography.https://iee explore.ieee.org/abstract/document/7045306/s
9. Uludag U, Pankanti S, Prabhakar S, Jain AK Biometric cryptosystems: issues and challenges. https://scholar.google.com.pk/scholar?q=Biometric+Cryptosystems:+Issues+and+Challe nges+by+Umut+Uludag,+Sharath+Pankanti,+Salil+Prabhakar+and+Anil+K.+Jain&hl=en& as_sdt=0&as_vis=1&oi=scholart
10. https://cryptobook.nakov.com/mac-and-key-derivation/pbkdf

The Application of Technology in Combating Human Trafficking

Reza Montasari and Hamid Jahankhani

Abstract Human trafficking is a complex, burgeoning crime with a global foothold that impacts an estimated 40.3 million people worldwide. Currently, the number and the scale of innovation and technology tools do not correspond with the magnitude of the problem. There is little awareness about existing digital innovations and technology initiatives within the field of anti-trafficking. This adds to the danger of fragmented and disjoined development and the application of technology-based tools. Therefore, considering its importance, this paper aims to provide an analysis of the current landscape of technology tools used to combat HT. The multi-pronged contributions of this study is: (1) to enable antitrafficking stakeholders to engage with technology more effectively and (2) to raise awareness about tools to assist their work.

Keywords Human trafficking · Modern slavery · Technology · Innovation · Digital forensics · Cyber crime · Cyber terrorism · Cyber threats

1 Introductions

Human Trafficking (HT) covers a plethora of crimes and exploitations spanning from withheld wages or documents, child labour in mining or natural resource exploitation to sex-trafficking, domestic service and car-washing [1]. Being a multidimensional challenge, HT affects all societies around the world. It is a global conundrum that has transformed into a $150 billion industry as attested by estimates published by the International Labour Organization (ILO). According to another report published by Oceana, non-profit ocean conservation organization, HT has become one of the fastest growing transnational criminal enterprises worldwide. Child labour exists in

R. Montasari (✉)
Hillary Rodham Clinton School of Law, Swansea University, Swansea S2 8PP, UK
e-mail: reza.montasari@swansea.ac.uk

H. Jahankhani
Northumbria University London, 110 Middlesex Street, London 1 7HT, UK
e-mail: hamid.jahankhani@northumbria.ac.uk

© The Author(s), under exclusive license to Springer Nature Switzerland AG 2021
H. Jahankhani et al. (eds.), *Cybersecurity, Privacy and Freedom Protection in the Connected World*, Advanced Sciences and Technologies for Security Applications, https://doi.org/10.1007/978-3-030-68534-8_10

places such as chocolate, coffee and tea supply chains or in places such as fishing and shrimping boats. Being a pervasive and a highly profitable crime, HT is the third-largest criminal activity globally, behind drug trafficking and counterfeiting [2]. With a little cost and yet large profits to be made, combined with a low risk of being identified, prosecuted and convicted, traffickers are often highly motivated to remain involved in this odious crime. According to [3], between the period 2015 and 2018, the prosecution rate of traffickers decreased by 42% worldwide. In Europe, this rate dropped by 52%. 2. This highlights that criminals are operating with impunity and there is little challenge from the technology systems that are in place to bring them to justice.

However, despite its devastating impacts, tackling HT is a difficult undertaking that has become even more challenging through the advancements in technology constantly exploited by the criminals and their accomplices at various stages of HT (such as recruitment, movement, control, advertising and abuse of victims). Such misuse of technology is a key factor contributing to the increase in HT profitability and the impunity of traffickers [3]. For instance, the members of HT ring exploit technologies such as instant and secure communication, remote control of victims using GPS location apps, or receiving and moving illegitimate profits through cryptocurrency [3]. While the misuse of technology tools presents many challenges to the anti-trafficking stakeholders worldwide, their constructive use could also play a critical role in the fight against HT and modern slavery [1]. For instance, such tools could assist in detecting slave-owners, providing satellite imageries to search for illegal factories to secure blockchain databases that trace fashion back to the source, or providing simple apps offering advice and contacts to victims or vulnerable workers. This, in turn, could improve prosecutions, raise awareness and offers support and services to the victims, and elucidate the structure and tactics of webs of human traffickers [1].

Despite the positive roles that technology could play in combating HT, little consideration has been given, or few resources have been assigned, to investigate the manner in which technology can be leveraged to address HT, in particular in the contexts of prevention and protection [3]. We believe that any effective measures to tackle HT necessitate the fusion of a wide range of factors, which together could produce the desired impact. Thus, being one of such effective measures, technology must be considered as a critical factor and incorporated as a fundamental component into any anti-trafficking strategy. We also contend that any future success in eliminating HT in its various forms will depend not only upon the social and legal aspects of HT but also on technology tools and how nation states are prepared and willing to harness such tools in their efforts to combat this phenomenon [3]. In line with our assertion, various other official bodies have highlighted the importance of technology and communication network in combatting HT [4]. However, notwithstanding numerous calls for incorporating technology into any anti-trafficking interventions, existing studies and anti-trafficking solutions have not drawn upon technology tools or communication networks in their proposed interventions.

Therefore, in line with these pronouncements, this paper seeks to address the current research gap by analysing the impact of the misuse and use of technology tools in both facilitating and combating HT, respectively. The paper also offers recommendations to various anti-trafficking stakeholders on how to maximize the value of technology-based solutions. These stakeholders could be: law enforcements, governments, NGOs, businesses, technology companies and academia, etc. The paper is intended to provide the stakeholders with knowledge and information that could be useful in leveraging the power of technology to combat the exploitation of victims.

2 Background

The main purposes of HT are exploitation and enslavement. Trafficking can occur among countries or in areas within a country. Akin to sexual assault or domestic violence, HT can affect any individual irrespective of age, gender, religion, citizenship or immigration status or nationality [5–7]. At its core, HT is the violation of human rights through which individuals are exploited for financial gains. The economic sectors that profit most from HT consist of agriculture, restaurants, manufacturing, domestic work, entertainment, hospitality, and the commercial sex industry. Although any individual can become a victim of HT, it is often those within poor, marginalised and underprivileged communities who are at greater risks of HT. While there exist numerous types of HT, there are three common forms including: the sex trade, forced labour, and domestic servitude. Many other forms of exploitation are thought to be under-reported often due to language barriers, fear of traffickers or fear of law enforcement. These could include forced marriage, organ removal, the exploitation of children in begging and warfare. While each incident of HT varies in specifics, all have three constituent elements consisting of: acts, means and purpose. These elements can be formulated as: *the action of a trafficker (what is done) + by means of (how it is done) + for the purposes of (why it is done) = HT (the result)*. To determine whether a specific circumstance indicates HT or not, one should consider both the definition of HT as well as the constituent elements of the crime as defined by relevant national and international legislations.

3 Technology

3.1 Misuse of Technology

Due to the advancements in technology, sex traffickers are no longer limited to recruit their victims in person. Technology enables traffickers to have a wider access to an unlimited number of potential victims. Traffickers are increasingly becoming adept at exploiting technological advances to recruit and sell victims. For instance, traffickers exploit tools such as proxy servers that can mask the source of communication hence concealing identity; online means such as web applications, chat rooms and virtual currency to advertise; Global Positioning Systems in cell phones to track their victims; or use encrypted data to communicate with other criminals [8].

Operating systems offered by technology giants such as Appel and Google now encrypt user's data by default. Similarly, messaging systems such as WhatsApp, owned by Facebook, offer end-to-end encryption in which only the person sending the message and the person receiving the message can see the communication. Such technologies are important in that they increase privacy of user's data online, thus ensuring the protection of human rights. However, they simultaneously pose significant challenges to law enforcement agencies in relation to data extraction from devices containing such technologies. This has resulted in authorities voicing concerns that encryption technologies can threaten the public by blocking access to data exchanged between traffickers or other criminals [8]. There exist millions of child sexual exploitation images circulating online. These abuse images, that can be easily uploaded onto various online platforms, are created through various means such as capturing screenshots of live streaming abuse or snapping exploitation photos on mobile devices, etc. Some of these photos have remained in circulation even many years after the initial incident of sexual exploitation. Addressing the issue would require partnerships from online platforms that have facilitated such a wide scale propagation [9].

3.2   Constructive Use of Technology

Notwithstanding its constant misuse and the many challenges that technology tools present, they also offer many benefits such as providing better insight, real time intelligence and linking together disparate data sets. Technology tools could be a highly valuable instrument at the disposal of stakeholders involved in combatting trafficking offences. This section describes some of the key technologies that can be utilised by the stakeholders to combat HT more effectively. While the individual description of each given tool is outside the scope of this paper, the key categories of these tools are discussed below.

Data Aggregation and Analysis The online world is infinite, and there exist numerous web applications, social media, chatrooms, applications, and on-line video games that can be linked to the online activities of both the victims' and human traffickers'. This has resulted in large volumes of data (Big Data), the collection and analysis of which pose significant challenges to law enforcements and other stakeholders involved in the fight against HT. For instance, these challenges could concern the quality of data—including: accuracy, validity, reliability, timeliness, relevance and completeness—about the distribution of victims, traffickers and buyers of the victims' services. Furthermore, this data is often spread over many sources in different jurisdictions, creating additional issues such as data ownership, privacy, unwillingness to share or a lack of knowledge concerning what data are available. Considering this large volume of data, combined with the stated challenges, it would be impossible for analytics practitioners to screen and analyse this quantity of online data. As a result, data analysis tools could be utilised to aggregate and synthesise only pertinent data into useful reports hence saving time and valuable resources.

Machine Learning and Deep Learning Anti-trafficking stakeholders could also employ the computational power of Machine Learning (ML), Deep Learning (DL) and Pattern Recognition (PR), three sub-sets of Artificial Intelligence (AI), to address HT more efficiently. AI technologies are capable of making predictions, recommendations or decisions independently and without human involvement [9]. For instance, in case of HT, an AI-based automated age-progression software can be deployed to establish how a child victim of sex trafficking would age and look when she/he is an adult. This is important as it can facilitate independent device communication with buyers of services from trafficking victims. This, in turn, will enable investigators to specify, for instance the characteristics of physical space in which the child is being held, and to determine financial transactions that could demonstrate signs of HT networks. ML could be a key asset in combating HT. It can be used to automate manual assignments and enable difficult tasks to become less complex given the ease with which it can sift through large quantity of data and extract patterns. For instance, ML techniques could be utilised to develop child finder desktop and mobile applications that will be capable of matching pictures of exploited children against a given country's database of missing children. With this type of application, the law enforcement's reaction time, a key factor to assist an exploited child, could potentially be reduced from days to seconds [10].

Blockchain for Traceability and Provenance Businesses operating in high-risk sectors, such as fishing, mining, construction or textile industries, are especially susceptible to the risk of forced labour within their large, complex supply chains. Given that these businesses, often multinational corporations, have numerous suppliers dispersed worldwide, observing their supply chains could be an intricate undertaking. Therefore, in order to determine threats of HT and forced labour, these businesses could adopt the use of blockchain technologies which facilitate tracking the manufacturing of the merchandise from their origin to the end use to ensure transparency and assist in practising both duty of care and due diligence [9]. Through the use of blockchain, these corporations will be able to collect vital information, analyse their supply chains and third-party networks, detect signs of HT and forced labour, and prevent it in the future. For instance, the Global Fishing Watch mapping platform, developed by Google, can be employed to examine vessels with histories of HT and forced labour. Similarly, the International Air Transport Association (IATA), a trade organisation for the world's airlines, has established a global awareness and training program called Eyes Open International to train flight attendants, gate agents and other airline personnel to notice signs of HT, and also to increase public awareness of this crime [10].

Facial Recognition The computational power of Facial Recognition technology and visual processing software can be leveraged in web crawling to search for photos and videos of victims of trafficking for sexual exploitation. These types of technology can also assist investigators in examining large quantity of photographs and videos to detect content that can be ascribed to a specific person of interest [10].

Image and Video Fingerprinting Software that enable image and video finger-printing techniques could also be utilised to identify, extract and then summarise the characteristic elements of a video recording allowing that to be uniquely identified by its resultant fingerprint. This technology could be effective at identifying child sexual abuse material and comparing digital video data. For instance, such techniques could enable investigators to detect the digital fingerprint of an image or video that law enforcement have classified as child sexual exploitation material.

URL Blocking URL Blocking is another technology that could allow internet service providers (ISPs) to block any part of a URL that includes known child sexual exploitation material. As a result, this would restrict the propagation of such material and prevent the cycle of re-victimization.

Existing Initiatives Technology could also be utilised to monitor HT and raise public awareness. There exist a number of apps such as GoodGuide that enable conscious consumers to be cognizant of the environmental and social impacts of their purchases. GoodGuide ranks a wide range of products and scores them based on health, environmental, and social aspects. Red Light Traffic is an app that enables the public to report suspected cases of HT anonymously. It also informs people of the "red flags" of HT, so it can be identified and reported [3].

The United States' Defence Advanced Research Project Agency (DARPA) have launched the Memex program, an initiative that seeks to develop software that enhances online search capabilities beyond the existing technologies [11]. Currently, Memex [12] can be employed to solve complex search problems such as HT, court documents, and research papers. Owing to its domain-specific indexing and search capabilities, Memex could enable investigators to engage with advanced content discovery, information extraction, information retrieval, user collaboration and other key search functions. For instance, Memex can be used to search and scan the most hidden contents of the deep web, the dark web (such as advertisements for sex labour), and non-traditional content (such as multimedia) that cannot be discovered or linked together with standard search engines [11].

Furthermore, "Made in a Free World", a not-for-profit organisation, have created the Slavery Footprint website that is claimed to enable the users to visualise how their consumption habits are related to modern slavery [13]. Likewise, LexisNexis, an Information Technology corporation, provides legal, regulatory and business infor-mation and analytics. Headquartered in New York City and serving customers in more than 130 countries worldwide, LexisNexis provides

technology tools to combat HT. It combines legal and business information with analytics and technology to provide powerful new decision tools by utilising Machine Learning, Natural Language Processing, Visualisation, and Artificial Intelligence to their worldwide legal database [14]. LexisNexis have designed and implemented a supply chain monitoring product entitled Smartwatch. According to LexisNexis [15], as of 2020, it searches 40,000 premium and online news archives and business sources and a comprehensive archive dating back 40 years. As well as news articles and select social media sources, users are provided with access to 80 million companies over 1,000 industry sources, and 75 million executive and biographical sources, in addition to important regulatory, legal, and public records content.

Unseen UK operates the UK Modern Slavery Helpline and Resource Centre. Its mobile application, the Unseen UK App, facilitates reporting to the UK Modern Slavery Helpline easier. It can assist the police and modern slavery groups in understanding understand how to respond to cases or suspected cases of HT, create greater awareness on the issue, inform individuals on how to identify signs of exploitation, and report situations of concern [16]. The Child Safety Hackathon, organised by Thorn and Facebook, is a 48-hour collaborative event that occurs once a year. This event brings together experts from the technology industry (such as Google, Microsoft, Amazon, Twitter, Pinterest, Intel, Facebook, and more) to hack on developing advanced solutions that will facilitate locating victims faster, deter predatory behaviour and make platforms safer [17].

4   Conclusion

Technology tools can be both misused and used to facilitate or to assist with combatting HT, better detecting and protecting victims. Although criminals are becoming more skilled in technology and able to exploit it successfully, the same might not be necessarily true of the actors accountable for combating HT. Considering this, any future success in eliminating HT in its various forms will be contingent on the manner in which nation states and civil societies are primed to develop, utilise and regulate technology in their responses to this crime [10]. This approach is of paramount importance, necessitating a collective and strong commitment and cooperation amongst nation states. This collective efforts are often a long process that necessitates vital improvements of both human and financial resources for monitoring, identifying, investigating, sentencing and disrupting all methods of HT enabled by information communication technologies especially by the Internet.

References

1. Melson C (2018) The role of tech in stopping human trafficking and modern slavery. Waterman MS, Identification of common molecular subsequences. https://www.techuk.org/insights/news/item/13574-the-role-of-tech-in-stopping-human-trafficking-modern-slavery. Accessed 19 Sept 2020
2. Ecpat (2018) The trafficking of children for sexual purposes: one of the worst manifestations of this crime. https://www.ecpat.org/news/trafficking-the-third-largest-crime-industry-in-the-world/. Accessed 22 Sept 2020
3. OSCE Office of the Special Representative and Co-ordinator for Combating Trafficking in Human Beings and Tech Against Trafficking (2020) Leveraging innovation to fight trafficking in human beings: a comprehensive analysis of technology tools. Vienna, Austria
4. United Nations Human Rights Office of the High Commissioner (2014) Special Rapporteur on trafficking in persons, especially women and children. https://www.ohchr.org/en/issues/trafficking/pages/traffickingindex.aspx. Accessed 25 Sept 2020

5. United Nations Office on Drugs and Crime (Unodc) (2020) Human trafficking. https://www.unodc.org/unodc/en/human-trafficking/what-is-human-trafficking.html. Accessed 21 Sept 2020
6. General Assembly resolution 55/25 (2000) Protocol to prevent, suppress and punish trafficking in persons, especially women and children, supplementing the United Nations convention against transnational organized crime. In: United Nations Office on drugs and crime, United Nations Convention against transnational organized crime and the protocols thereto
7. Modern Slavery Act 2015 (2015) c. 30. United Kingdom Legislation [online]. https://www.legislation.gov.uk/ukpga/2015/30/contents/enacted. Accessed 21 Sept 2020
8. Wulfhorst E (2017) Latest technology helps sex traffickers recruit, sell victims—FBI. Thomson Reuters Foundation
9. Ferrer C (2017) Machine learning can help find kids faster. https://www.thorn.org/blog/machine-learning-find-kids-faster/. Accessed 17 Sept 2020
10. Inter-Agency Coordination Group against Trafficking in Persons (2019) Human trafficking and technology: trends, challenges and opportunities. ICAT. Issue brief (7):1–6
11. United States Department of Defense (2017) Defense advanced research projects agency project. DARPA program helps to fight human trafficking. https://www.defense.gov/Explore/News/Article/Article/1041509/darpa-program-helps-to-fight-human-trafficking/. Accessed 25 Sept 2020
12. United States Department of Defense (2017) Memex. Defense advanced research projects agency project. https://www.darpa.mil/about-us/timeline/memex. Accessed 25 Sept 2020
13. Made in a Free World (2020) Using the free market to free people. https://slaveryfootprint.org/miafw/about.php. Accessed 25 Sept 2020
14. Jefferson A (2017) LexisNexis launches Lexis answers, infusing new artificial intelligence capabilities into the company's flagship legal research platform, Lexis advance. https://www.lexisnexis.com/community/pressroom/b/news/posts/lexisnexis-launches-lexis-answers-infusing-new-artificial-intelligence-capabilities-into-the-company-s-flagship-legal-research-platform-lexis-advance. Accessed 25 Sept 2020
15. LexisNexis (2020) Get insights from 40,000+ premium, licensed and web sources. https://www.lexisnexis.com/en-us/professional/nexis/nexis.page. Accessed 25 Sept 2020
16. Unseen (2020) Modern slavery helpline. https://www.unseenuk.org/. Accessed 25 Sept 2020
17. Thorn (2016) Child Safety Hackathon (2020) Child Safety Hackathon Brings silicon valley together. https://www.thorn.org/blog/child-safety-hackathon/. Accessed 25 Sept 2020

Countering Adversarial Inference Evasion Attacks Towards ML-Based Smart Lock in Cyber-Physical System Context

Petri Vähäkainu⬤, **Martti Lehto**⬤, **and Antti Kariluoto**⬤

Abstract Machine Learning (ML) has been taking significant evolutionary steps and provided sophisticated means in developing novel and smart, up-to-date applications. However, the development has also brought new types of hazards into the daylight that can have even destructive consequences required to be addressed. Evasion attacks are among the most utilized attacks that can be generated in adversarial settings during the system operation. In assumption, ML environment is benign, but in reality, perpetrators may exploit vulnerabilities to conduct these gradient-free or gradient-based malicious adversarial inference attacks towards cyber-physical systems (CPS), such as smart buildings. Evasion attacks provide a utility for perpetrators to modify, for example, a testing dataset of a victim ML-model. In this article, we conduct a literature review concerning evasion attacks and countermeasures and discuss how these attacks can be utilized in order to deceive the, i.e., CPS smart lock system's ML-classifier to gain access to the smart building.

Keywords Adversarial machine learning · Defensive mechanisms · Evasion attacks · Cyber-physical system

1 Introduction

A cyber-physical system (CPS), such as a smart building, utilizes technology aiming to create a safe and healthy environment for its occupants. Buildings are pertinent to smart cities, but as the number of buildings grows, it also increases security risks. Smart building technology is still in the early stages of growth and adoption increases moderately and is steadily becoming a significant business around the world. Actors and industries interested in implementing intelligent building solutions are, for example, airports, factories, hospitals, military bases, residential buildings, etc. Adoption of smart building technology brings in associated security threats consumers are usually unaware of. In the smart-building CPS context, for example,

P. Vähäkainu (✉) · M. Lehto · A. Kariluoto
University of Jyväskylä, Jyvaskyla 40100, Finland
e-mail: petri.vahakainu@jyu.fi

© The Author(s), under exclusive license to Springer Nature Switzerland AG 2021   157
H. Jahankhani et al. (eds.), *Cybersecurity, Privacy and Freedom Protection in the Connected World*, Advanced Sciences and Technologies for Security Applications, https://doi.org/10.1007/978-3-030-68534-8_11

there exists a chance that, with e.g., adversarial attack, a perpetrator could fool the ML-model and gain entry to a building causing significant security threats.

A cyber-physical system (CPS) is an interconnected network of sensors and actuators, which are controlled by a program within a cloud. The program accepts data from the sensors and calculates, based on previous events, how the actuators should be adjusted to implement changes to the flow of the system. A smart building is an example of a CPS. Cyber-physical systems tend to produce and gather an abundance of data in real-time.

Artificial intelligence (AI) is known to be an algorithm that can mimic human behavior to an extent. It is used in multiple industries and to solve many different tasks autonomously. The use requires data for training and for the necessary testing in order to validate the desired functioning of the AI. Utilizing artificial intelligence or machine learning (ML) within the cloud program of the CPS can help to improve the operation of the CPS control cycle by optimizing, for example, the energy consumption usage of a smart building. Later, we introduce a ML-assisted smart lock as access control to an imaginary smart building (CPS) functioning as an example.

The concept of cybersecurity is extensively used and remains vague and intricated. Its goal is to enable operations in cyberspace without risk of physical or digital harm [7]. It can be applied to various contexts, from business operations to ICT technologies. According to De Groot [5], cybersecurity can be defined as "the body of technologies, processes, and practices designed to protect networks, programs and data from attack, damage or unauthorized access. Cyber security may also be referred to as information technology security." Gartner defines cybersecurity as follows: "Cybersecurity is the combination of people, policies, processes and technologies employed by an enterprise to protect its cyber assets" [8].

Cybersecurity is an important viewpoint for the running of the CPS, especially when concerning the machine learning algorithms. These algorithms are probabilistic, involve a vast amount of data, and their training can take a long time as well as be costly. Since ML models are also used for various tasks ranging from non-demanding choices to highly critical decisions, the protection of the models against perpetrators is indispensable.

In this article, the authors have showcased the need for artificially intelligent systems and the importance of cybersecurity against the threats involved with both AI and CPS. In Sect. 2, the authors explain artificial intelligence and machine learning in more detail. In Sect. 3, the authors describe adversarial machine learning, where Sect. 3.1 showcases the ML testing-phase adversarial inference evasion attacks, and Sect. 3.2 reviews the defenses against evasion attacks. In Sect. 4, we discuss the application of evasion attacks and defenses against them in a smart building context. Lastly, Sect. 5 concludes the paper.

2 Artificial Intelligence and Machine Learning

AI is a mathematical approach to estimate a function, and it can be expressed mathematically as f(x): $Rn \rightarrow Rm$, where f(x) is the function to be modeled, Rn represents the real input values and Rm represents the possible real output values. ML research field is needed to make AI models and systems more capable of handling new situations [11], because resources might have been limited during initial training, and the new situation might be from outside the original input or output domain that was used for training of the model.

Deep Learning (DL) is a subfield of ML, where the learning is done with models that have multiple layers within their structure. The additional depth can help the models to learn more complex associations within the given data than regular AI models [12]; hence DL models are called deep. AI is a very enticing choice for many different use cases, where the function to be estimated is either unknown or difficult to implement in practice, such as machine translations. In practice, the quality and quantity of data, the structure of the model, and training time, as well as the training method, affect how any AI learns to make its choices.

Neural network (NN) is a popular base model used in the development of AI solutions. The model has three layers: an input layer, hidden layer, and output layer, where data flows from the input layer through the hidden layer consisting of multiple layers, and the result is produced to the output layer. NNs are a collection of structured, interjoined nodes whose values are comprised of all the weights of the connections coming to each node. Every value of a node is inputted to an activation function, such as a rectified linear unit (ReLu). The activation function is typically the same for all the nodes in the same layer.

Long-Short Term Memory neural network (LSTM) is a special case of Recurring Neural Network (RNN) [13], which retains output information from previous timesteps as part of the input information. The extra information can be helpful i.e. when forecasting with sequential data. Because NNs can suffer from the problems of vanishing and exploding gradients, which likely will increase with the growth of sequence size, LSTMs have three gates within each node that are used to control the information going through them [13]. These logical gates use sinh and tanh activation functions to control the flow and size of internal representations of the inputs and outputs.

3 Adversarial Machine Learning

In recent years, the utilization of the ML approach has been flourishing and seen an expeditious increase. ML has been used to identify relevant patterns in the information and make even more precise predictions as a function of time. ML methods have been used to, i.e., malware detection, facial and speech recognition, robotics, autonomous cars, etc. The benefits come with the disadvantages, and ML models

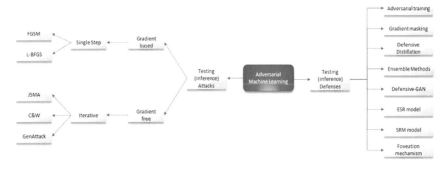

Fig. 1 Inference evasion attacks and defenses in adversarial machine learning

may be vulnerable to exploitation and manipulation. This kind of risk can be considered as "AML" as systems can be fooled by perpetrators through malicious input to cause a malfunction in ML models to make incorrect assessments. During the AML attacks (Fig. 1), inaccurate or misrepresentative data can be injected into an ML model at the model training phase, or malicious data may be used to swindle a trained model to make it behave abnormally and provide false predictions [10].

The field of AML has emerged to study vulnerabilities of the ML approach in adversarial settings and to develop techniques to make learning robust to adversarial manipulation [32]. AML is about learning in the presence of adversaries, and the learning can happen, i.e., in an exploratory way in testing (inference time), when the attacker aims to confuse the decision of the ML model after it has been learned [29]. Adversarial attacks have been under extensive research recently, and one prominent finding by Szegedy et al. [30] was in the field of computer vision, revealing that a small perturbation in the form of carefully crafted input could confuse a deep neural network (DNN) to misclassify an image object. After the study, adversarial attacks have been more widely explored beyond image classification.

3.1   ML Testing-Phase Adversarial Inference Evasion Attacks

An evasion attack occurs when the neural network receives an adversarial example as an input, which can be considered as a deliberately perturbed appearing unmodified and precisely the same to human eyes, but is able to deceive the classifier [20]. Attacks against ML can be traced back to the year 2004 when Battista Biggio published the paper: "Wild Patterns" concerning the rise of adversarial machine learning. Research in the field of adversarial examples has been conducted to a great extent in computer vision, but they are basically applicable to any type of ML systems. For example, Sharif et al. [27] managed to impersonate as another person by wearing particularly designed adversarial glasses. The use of glasses embodies that perturbations can occur due to other incidents than modifying images with proper adversarial techniques.

Evasion attacks can be roughly divided into gradient-based and gradient-free classes. Gradient-based attacks can be further divided into a single step and iterative attacks. Fast Gradient Sign Method (FGSM), proposed by Goodfellow et al. [9], is a well-known single-step white-box type of an attack utilizing gradients of neural network in generating an adversarial example. In the case of adversarial images, the aim is to create an image that maximizes the loss. This can be expressed with the following equation: $adv\_x = x + \varepsilon * sign(\nabla_x J(\Theta, x, y))$ in which adv_x = adversarial example, x = original input image, y = original input label, ε = multiplier to ensure the perturbations are small, Θ = model parameters and J = loss. (TensorflowCore) [31] FGSM attack can be used to counter any kind of ML algorithms making use of gradients and weights. The backpropagation method can be used in calculating gradients. FGSM provides low computational cost and can be effective to run if weights and learning architecture is known. FGSM applies well for crafting adversarial examples with major perturbations, but as a downside, it is easier to detect than other methods, such as L-BFGS and JSMA [3].

Jacobian-based saliency map algorithm (JSMA) was presented by Papernot et al. to optimize L_0 distance. JSMA attack can be used for deceiving classification models, for example, neural network classifiers, such as DNNs in image classification tasks. The algorithm is able to induce the model to misclassify the adversarial image concerned as a determined erroneous target class [33]. JSMA is an iterative process, and in each iteration, it saturates as few pixels as possible by picking the most important pixel on the saliency map in a given image to their maximum or minimum values to fool the classifier [24]. Even though the attack alters a small number of pixels, the perturbation is more significant than $L\infty$ attacks, such as FGSM [19]. The method is reiterated until the network is cheated, or the maximal number of altered pixels is achieved. JSMA can be considered as a greedy attack algorithm for crafting adversarial examples, and it may not be useful with high dimension input images, such as images from ImageNet dataset [19].

The perturbation generated by JSMA is the gradient direction of the predicted value of the target class label, a forward derivative. The forward derivative is composed of the partial derivative value of the target class for each pixel [14]. To craft an adversarial example from a given input X (example point), JSMA first computes the gradient $\nabla F(x)$ in which F denotes feedforward neural network [17]. The dimensions for the model output (number of classes) and the inputs are M and N, respectively. The Jacobian is computed by:

$$\nabla F(X) = \frac{\partial F(X)}{\partial X} = \left[\frac{\partial F_j(X)}{\partial F_j(X)} \right]_{i \in 1...N, j \in 1...M} \tag{1}$$

The next step is constructing a saliency map used to choose the most relevant component to perturb. The goal is to maximize the output for the target class c, $F^c(X)$, and minimize the output for the other classes $j \neq c$ [34]. This can be reached by utilizing the adversarial saliency map:

$$S(X, c) = \begin{cases} 0, & \text{if} \frac{\partial F_c(X)}{\partial X} < 0 \text{ or } \sum_{j \neq t} \frac{\partial F_j(X)}{\partial X} > 0 \\ \frac{\partial F_c(X)}{\partial X} \left| \sum_{j \neq t} \frac{\partial F_j(X)}{\partial X} \right|, & \text{otherwise} \end{cases}. \tag{2}$$

The adversarial attack can be created by starting from a selected example point and to iteratively perturb the example point in the direction of $S(X, c)$ by minor steps until the predicted label changes. For an untargeted attack, the prediction score is minimized for the winning class in a similar fashion [34].

The Limited-memory-Broyden-Fletcher-Goldfarb-Shanno (L-BFGS) is a popular non-linear box-constrained gradient-based quasi-Newtonian numerical optimization algorithm using a limited amount of memory for adversarial examples generation [33]. The algorithm can be utilized for solving high-dimensional minimization problems in where both the objective function and its gradient can be computed analytically [4]. Due to the costly linear search method is used to find the optimal value, especially for complex DNN networks, the algorithm is considered time-consuming and therefore impractical [36].

According to Krzaczynski, L-BFGS algorithm is for finding local extrema of functions based on Newton's method of finding stationary points of functions. The second degree of approximation can be utilized to find minimum function f(x) using the Taylor series as follows:

$$f(x_0 + \Delta x) = f(x_0) + \nabla f(x_0)^T \Delta x + \frac{1}{2} \Delta x^T \cdot H \Delta x. \tag{3}$$

In the formula, H denotes hessian matrix ($H = B^{-1}$), $f(x_0)$ is a locally modelled f at point x_0 at each algorithm iteration, $\nabla f(x_0)$ is a gradient of the function. The minimum can then be solved from the following equation:

$$\nabla f(x_0 + \Delta x) = \nabla f(x_0) = H \Delta x \Rightarrow \nabla f(x_0 + \Delta x) = 0 \Rightarrow \Delta x_0 = -B^{-1} \cdot \nabla f(x_0). \tag{4}$$

According to Okazaki [22], the computational cost of the inverse Hessian matrix used in the L-BFGS algorithm is high, specifically when the objective function takes a significant number of variables. L-BFGS algorithm iteratively looks for a minimizer by using approximation of the inverse Hessian matrix by information from last m iterations. The process mitigates computational time in solving large-scale problems, and saves the memory storage. However, the L-BFGS algorithm solves the minimization problem only if the objective function F(x) and its gradient G(x) are computable.

Carlini and Wagner (C&W attack) has been presenting one of the most powerful iterative gradient-based attacks towards Deep Neural Networks (DNNs) image classifiers due to its ability to break undefended and defensively distilled DNNs on which, for example, L-BFGS and DeepFool attacks fail to find the adversarial samples. In addition, it can reach significant attack transferability. C&W attacks are optimization-based adversarial attacks, which can generate L_0, L_2 and L_∞ norm measured adversarial samples, also known by CW_0, CW_2, and CW_∞. The attack attempts to minimize the distance between a valid and perturbed image while still causing the perturbed image to be misclassified by the model [28]. In many cases, it can decrease classifier accuracy near to 0%. According to Ren et al. [25], C&W attacks reach a 100% success rate on naturally trained DNNs for image datasets, such as MNIST, CIFAR-10, and ImageNet. C&W algorithm is able to generate powerful adversarial examples, but computational cost is high due to the formulation of the optimization problem. The C&W attack formulates the following optimization objective:

$$\min_{\delta} D(x, x + \delta) + c \cdot f(x + \delta), \text{ where } x + \delta \in [0, 1] \text{ and } f(x + \delta) \leq 0. \quad (5)$$

In the optimization formula, δ signifies the adversarial perturbation, D means L_0, L_2 or L_∞ distance metric, and $f(x + \delta)$ denotes customized adversarial loss. The condition for function $f(x + \delta) \leq 0$ is valid only if DNN's prediction is targeted by attack [25]. This attack is to search for the smallest weighted perturbation by norms concerned in order to simultaneously force network to improperly classify the image. In the formula, c stands for a hyperparameter to balance the two parts of equation, and $f(x + \delta)$ is the loss function to measure the distance between the input image and the adversarial image [15, 16].

Gradient-based optimization and computation can solely be carried out in situation (white-box setting) in which a perpetrator possess full information of the model architecture and weights and in addition, full access and control over a targeted DNN. As this is not likely a real-world case, adversarial attacks in the black-box settings are considered. Gradient-based approaches usually are computationally expensive; therefore gradient-free optimization can be considered as a practical alternative. Alzantot et al. [1] introduced GenAttack, which bases on genetic algorithms being population-based gradient-free optimization strategies. GenAttack is robust to defenses executing gradient masking or obfuscation. It is also able to craft perturbations in the black-box setting to override some gradient-altering defense methods. The algorithm can conduct successful targeted black-box attacks by querying the target model remarkably less than other comparable methods, even against large-scale high-dimensional ImageNet models, which earlier methods have had difficulties to scale to.

3.2 Defense Mechanisms Against Evasion Attacks

Several defenses have been presented in the literature to reduce the effect of adversarial attacks, such as C&W, FGSM, JSMA, and L-BFGS. The network's robustness can be improved by continuously training the model with new types of adversarial samples to make the classifier more robust against forthcoming attacks. According to Moosavi-Dezfooli et al. [21], the number of samples to train the model doesn't solve the problem as novel types of adversarial samples emerge at all times. Luo et al. [18] presented the 'foveation' mechanism, which can be used in defending against adversarial samples generated by L-BFGS and FGSM. It is assumed that by training a significant number of data sets based on the DNN classifier can be considered to be robust to image scaling and transformation changes. Confrontation mode does not have this feature.

Many potential defense mechanisms can be thought of as belonging to the group of gradient masking. These techniques generate a model without useful gradients, for example, by using the nearest neighbor classifier instead of DNN. Nearest neighbor, though, has been shown to be vulnerable to attacks based on transferring adversarial examples from smoothed nearest neighbors. Papernot et al. [23] Gradient masking is used because most white-box attacks operate by calculating the gradient of the DNN model [14]. Therefore, if the efficient gradient cannot be calculated, the attack will not be successful. Gradient masking's primary aim is to make the gradient useless. Yanagita and Yamamura [35] states that gradient masking is able to eliminate the valuable gradient for perpetrators, but adversarial perturbations easily transfer over most models. The models concerned can be fooled by adversarial examples crafted based on other models. A Black-box type of attack can be then utilized to overcome gradient masking defenses.

Defensive distillation can be counted to one of the adversarial training techniques providing flexibility to an algorithm's classification process, making the model less prone to exploitation. According to Carlini and Wagner [2], it can take an arbitrary neural network, increase its robustness, and mitigate the ability to find adversarial examples from 95 to 0.5%. It was originally introduced to transfer learned information from one NN to another (defensive technique). Its feasibility to defend against FGSM and JSMA was demonstrated [35] presented a defense distillation method that can reduce the input variations making the adversarial crafting process more difficult, helping DNN to generalize the samples outside the training set and reducing the effectiveness of adversarial samples on DNN. The distillation method transfers the knowledge from one architecture to another by decreasing the size of DNN. In distillation adversarial training, one model can be trained to predict the output of probabilities of another model trained on an earlier baseline standard. Defense distillation provides the advantage of being compliant with yet unknown threats. Usually, the most efficient adversarial defense training methods demand interminable input of signatures of known vulnerabilities and attacks into the system. The distillation provides a dynamic method requiring less human intervention.

As a drawback, if a perpetrator has a lot of computing power available and the proper fine-tuning, she can utilize reverse engineering to find fundamental exploits. Defense distillation models are also vulnerable to poisoning attacks in which a malicious actor corrupts a preliminary training database. (DeepAI) [6] Defensive distillation can be evaded by the black-box approach [23] and also with optimization attacks [30]. Carlini and Wagner [2] proved that defensive distillation failed against their L_0, L_2 and L_∞ attacks. These new attacks succeed in finding adversarial examples for

100% of images on defensively distilled networks. Previously known weaker attacks can be stopped by defensive distillation, but it cannot resist more powerful attack techniques.

Empirical experiments indicate it is challenging to detect adversarial examples generated by the C&W method. However, C&W attacks can be detected by a relatively high 93% accuracy on ImageNet-1000 by utilizing the Enhanced Spatial Rich Model (ESRM), which also provides high detection accuracy against weaker single step gradient-based FGSM attacks. The computational time of ESRM is long due to high-dimensional features. ESRM is an extended version of SPAM (Spatial Rich Model), which extracts residuals from images. Residuals can be seen as the image noise components gained by subtracting from each pixel its estimate received using a pixel predictor from the pixel's neighborhood. SPAM doesn't provide means to consider the location of modified pixels caused by adversarial attacks; therefore, estimation of the relative modification probability of each pixel is required to be done. This can be done by utilizing MPM (Modification Probability Map), which is the matrix of all pixel's modification probabilities. ESRM provides a new Markov transition probability estimation based on MPM [15, 16].

Defense-GAN (Generative Adversarial Networks) is a defense strategy providing sophisticated defense methods against white-box and black-box adversarial attacks used to threaten classification networks. Defense-GAN is trained to model the distribution of unperturbed images, and at inference time, it finds a close output to a given image not containing adversarial changes. Prior to sending the image to the classifier, it is projected onto the generator by minimizing the reconstruction error $||(G(z)-x||_2^2$, and the resulting reconstruction G(z) will then be passed to the classifier concerned. Training the generator to model the unperturbed training data distribution reduces potential adversarial noise [26].

Defense-GAN can be utilized jointly as an add-on with any classifier without modifying the classifier structure. In addition, re-training the classifier ought not to be required, and mitigation of performance should not be prominent. The defense method concerned can be utilized to defend against any attack as it does not presume an attack model but takes advantage of the generative efficiency of GANs to reconstruct adversarial examples. Due to the GD (Gradient Descent) loop and non-linear nature of Defense-GAN, a white-box type of attacks are challenging to conduct [26]. The authors mentioned a conducted C&W L2-norm attack under white-box setting against a convolutional neural network classifier. Under a white-box setting, with no attack and for most of the target classifiers, accuracy was higher than 0.992, utilizing the MNIST image dataset. By conducting the attack (both FGSM and C&W) and taking advantage of the Defense-Gan defense strategy method, accuracy decreased only less than 1%.

Defense-GAN overcomes adversarial training as a defense method, and when conducting adversarial training using FGSM in generating adversarial examples against, for example, the C&W attack, adversarial training efficiency is not sufficient. In addition, adversarial training does not generalize well against different attack methods. Increased robustness gained by using adversarial training is reached when the attack model used to generate the augmented training set is the same as that used by the perpetrator. Hence, as mentioned, adversarial training endures inefficiently against the C&W attack; therefore, a more powerful defense mechanism should be utilized. Training GANs is a remarkably challenging task, and if GANs are not trained correctly, and hyperparameters are chosen incorrectly, the performance of the defensive mechanism may significantly mitigate [26].

4   ML-Based Smart Lock System in Smart Buildings

Figure 2 showcases an example case of a smart lock system integrated into a smart building under evasion attacks. The intended use is such that in order to control access to the building, the building user is recorded on video surveillance, and each image is sent to the CPS's ML model that uses pre-taken and confirmed images from the database to compare the current user's image to the confirmed ones. A third-party service is used in the background for the comparisons, and each image before sending it to the cloud-based service is preventatively classified and perturbed. The figure also includes the API, and it is intended for users to register, sign-in, and update their own profiles, including their face images. Therefore, there are two ways a perpetrator might attack the CPS's ML model, either by means of physical adversarial evasion attacks or by using the API to send the malicious image as input to the ML model to achieve the evasion attacks.

To protect other inhabitants of the building, the ML model ought to be trained against different adversarial attacks e.g. using a model that exhibits a structure of stacked ensemble. The stacked ensemble should firstly be able to take in the image as input and use defensive distillation to reduce the number of malicious pixels' effects in the image since defensive distillation is a valuable technique to combat unknown attacks. Then, the "new" image can be given to a group of neural networks that have different structures and have been trained with different data against multiple

Fig. 2 Evasion attacks towards ML-based smart lock system in smart building's context

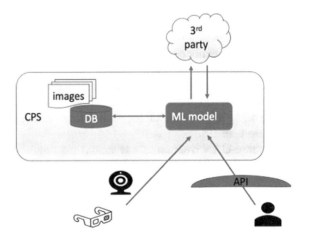

attacks. For example, it might be beneficial to have a CNN or LSTM that has been trained against FGSM and JSMA based adversarial attacks and to have another DNN trained with an enhanced spatial rich model to find any remaining potential C&W attack generated pixel combinations. Lastly, the second to last model of the main ML model combines the classification results and the corresponding confidence values and outputs the ruling of the other (sub)NN models. If no malicious inputs have been detected, then the input and the likeliest images from the database are sent to be compared in the cloud by the third-party vendor(s). The third-party vendor's results are used to ascertain which of the known users is in the image. This result gets inputted for the last model in the ML model, and it creates the access-allowing commands for the user to the smart building.

5 Conclusion

In this paper, the authors reviewed the concept of adversarial machine learning and related evasion attacks. These attacks include gradient-based and gradient-free techniques, such as FGSM and C&W, respectively. The attacks can perturb the input data in such a way that the inputs seem valid for a human but mess maliciously with an artificial intelligence model.

The defenses against the adversarial evasive attacks showcased here were defensive distillation, defense-GAN, foveation, nearest neighbor clustering, Spatial Rich Model (SPAM), and enhanced spatial rich model (ESRM). To detect and prevent the adversarial attack effects, it is recommended to utilize a multitude of defensive tragedies, such as to maintain role-based access control to the models' APIs and functions. In addition, it is vital to train the desired ML models against different attacks. Unfortunately, none of the defensive methods were completely impenetrable. Thus an ensemble of AI models is recommended, even though training NNs against these attacks is time-consuming, and it costs both money and resources.

References

1. Alzantot M, Sharma Y, Chakraborty S, Zhang H, Hsieh C-J, Srivastava MB (2019) GenAttack: practical black-box attacks with gradient-free optimization (2019). arXiv:1805.11090v3 [cs.LG]
2. Carlini N, Wagner D (2017) Towards evaluating the robustness of neural networks. arXiv:1608.04644v2 [cs.CR]
3. Co KT (2018) Bayesian optimization for black-box evasion of machine learning systems. Imperial College London, Department of Computing
4. Coppola A, Stewardt BM (2014) LBFGS: efficient L-BFGS and OWL-QN optimization in R. http://cran.csiro.au/web/packages/lbfgs/vignettes/Vignette.pdf. Accessed 26 Aug 2020
5. De Groot J (2020) What is cyber security? Definition, best practices & more. Data Insider. https://digitalguardian.com/blog/what-cyber-security. Accessed 17 Aug 2020

6. DeepAI, What is defensive distillation? https://deepai.org/machine-learning-glossary-and-terms/defensive-distillation. Accessed 9 Oct 2019

7. Dewar RS (2014) The "Triptych of Cyber Security": a classification of active cyber defence. In: 6th international conference on cyber conflict. NATO CCD COE Publications, Tallinn. https://www.ccdcoe.org/uploads/2018/10/d1r1s9_dewar.pdf. Accessed 17 Aug 2020

8. Gartner: Cybersecurity (2020). https://www.gartner.com/en/information-technology/glossary/cybersecurity. Accessed 2 Sept 2020

9. Goodfellow I, McDaniel P, Papernot N (2018) Making machine learning robust against adversarial inputs. Commun ACM 61(7):56–66

10. Ibitoye O, Abou-Khamis R, Matrawy A, Shafix MO (2019) The threat of adversarial attacks on machine learning in network security—a survey. arXiv:1911.02621v1 [cs.CR]

11. Jordan MI, Mitchell TM (2015) Machine learning: trends, perspectives, and prospects. Sciencemag.org. Science 349(6245)

12. LeCun Y, Bengio Y, Hinton G (2015) Deep learning. Nature 521(7553), 436–444

13. Lipton ZC, Berkowitz J, Elkan C (2015). A critical review of recurrent neural networks for sequence learning. arXiv preprint arXiv:1506.00019

14. Liu C, Ye D, Shang Y, Jiang S, Li S, Mei Y, Wang L (2020) Defend against adversarial samples by using perceptual hash. Comput Materials Continua (CMC) 62(3):1365–1386. https://doi.org/10.32605/cmc.2020.07421

15. Liu J, Zhang W, Yu N (2018) CAAD 2018: iterative ensemble adversarial attack. arXiv:1811.03456 [cs.CV]

16. Liu J, Zhang W, Zhang Y, Hou D, Liu Y, Zha H, Yu N (2018) Detection based defense against adversarial examples from the steganalysis point of view. arXiv:1806.09186v2 [cs.CV]

17. Loison A, Combey T, Hajri H (2020) Probabilistic Jacobian-based saliency maps attacks. arXiv:2007.06032 [cs.CV]

18. Luo Y, Boix X, Roig G, Poggio T, Zhao O (2015) Foveation-based mechanisms alleviate adversarial examples. arXiv:1511.06292

19. Ma S, Liu Y, Tao G, Lee WC, Zhang X (2019) NIC: detecting adversarial samples with neural network invariant checking. In: Network and distributed systems security (NDSS) symposium 2019, San Diego, CA, USA. https://doi.org/10.14722/ndss.2019.23415

20. Moisejevs I (2019) Evasion attacks on machine learning (or "adversarial examples"). In: Towards data science. https://towardsdatascience.com/evasion-attacks-on-machine-learning-or-adversarial-examples-12f2283e06a1. Accessed 30 July 2020

21. Moosavi-Dezfooli SM, Fawzi A, Fawzi O, Fossard P (2017) Universal adversarial perturbations. In: Proceedings of the IEEE conference on computer vision and pattern recognition, pp 1765–1773

22. Okazaki N (2014) libLBFGS: a library of limited-memory Broyden-Fletcher-Goldfarb-Shanno (L-BFGS). http://www.chokkan.org/software/liblbfgs. Accessed 26 Aug 2020

23. Papernot N, McDaniel P, Goodfellow I, Jha S, Celik ZB, Swami A (2016) Practical black-box attacks against machine learning. arXiv:1602.02697 [cs.CR]

24. Pawlak A (2020) Adversarial attacks for fooling deep neural networks. https://neurosys.com/article/adversarial-attacks-for-fooling-deep-neural-networks. Accessed 31 July 2020

25. Ren K, Zheng T, Qin Z, Liu X (2020) Adversarial attacks and defences in deep learning. Engineering 6(3):346–360. https://doi.org/10.1016/j.eng.2019.12.012

26. Samangouei P, Kabhab M, Chellappa R (2018) Defense-GAN: protecting classifiers against adversarial attacks using generative models. arXiv:1805.06605v2 [cs.CV]

27. Sharif M, Bhagavatula S, Bauer L, Reiter MK (2016) Accessories to a crime: real and stealthy attacks on state-of-the-art face recognition. https://doi.org/10.1145/2976749.2978392

28. Short A, Pay TL, Gandhi A (2019) Defending against adversarial examples. In: Sandia report, SAND 2019-11748. Sandia National Laboratories

29. Song D (2019) Plenary session—toward trustworthy machine learning. In: Proceedings of a workshop of robust machine learning algorithms and systems for detection and mitigation of adversarial attacks and anomalies, pp 35–38

30. Szegedy C, Zaremba W, Sutskever I, Bruna J, Erhan D, Goodfellow I, Fergus R (2013) Intriguing properties of neural networks. arXiv:1312.6199
31. TensorflowCore: Adversarial Example Using FGSM (2020). https://www.tensorflow.org/tutorials/generative/adversarial_fgsm. Accessed 30 July 2020
32. Vorobeychik Y, Kantarcioglu M (2018) Adversarial machine learning synthesis lectures on artificial intelligence and machine learning. https://doi.org/10.2200/S00861ED1V01Y201806AIM039
33. Wiyatno R, Xu A (2018) Maximal Jacobian-based saliency map attack. arXiv:1808.07945v1 [cs.LG]
34. Wu H, Wang C, Tyshetskiy Y, Docherty A, Lu K, Zhu L (2019) Adversarial examples on graph data: deep insights into attack and defense. arXiv:1903.01610 [cs.LG]
35. Yanagita Y, Yamamura M (2018) Gradient masking is a type of overfitting. Int J Mach Learn Comput 8(3):203–207. https://doi.org/10.18178/ijmlc.2018.8.3.688
36. Yuan X, He P, Zhu Q, Bhat R, Li X (2017) Adversarial examples: attacks and defences for deep learning. arXiv:1712.07107 [cs.LG]

Software License Audit and Security Implications

Wee Kiat Yang, Amin Hosseinian-Far, **Luai Jraisat**,
and Easwaramoorthy Rangaswamy

Abstract The typical purpose of software auditing is to assess the conformant of the developed software with the original plans, procedures, relevant regulations. Every audit involves several people with various roles in the auditing processes. The audit itself entails a number of preferable characteristics. In any audit engagement, the perceptions of the audit quality are directly related to the perceived reputation, credibility and objectivity of the auditor. This paper highlights and critically reviews the research works related to quality in audit, in particular, software license audit from the perspective of internal control and security. The paper examines existing studies in the field with a view to identifying future research opportunities in relation to software license auditing. Moreover, security implications and challenges in the context of software auditing, and a set of recommendations are provided.

Keywords License audit · Software audit · Compliance · Security

1 Introduction

In every software company, survival depends on aggressive development and protection of its intellectual property [1, 2]. Other than protection from illegal installation of products, compliance is another important area of software licensing. This is especially so in relation to consumption and proper licensing of the software products and ensuring their customers software assets are managed effectively [3].

W. K. Yang · A. Hosseinian-Far (✉) · L. Jraisat
Faculty of Business & Law, University of Northampton, Northampton NN1 5PG, UK
e-mail: amin.hosseinian-far@Northampton.ac.uk

W. K. Yang
e-mail: wee.yang@Northampton.ac.uk

L. Jraisat
e-mail: luai.jraisat@Northampton.ac.uk

E. Rangaswamy
Amity Global Institute, Singapore 238466, Singapore
e-mail: moorthy@singapore.amity.edu

© The Author(s), under exclusive license to Springer Nature Switzerland AG 2021   171
H. Jahankhani et al. (eds.), *Cybersecurity, Privacy and Freedom Protection
in the Connected World*, Advanced Sciences and Technologies for Security
Applications, https://doi.org/10.1007/978-3-030-68534-8_12

In an audit engagement, the perceptions of the audit quality are directly related to the perceived reputation, credibility and objectivity of the auditor; in essence, the quality and experience of an auditor [4].

Considering the plethora of software development activities in today's digital world, license auditing process entail a number of challenges. Some of these challenges are:

- Inappropriate licensing model for customer
- Ineffective software asset management process
- Ineffective users' access management
- Ineffective authorization management
- Engaging software customers in the audit process
- Third party audit firms that do not understand or abide by the licensing model and measurement methodology of software vendors
- Inaccurate information or advice provided by third party audit firms

Considering these challenges, this paper attempts to provide a concise overview of the software license auditing process. It also attempts to provide a brief discussion on security implications in the context, and offer a number of recommendations.

The rest of the paper is structured as follows. Section 2 provides a literature review on software licensing, contributing factors on audit quality, and relevant concepts within the field. Section 3 provides a brief discussion on auditing challenges and security implications. A brief discussion, the future of software license auditing, and a number of recommendations are provided in Sect. 4. The paper is concluded in Sect. 5.

2 Literature Review

Audit is defined as "an official examination of business and financial records to see that they are true and correct" or "an official examination of the quality or standard of something" [5] and License Auditing is the official examination of the licensees' deployment and utilization of the software to ensure proper Software Asset Management (SAM). Most of the existing studies are conducted in the area of financial auditing and are focused on the financial examinations.

Other than the qualification of the auditor, audit quality is an important aspect of a successful audit. Audit quality, as defined by DeAngelo [6] "..to be the market-assessed joint probability that a given auditor will both (a) discover a breach in the client's accounting system, and (b) report the breach.". It is also determined between low-to-high quality involving various factors [7].

In the area of financial auditing, failure in an audit often occur when the auditor does not enforce the proper audit principles. This is known as the Generally Accepted Accounting Principle (GAAP) failure. Another circumstance when an audit failure can occur is when the auditor fails to produce a qualified report for the engagement known as audit report failure [8].

Fig. 1 Framework adopted from Francis's Audit Quality Assessment Framework [7]

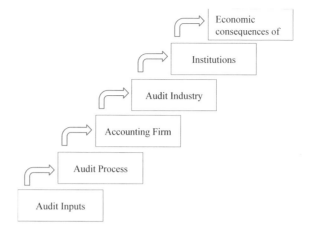

To date, several research works have studied audit quality and the factors that are affecting the audit quality. Some of the notable factors include: Audit firm's reputation [9, 10], Audit firm's size [6, 7, 11–14], Non-Audit services [15], Task complexity [16], Auditor's rotation [16, 17] and Auditor's independence [16, 18].

To effectively study the audit quality, Francis [7] developed a framework that assesses some of the factors that are impacting an audit quality. The framework consists of 6 levels namely Audit inputs, Audit process, Accounting firm, Audit industry, Institutions and Economic consequences of audit outcome (Fig. 1).

Audit testing procedures and team personnel are the inputs to the framework. The inputs are then transited to the next level of the framework for the decisions to be made by the engagement team through the audit process. The assumption of Francis was that auditing occurs in the context of the audit firm [7]. The audit firm is also the institution that hires, trains and develops the auditors. The audit report is also cleared and issued under the name of the firm. It is this collection of audit forms that has formed the Audit Industry and the audit market which is in turn governed by various certification bodies and institution. Therefore, these activities affect clients' companies in terms of report outcome. It was also noted by Francis that if an audit is carried out by competent auditors, the audit result will be of a higher quality. However, there are no evidence to support this claim. On the contrary, performance can be affected by many factors [11–13]. It was also perceived that larger firms have more in-house expertise and therefore have a greater opportunity to produce higher quality audits. While there were insufficient data to link audit quality and size of audit form [7], it is somewhat true in terms of expertise. In the context of principle software vendors, the auditors, who are the employees of the principal software vendors, usually have the expertise and expert knowledge in their own software products. Additionally, they are supported by a large number of in-house resources that are experts in different dimensions of the software product. Thus, the quality of audits conducted by principle software vendors are perceived to be of a higher quality compared to external audit firms. Other factors should be considered. There are instances that accounting audits in their audit tenure conduct non-audit services, and this may influence the level of specialty required for nonfinancial audits.

There are no firm evidence pointing to the loss of audit quality for firms registered with the Securities and Exchange Commission (SEC) when performing non-audit services. Instead, such registration is observed to be positively associated with the audit quality, and therefore it is of the opinion that it is actually the result of auditors providing higher quality audit [19]. In a separate study conducted by Boskou, Kirkos and Spathis [20], information from publicly available annual reports were used to develop a classification of audit quality. Using machine learning techniques to perform text mining from a company's annual report also yielded the conclusion that there is a positive relationship between financial, operational and strategic risks [21]. Boskou, Kirkos and Spathis used models and natural language processing to review different models and developed a classification that can be used to predict internal audit quality, by enabling the auditor to effectively assess the risks and better plan audit procedures [20].

In order to provide inputs relating to the attributes that are important to audit quality, around 4,600 audit professionals in IT and Finance were surveyed by Stoel et al. Accounting skill and Audit skills, Business process knowledge and experience were rated higher by the Financial auditors while Auditor Experience with Auditee, IT and Controls Knowledge, and Planning and Methodology were rated higher by IT [22]. This information provides greater insights to audit leads when planning for an audit that can contribute positively towards improving the quality of audit.

In a separate study conducted by Kilogre and Bennie [23] on Australian auditors, the audit quality attribute that was perceived to be the most important was reported to be 'Audit firm size' (Table 1).

"Whether or not they are IT auditor or Financial auditors, it was observed that there was a positive correlation between auditors in big-N firms and audit quality that is referred to as the "Big N effect" [14]; therefore, it seems that the higher the quality of the audit, the greater the assurance of report quality [24]. However,

Table 1 Categorization of attributes and their relative ranking, adopted from [23]

Rank	Attributes	Category
1	Audit firm size	Independence/Competence
2	Partner/Manager attention to audit	Competence/Interaction
3	Communication between audit team and client management	Interaction
4	Knowledgeable audit team	Competence
5	Audit firm industry experience	Competence
6	Partner knowledge of client industry	Competence
7	Senior manager/manager knowledgeable—client industry	Competence
8	Provision of non-audit service	Independence
9	Audit quality review	Competence
10	Audit partner tenure	Independence

it is challenging to compare the audit quality between firms that assign the audits in Big-N and non-Big-N as the audit firms choose their auditors based on firm or the characteristics of the auditor. Nevertheless, larger firms have more resources and technology support over non-Big-N firms. Moreover, Big-N auditors are known to be "generalist" [25, 26]. That being said, firms with poor performance are more likely to change auditors [27].

Apart from the studies involving attributes contributing to audit quality, some research works have attempted to adopt text mining to examine audit report and evaluate the quality of audits. A company's annual report is a useful tool where the information is available publicly. This information can be analyzed textually using classification methods. Machine learning and natural language processing tools can also be used to effectively assess risks and provide recommendation to auditor to improve audit procedures [20].

Generally, in an audit practice, auditors collect artefacts and supporting documents through the use of an audit procedure to detect materials that may demonstrate misrepresentation in financial statements. When such misinterpretation is detected by auditors, such misconceptions are usually communicated to auditees' management to adjust the misstatements. This is similar with license auditing where the license auditor communicates with customers (or management) to understand the cause of such misinterpretation. These misinterpretations, however, can also be due to the internal processes that result in certain findings. For instance, suppose a customer has User Licenses created in the System that exceeds his/her entitlement. Discussions are carried out with the customer to understand the issue, and where necessary, User number is adjusted downwards based on the assessment and feedback; Also, another underlying reason may be due to the customer's failure to clean up or deactivate an unused User account, that was later verified using system data. In summary, the phases of determining audit quality are (i) Detecting (ii) Adjusting (iii) Reporting of misstatement/misrepresentation to achieve audit quality.

The observable audit objectives include audit adjustments, audit opinion and the quality of audited financial statements. Adjustment occurs when there is a misstatement in the pre-audit financial statements and is required by the auditor to correct the misstatement or issuing a modified audit opinion in response to the misstatement [28]. According to Xiao et al. [28], an audit adjustment appears to occur more frequently when the audit effort is greater. The increase audit effort means the auditor can perform more comprehensive audit verifications with the attainment of more artefacts to demonstrate certain system behavior or financial postings, thus improves the ability to detect misstatements. With enough artefacts and evidences, the auditor is in a better position to determine anomalies, and less likely to waive any audit adjustments proposed by the client's management. If an adjustment is required with the support of the evidence, auditors are in a better position to ensure detected misstatements can be corrected through adjustments. Therefore, higher audit effort does not naturally means that there are more issuances of modified audit opinions [28].

2.1 Licensing

In the software market, infringement of copyright includes the reverse engineering of software codes and unauthorized duplication and use of software. The Business Software Alliance estimated that, in 2011, that the illegal software market caused about US$63 billion in damages all over the world [29].

When we look at software licensing, the general categories are opensource, non-opensource and subscription based. There are well over 60 different types of opensource software [30] such as mSQL, Linux, opensource Office etc. In terms of non-Opensource, there are several major software vendors with products that are non-Opensource, for instance Microsoft^{TM}, SAP^{TM}, Oracle^{TM}, Adobe^{TM} etc. Subscription software examples are Salesforce^{TM}, Workday, SAP Cloud Solutions etc.

In terms of license auditing, in particular to SAP licensing, there are different licensing models namely perpetual, subscription and consumption based. SAP software are based on two components (i) Software and (ii) SAP Named Users. There are two perpetual license models—the Classic SAP software and SAP S/4HANA–branded software. Software provides the opportunity to support business functionality and is licensed in accordance with specific metrics. SAP Named Users provide the rights for individuals to use the software. The named users are further divided into a few types, each providing specific use rights. SAP uses the analogy of a house and the key where the Software is the "house" and the Named Users are the "keys" [3].

In a subscription model, the customer does not have perpetual use rights over the software, instead pays an annual subscription fee as part of a term contract. The fee includes all the Software as a Service (SaaS) components, including product support. Under the consumption-based model, and the customers pay the dues based on actual usage. There are also various types of database licensing options that need to be considered.

Compliance is an important aspect of SAP software licensing. When it comes to the consumption and proper licensing of SAP products, SAP's global audit and license compliance process protects SAP's core business assets and ensures that SAP's customers can manage their SAP assets effectively and manage any overuse of the software [3].

If an audit is initiated by a software vendor, organizations should cooperate with the software vendors. There are instances where customers are evasive and purposefully delay requests, provide inaccurate or wrong information, and are non-responsive to the request of information or organization's attempt to circumvent the software's built in monitoring mechanisms [31]. In such cases, there will be difficult conversations with the customer with possible escalations from both sides. At times, it may involve legal interventions to enforce the contractual rights. Therefore, it is important to properly plan an audit engagement. Planning the information systems audit must include all the stages necessary for the achievement of the objectives of the audit mission, namely documentation of the audited activity, the program or system under scrutiny, the establishment of the audit strategy, the establishment of the audit procedures and techniques, and the methods of synthesis, analysis and interpretation of the audit evidence [32].

3   Security Implications

Many software organizations are facing the situation where it is believed that their intellectual property (IP) rights are not used in a lawful manner causing revenue leakage and potential loss of control over the protection of IP rights and discouragement of infringements [33]. All usage of Intellectual Properties should be appropriately licensed depending on the usage and scenarios. In the factsheet published by the European Intellectual Property Rights (IPR) Helpdesk, license agreement is defined as "a contract under which the holder of intellectual property (licensor) grants permission for the use of its intellectual property to another person (licensee), within the limits set by the provisions of the contract" [34]. Without such an agreement, the use of the intellectual property would most likely result in an infringement.

Based on a survey conducted by Deloitte on consumer privacy, it was found that 91% of the 2,000 participants aged between 18–75 are willing to accept legal terms and conditions without reading them [35]. Not just in the consumer space, companies sometimes agree and sign the Software agreement without fully understanding the implications of the contract or the "fine prints". While some may say that contracts of larger and younger companies tend to be "pro-seller", there are no evidences to support that larger firms offer software terms that are worse than those that are offered to general public as compared to business and corporate users [36].

Proper software licensing and audit not just ensure company remains compliant to the agreement that they have signed with the Software vendor, but also ensure their internal control and governance are functioning properly. Internal control that we are discussing in this paper are the segregation of duties and unauthorized access. Segregation of duties is one of the fundamental elements of internal control [30]. Other than ensuring no single individual has control over the whole process, exposing the organization to risk. but also to ensure legal compliance [37]. Not just in Information Technology, this notion is also an important topic in accounting, given that the segregation of duties seeks the prevention of possible fraud through collusion, where there are conflict of interest [38]. In licensing terms, one scenario relating to the segregation of duties include sharing of User IDs within the organization. Some companies attempt to cut down the number of licensed Users by allowing Users to share a single User ID for various functions within the department. This results in the inability to identify the exact person who entered a request entry. This issue leads to the inability to identify the individual who approved the request. This also creates collusion between employees and potentially results in fraud cases within the organization.

With the sharing of User ID, an employee who is no longer with the department or organization may gain unauthorized access to the application software by using the shared User ID. This can be done by a disgruntled employee or an attacker with the aim of stealing information, resulting in unauthorized access to the protected or sensitive information. Eventually, this compromises the data integrity and availability within the firm [39]. Software license audit can potentially expose sharing of licenses by reviewing the Usage information and User management procedures. While protecting the intellectual property of the software vendors, it can also uncover malpractices and make known to the management for corrective measures.

4 Discussion and Recommendation

In every license audit scenario, the outcome is not just ensuring the intellectual property is protected, but it also provides a significant revenue stream to the company. Several existing research studies have been conducted on audit quality. However, more remains to be done in the area software auditing, in particular, software license auditing and how it will inform information security and compliance procedures.

Some security implications relating to license management are:

(a) The issue of unauthorized access; In addressing the issue on the sharing of User IDs, proper management of User licenses reduce sharing of User ID thus cutting down unauthorized access incidents to the system. Preventive measures include:

- Ensuring that every individual is assigned to their own User ID; that way, any access and transactions can be traced back to the individual;
- Ensuring User Management policy is in place. For instance, if an individual has not logged in for 30 days, the User ID should be locked and if the individual has not logged in for next 30 days, the ID will be invalidated;
- Authorization assigned to individual IDs should have an expiry date and set access levels that should be approved periodically;

(b) Sharing of User IDs potentially leads to the risk of segregation of duties; User ID sharing between a few individuals performing different roles leads to the User ID being authorized in excess of the requirement. For example, an individual may submit a purchase requisition and uses the same ID; the requisition can be approved by the same individual thus making unauthorised purchases through the system that are challenging to uncover by examining the existing system data.

Addressing such licensing concerns, provides the opportunity to organizations to minimize risks related to internal controls.

5 Conclusion and Future Work

License auditing entails a number of processes and a good quality audit is typically defined by several characteristics. There are security implications in the software license auditing sector. This paper highlighted the importance of audit quality, and discussed some of the security implications within the context.

Moreover, the paper explored other research opportunities. One of the future areas of research is to assess and identify further factors that affect the quality of an audit. This could be achieved, for instance, by the application of information seeking behavior and foraging theories and practices, which may also inform other security implications specific to license auditing. In this context, various categories of information, i.e. information as process, information as knowledge and information as thing will need to be analyzed considered, and 'information as thing' as a notion will need to be used to assess data in the license auditing process [40]. Data in the context of future research refers to the facts and statistical representation residing in the information system that can be downloaded for analysis.

References

1. Bader MA (2006) Intellectual property management in R&D collaborations: the case of the service industry sector. Physic-Verlag-A Springer Company
2. Perleberg GB (2015) Technology licensing, pp 27–30
3. SAP (2018) SAP licensing guide.pdf
4. Soobaroyen T, Chengabroyan C (2006) Auditors' perceptions of time budget pressure, premature sign offs and under-reporting of chargeable time, vol 218, pp 201–218
5. Oxford Learners D (2020) Definition of audit. https://www.oxfordlearnersdictionaries.com/definition/english/audit_1?q=audit. Accessed 26 June 2020
6. DeAngelo LE (1981) Auditor size and audit fees. J Acc Econ 3(May):183–199
7. Francis JR (2011) A framework for understanding and researching audit quality. Auditing 30(2):125–152. https://doi.org/10.2308/ajpt-50006
8. Francis JR (2004) What do we know about audit quality? Br Acc Rev 36(4):345–368. https://doi.org/10.1016/j.bar.2004.09.003
9. Skinner DJ, Srinivasan S (2012) Audit quality and auditor reputation: evidence from Japan. Acc Rev 87(5):1737–1765. https://doi.org/10.2308/accr-50198
10. Aronmwan EJ, Ashafoke TO, Mgbame CO (2013) Audit firm reputation and audit quality. Eur J Bus Manag 5(7):2222–2839
11. Castka P, Bamber CJ, Sharp JM (2001) Factors affecting successful implementation of high performance teams. Team Perform Manag Int J 7(7–8):123–134. https://doi.org/10.1108/13527590110411037
12. Caillier JG (2010) Factors affecting job performance in public agencies. Public Perform Manag Rev 34(2):139–165. https://doi.org/10.2753/pmr1530-9576340201
13. Ariffin ZZ (2012) Do industry affiliations affecting corporate tax avoidance. In: International conference on management, economics and finance (Icmef 2012) proceeding, no October, pp 784–796
14. DeFond M, Erkens DH, Zhang J (2015) Do client characteristics really drive the Big N audit quality effect? Manag Sci
15. Beck PJ, Wu MGH (2006) Learning by doing and audit quality. Contemp Acc Res 23(1):1–30. https://doi.org/10.1506/axu4-q7q9-3yab-4qe0

16. Agus Wijaya I, Yulyona MT (2017) Does complexity audit task, time deadline pressure, obedience pressure, and information system expertise improve audit quality? Int J Econ Financ Iss 7(3):398–403 [Online]. http://www.econjournals.com
17. Knechel WR, Vanstraelen A (2007) The relationship between auditor tenure and audit quality implied by going concern opinions. Auditing 26(1):113–131. https://doi.org/10.2308/aud.2007.26.1.113
18. Tepalagul N, Lin L (2015) Auditor independence and audit quality: a literature review. J Acc Audit Financ 30(1):101–121. https://doi.org/10.1177/0148558x14544505
19. Bell TB, Causholli M, Knechel WR (2015) Audit firm tenure, non-audit services, and internal assessments of audit quality. J Acc Res 53(3):461–509. https://doi.org/10.1111/1475-679X.12078
20. Boskou G, Kirkos E, Spathis C (2019) Classifying internal audit quality using textual analysis: the case of auditor selection. Manag Audit J 34(8):924–950. https://doi.org/10.1108/MAJ-01-2018-1785
21. Yang R, Yu Y, Liu M, Wu K (2018) Corporate risk disclosure and audit fee: a text mining approach. Eur Acc Rev 27(3):583–594. https://doi.org/10.1080/09638180.2017.1329660
22. Stoel D, Havelka D, Merhout JW (2012) An analysis of attributes that impact information technology audit quality: a study of IT and financial audit practitioners. Int J Acc Inf Syst 13(1):60–79. https://doi.org/10.1016/j.accinf.2011.11.001
23. Kilgore A, Bennie NM (2014) The drivers of audit quality: auditors' perceptions. Assoc Chart Certif Acc 3–15
24. DeFond M, Zhang J (2014) A review of archival auditing research. J Acc Econ 58(2–3):275–326. https://doi.org/10.1016/j.jacceco.2014.09.002
25. Sirois L-P, Simunic DA (2012) Auditor size and audit quality revisited: the importance of audit technology. SSRN Electron J. https://doi.org/10.2139/ssrn.1694613
26. Jiang J, Wang IY, Philip Wang K (2019) Big N auditors and audit quality: new evidence from quasi-experiments. Acc Rev 94(1):205–227. https://doi.org/10.2308/accr-52106
27. Landsman WR, Nelson KK, Rountree BR (2009) Auditor switches in the pre- and post-enron eras: risk or realignment? Acc Rev 84(2):531–558. https://doi.org/10.2308/accr.2009.84.2.531
28. Xiao T, Geng C, Yuan C (2020) How audit effort affects audit quality: an audit process and audit output perspective. China J Acc Res 13(1):109–127. https://doi.org/10.1016/j.cjar.2020.02.002
29. Arai Y (2018) Intellectual property right protection in the software market. Econ Innov New Technol 27(1):13. https://doi.org/10.1080/10438599.2017.1286734
30. Peng G, Di Benedetto CA (2013) Learning and open source software, vol 44, no 4, pp 619–643
31. Bigler M (2003) The high cost of software piracy. Intern Audit
32. Lenghel RD, Vlad MP (2017) Information systems auditing. Quaestus Multidiscip Res J 11:178–182. https://doi.org/10.1016/0378-7206(84)90006-5
33. European IPR Helpdesk (2015) Fact sheet defending and enforcing IP. European IP Helpdesk. http://www.iprhelpdesk.eu/sites/default/files/newsdocuments/Fact-Sheet-Defending-and-Enforcing-IP.pdf. Accessed 29 July 2020
34. European IPR Helpdesk (2015) Commercialising intellectual property : license agreements. European Intellectual Property Rights. https://www.iprhelpdesk.eu/sites/default/files/newsdocuments/Fact-Sheet-Commercialising-IP-Licence-Agreements.pdf. Accessed 29 July 2020
35. Deloitte (2017) Global mobile consumer survey: US results
36. Marotta-Wurgler F (2007) What's in a standard form contract? An empirical analysis of software license agreements. J Empir Leg Stud 4(4):677–713. https://doi.org/10.1111/j.1740-1461.2007.00104.x
37. Ferroni S (2016) Implementing segregation of duties. A practical experience based on best practices. ISACA J 3:1–9 [Online]. www.isaca.org
38. Kim R, Gangolly J, Ravi SS, Rosenkrantz DJ (2020) Formal analysis of segregation of duties (SoD) in accounting: a computational approach. Abacus 56(2):165–212. https://doi.org/10.1111/abac.12190

39. Sen R, Borle S (2015) Estimating the contextual risk of data breach: an empirical approach. J Manag Inf Syst 32(2):314–341. https://doi.org/10.1080/07421222.2015.1063315
40. Buckland MK (1991) Information as thing. J Am Soc Inf Sci 42(10):758–758. https://doi.org/10.1002/(SICI)1097-4571(199112)42:10%3c758:AID-ASI11%3e3.0.CO;2-4

The Deployment of Autonomous Drones During the COVID-19 Pandemic

Usman Javed Butt, William Richardson, Maysam Abbod, Haiiel-Marie Agbo, and Caleb Eghan

Abstract Drones are being utilised in diverse domains all over the world. The latest employment is the utilisation of drones during the global pandemic for crowd dispersal, infection monitoring, facial recognition, and logistical roles. Drones with artificial intelligence are unique devices that enhance the economy and are capable of completing tasks that humans cannot. The issue is that companies have disregarded the security of these machines, making them vulnerable to cyberattacks. Furthermore, a compromised drone can be used as a malicious offensive weapon system for tasks such as illegal imagery, videography, and as a functioning autonomous weapon system. This paper critically analyses the amalgamation of drones and artificial intelligence, reviews the current threat landscape, and studies drone vulnerabilities in regard to attack methods. Finally, a robust framework with secure countermeasures is proposed and therefore recommended for the drone industry to adopt and implement.

Keywords Drones · Cyberattacks · Cyberspace · Cyber Criminals · Artificial intelligence · Autonomous

U. J. Butt (✉) · M. Abbod
Department of Electronic and Computer Engineering, Brunel University, London, UK
e-mail: usman.butt@brunel.ac.uk

M. Abbod
e-mail: maysam.abbod@brunel.ac.uk

W. Richardson · H.-M. Agbo · C. Eghan
The Department of Computer and Information Sciences, Northumbria University, London, UK
e-mail: william.j.richardson@northumbria.ac.uk

H.-M. Agbo
e-mail: haiiel-marie.agbo@northumbria.ac.uk

C. Eghan
e-mail: caleb.eghan@northumbria.ac.uk

1   Introduction

The interesting concept of artificial intelligence (AI) and drones is becoming a fast and growing form of unique amalgamation. This is due to its low costs, manoeu-vrability, and speed, especially when concerning logistical issues with transferring packages and equipment that need to be distributed in hasty situations.

Recently, during these unprecedented times AI and drones have been combined for a rapid distribution of emergency medical supplies, personal protective equipment (PPE) and COVID-19 tests equipment to people who need them in undulating terrain: where these are logistically hard to get to, or whom need, rapidly distributed.

This has already been utilised in the south of England, the Isle of Wight, and Scot-land. Scotland, for example, has been supplying arduous areas in derelict locations such as the Isle of Mull where teams have desperately needed PPE. Further to this, globally, this concept is been rendered successful in countries such as Africa and the United States of America. In Africa, these drones are providing critical assistance to communities who have been viciously impacted by COVID-19 and these journeys have been successfully utilising this as far as ninety kilometres away. The journey by vehicles would normally take hours due to the terrain and quality of vehicles, making drones a unique alternative.

China has utilised their drones differently from others during the pandemic; they have used drones to fly at high elevations and disinfect buildings in urban areas which is said to be a lot less time - consuming than manual labour, and more effective. This is because they can spray up to 600,000 square metres a day and this is something the United Kingdom (UK) government is considering for UK cities. Furthermore, coun-tries have used drones in more diverse ways, the Netherlands and Malaysia are using them to enforce lockdown measures. These countries are using AI to visually identify people who are not complying with these measures, and consequently enabling the police force to follow the crimes up with visual identification as evidence.

These examples of the deployment and utilisation of autonomous drones have been successful and effective during these unprecedented times; however, the security of these machines has been overlooked. This is a problem since attacks such as wireless attacks, hijacking, and man-in-the-middle attacks could be performed against these machines. This leaves medical supplies vulnerable, data insecure, and equipment exposed. Further, this could leave these machines vulnerable to compromise and used maliciously by stealing sensitive data such as facial recognition data and voice recognition data which could then be used for personal authentication methods. This would be detrimental to the public and to the governments using AI and drones, which is why the enhancement of security is needed.

This paper will contribute to the enhancement of security for autonomous drones by exploring their architecture, locating their vulnerabilities, and rectifying these by recommending secure protocols to defend against malicious attacks.

2 Autonomous Drones and Their Role in the COVID-19 Pandemic

Autonomous drones have been employed for various roles since the pandemic erupted in early 2020. The roles that drones have been employed in have been in a positive light, since they have engaged with the population during critical situations. The most notable role during the pandemic is delivering emergency medical supplies to locations that are in derelict locations, or, for people who cannot reach the intended location to pick-up these supplies. The reason the United Kingdom (UK) employed these machines was because the UK Government enforced minimal movement to mitigate the virus spreading and decrease contact with other humans, this was a safer alternative to mitigate the spread of the disease. At the time of writing, there are more than sixteen million cases confirmed globally linked to COVID-19 [1]. The profound effects of the pandemic are expected to plunge vast amounts of countries into recession and the employment of autonomous drones are an alternative to ignite economies again.

At the beginning of the pandemic, the UK was not fully prepared, and for that reason, there was a shortage of personal protective equipment (PPE) throughout industries [2]. This issue, along with the high risk of spreading the disease through active movements, presented unprecedented challenges. For this reason, drones were utilised quicker than expected in the industry, and proved to provide a monumental advantage when supplying critical facilities such as hospitals, key workers, and vulnerable people.

Additionally, except for being used logistically for medical supplies, drones are being employed to sterilise urban terrain to mitigate the spread of disease [3]. Further, when employed in these urban areas, in Wuhan China, using artificial intelligence (AI) capabilities, the drones were able to identify humans so that they could be identified, they could then disperse crowds easier and therefore told to isolate. Furthermore, if these people did not comply, they were then able to identify for prosecution by using surveillance technology, along with the capability to be told to comply through automated speech [4].

In addition to medical supplies being logistically utilised by drones, e-commerce industries have also taken advantage of autonomous drones. In the United States of America (USA), the company "Wing" owned by google has conducted thousands of automated flights carrying various packages across the USA using drones that fly around sixty-five miles per hour [5]. These drones have proved invaluable when conducting logistical tasks, especially because this minimises person-to-person contact. The journeys have been anything from two to twelve-minute journeys with the furthest being twelve miles. Wing, call this their drones-as-a-service (DaaS) platform and they report that this has solved many isolated family's issues when needing crucial household items such as food, toilet paper and bathroom accessories [6]. Amazon is also in the early stages of their prime air platform, which is the same concept as Wing, however, Wing is leading the way in drone delivery innovation.

In Australia, a group of academics teamed up with Dragonfly to develop drones that can test human temperatures using artificial intelligence to detect COVID-19 symptoms and infectious diseases. The drones can establish symptoms from just ten metres away from the intended person and transfer this data in real-time back to the required personnel. Its main objective is said to track and trace the disease as fast as possible in order to contain and minimise possible contact between humans [7].

Figure 1 illustrates the sectors utilising drones: augmented intelligence is being used to identify people to comply with lockdown procedures. It is also being enabled to identify criminals and missing persons by UK police force [8]. Population management is being mitigated by drones to control crowds of people not complying with rules. These people are being identified and arrested in China if they do not comply with government rules [3]. Disinfection by drones are being utilised in urban areas to sanitise areas in built-up high populated areas. Drones in urban areas are significant, since they can reach areas that humans cannot, or is not safe. Finally, PPE and food supplies are being distributed to key workers in critical need of resupply. Moreover, essential medical supplies are being distributed to vulnerable people in derelict locations, because of speed and versatility when mitigating terrain and weather.

At the time of writing, the data collected from [9] illustrates the industries that are utilising drones the most. In Fig. 2, we can see that construction is the largest sector followed by information communications and manufacturing. Drones can make industries objectives easier by using their advanced abilities such as elevated imagery, video-link, and live data feeds. This can render industries jobs easier as they can identify issues and rectify them faster than they would if not identified. They also make industries ultimately more cost-effective in the long-term, since the

Fig. 1 Sectors utilising drones

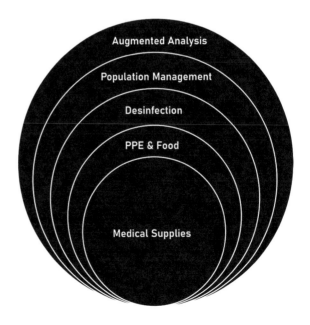

Fig. 2 Sectors mostly
utilising drones

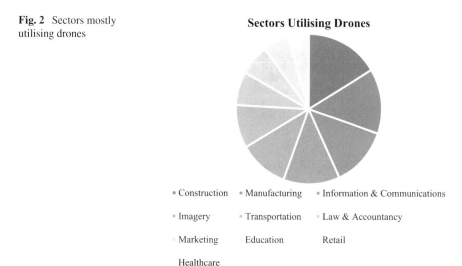

advanced skills, faster utilisation, and more accurate data create a significant return
on investment (ROI).

Drones are becoming a main driver in countries and more so in smart cities. This is
because of the lucrative revenue they accumulate; they enhance the economy making
them valuable assets by boosting digital transformation through employment, along
with valuable data to industries. Figure 3 shows the commercial industry projections
from 2016 to 2025. The projection for 2025 is estimated to be 12,647.02 million U.S
dollars [10].

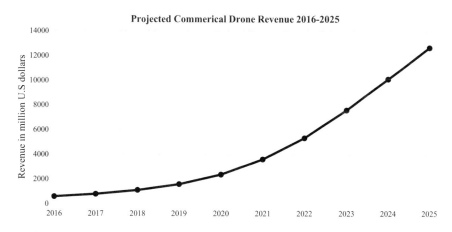

Fig. 3 Commercial drone projected revenue [10]

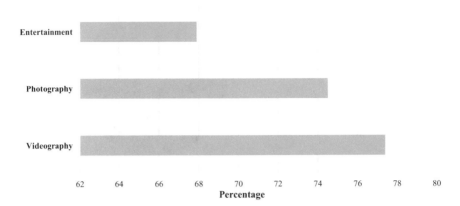

Fig. 4 Most common use of drones [9]

Figure 4 shows data collected from [9] on how drones are being utilised. The data indicate that videography is the largest and most popular function across industries and general hobbyists, followed by photography and entertainment. The data collected indicates that drones are as much popular with the public, as with industries. It also identifies the ease with which the public can operate such a complex machine with the many technical sensors they possess. Governments should seek to protect the public's privacy against the misuse of drones through offensive techniques. This should also be incorporated with national security when offensive drones can be misused with illegal imagery and videography.

3   A Review of Drones, Their Architecture and Vulnerabilities

3.1   Drone Types

There are three main types of commercial drones being used in industries and during the pandemic. They all come under the classification of unmanned aerial vehicles (UAV) since the term drone can be used for autonomous robots or even vehicles. The three types used largely are:

- **Fixed-Wing Drones**: These are usually launched from a runway so that speed can be produced before take-off. Not necessarily used in the commercial industry, they are used primarily within the military with large weapon systems attached. These tasks normally endure long range missions with surveillance and reconnaissance functions [11].

- **Hybrid-Wing Drones**: A combination of fixed and rotary winged type of drone. This is a new concept that can glide and land or take-off using its rotors which adds an extra advantage, in diverse terrain and environment. This is a fully autonomous drone used in the military; however, this could be a new concept in the commercial industry since its emphasis is on longevity flights and endurance [12].
- **Multi-Rotor Drones**: These drones are the most common in the commercial industry as they can take-off in compact areas, are easily manoeuvrable and can travel at high speed [13]. Additionally, this type of drone presents the advantage of hovering over objectives to conduct further tasks. Futhermore, multi-rotor drones are especially useful when using artificial intelligence since AI enables them to pinpoint locations, people, tangible objects and to send real-time data back to the ground control station (GCS).

3.2 Drone Architecture

In this section we look at the vulnerabilities in drone architecture, this is critical to understand since there has been little or no emphasis on the security of Personal Identifiable Information (PII), which could be compromised when the drone is deployed, or when stagnant. Drones have been rendered targets for cyberattacks by threat actors such as criminals, nation-state and even script kiddies. This is due to their inadequate security along with their wireless connection which is prone to be intercepted by Man-In-The-Middle (MITM) attacks, remote takeover, and Denial-Of-Service (DOS) attacks.

According to [13] a typical drone architecture consists of three elements:

- **Flight Controller**: This is the central processing element within the drone architecture [13]. This central point processes the data that is transmitted to the drone consequently interpreting this into concise information which is then relayed to the GCS. The sensors included within this element are all pertinent to the mapping and flight destination of the drones, with the main sensor attached being the global positioning system (GPS).
- **Ground Control Station**: This is the ground-based hardware incorporating software that allows the operators controlling the drones to communicate with payloads [14]. Note that for autonomous drones the operator controlling will set the flight path and the destination allowing the drone to intelligently travel without interaction. The GCS is designed to communicate with the drones from huge ranges, this makes them more susceptible to attacks because of this increased proximity which is more susceptible to their communication being compromised.
- **Data Links**: This enables the data to be transferred from the drone to the GCS so that commands can be performed [15]. The data links system is better categorised into three types:

1. **Line of Sight (LOS)**: This is controlled through wireless communications which are close in proximity and can operate by radio waves. Commercial drones will

use this method especially with the introduction of 5G which increases the bandwidth capabilities.

2. **Beyond Line of Sight (BLOS)**: Missions that are controlled by satellite communications since they are out of proximity, or they can be controlled by a relaying device [16]. This concept is used with autonomous drones as their proximity is increased from the GCS.

3. **Tactical Data Communication**: A communication link, which is associated with military drones, so that communication can be processed through multiple aircraft [15]. This is a secure method since military drones transport and utilises weapon systems during kinetic warfare.

Figure 5 illustrates how drones communicate with the GSC. This diagram shows the two types: by human interaction or by setting the flight path using artificial intelligence from the GCS. Most modern drones now use 2.4 GHz or 5.8 GHz radio frequency depending on the proximity from the GCS to the drone [17]. Attached to the drone are the sensors such as the GPS for mapping, along with the camera

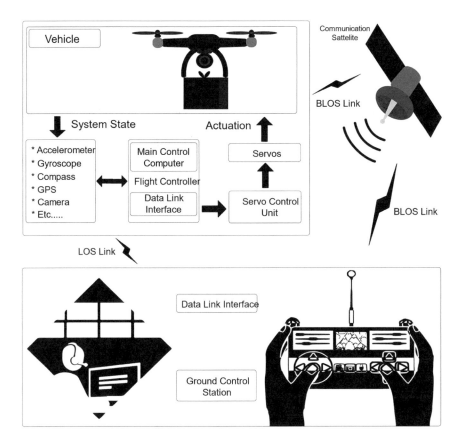

Fig. 5 Common drone architecture and communication link

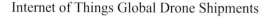

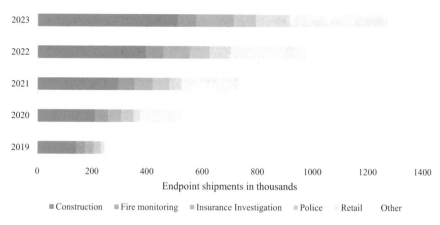

Fig. 6 Drone IoT global shipments [18]

for imagery, and intelligent voice sensors so that some drones can talk if needed. The drone architecture holds data regarding the flight path and Personal Identifiable Information (PII) if they are delivering to specific households. The attacker's main objective would be compromising the drone for valuable data and taking over the drone remotely to use as an offensive weapon system.

A report by Statista [18] shows current IoT sensor shipments for drones in 2019 and 2020, followed by the predicted shipments of IoT for the subsequent years up to 2023. The graph shown in Fig. 6 shows the profound increase of sensors being used to attach to drones from 2019. The coloured key indicates which industries have and are predicted to have the most IoT shipments. The graph suggests that construction is likely to obtain the most sensors for drones. This shows that surveying and mapping of urban areas are imperative to the growth and performance of a smart city, since drones in construction are used to assess and evaluate the potential and final constructions of large buildings. The drone industry is expected to surpass $141 billion by 2023 making this one of the world's fastest growing technology industries, this figure will also entice cybercriminals to attack drones, as the lucrative market increases.

3.3 Drone Vulnerabilities

It is common knowledge that commercial drones are not manufactured with security in mind. Drones primarily use a GPS for navigation that is not encrypted, which can be spoofed and thus controlled by an attacker. One of the most notable attacks on a drone was an attack by Iraq insurgents in 2009, these insurgents compromised the live feed of a US predator drone, and in doing so they were able to avoid lethal kinetic force by watching the video footage and successfully circumvent the attacks [19].

There are three fundamental ways that attackers could intercept the drone and maliciously compromise its confidentiality and integrity. Altawy and Youssef [16] categorise these into three capabilities:

Revelation: The attacker would have to perform eavesdropping into the data in real-time use. This could be easily conducted since commercial drones do not encrypt their data-in-transit. Performing such attacks would enable the attacker to compromise the wireless connection, this would establish a rogue connection between the attacker and the drone thus creating a secondary controller to control its commands.

Knowledge: The attacker would have knowledge on configuration and authentication information so that they could access the system with authenticated data. This would enable the attacker to gain PII from the drone software and control its architecture.

Disruption: The attacker conducts an attack by using a sophisticated technique such as a MITM attack where they could compromise the connection and maliciously take over the drone. Using the drone's autonomous software, the attacker could now use this as an offensive weapon by collecting live imagery from an intended target, intercept home or corporate networks using its powerful signal, and even used as an offensive kinetic weapon system by logistically carrying explosives used by terrorists.

Dey et al. [20] proved in their experimental study that drone architecture could be exploited through three methods when they tested against drone software: DJI Phantom 4 PRO and Parrot Bebop 2. In particular, the attacks performed were cracking authentication method which allowed modifications to the software, reverse engineering which allowed still images to be extracted, and GPS spoofing which allowed for the signal to be spoofed and thus controlled by a rogue controller. These two drones are some of the most sort after commercial drones and most likely used in the pandemic considering their credibility.

Autonomous drones are a new concept, with artificial intelligence software embedded within the architecture, this renders more avenues to attack and therefore more vulnerabilities to exploit. Astaburuaga et al. [21] performed a vulnerability assessment, they achieved this by using Nmap scanning software which revealed multiple ports were open. Their analysis showed that many vulnerabilities were found and anyone with adequate tools could connect to the drone and access their data along with other attached sensitive information. Accessing an autonomous drone would present even more open ports since the software used would require more connections from the hardware to the GCS. If an autonomous drone were then accessed, it could then be used maliciously by using offensive techniques by conducting intelligence gathering, facial recognition data, automated hacking, and speech impersonation data through manipulation of the program code. Securing autonomous drone architecture should be at the forefront of manufacturers incentive to help mitigate this threat.

This section reviewed the most significant usage of drones, their types, and their architecture. The paper then analysed the vulnerabilities between the GCS and the drone itself, with diagrams to illustrates drone to GCS architecture and imperative data to interpret the importance of drone security. This section signified the importance of drones with data to show the significant increase in demand and their significance on the economy.

4 The Amalgamation of Artificial Intelligence and Drones

4.1 Artificial Intelligence (AI)

Artificial Intelligence unequivocally enhances drone design, this is because of the unique sensors that can be attached to the drone to create a device that communicates with the GCS using real-time data. AI and drones can be used to enhance the design and assist with logistical, medical, intelligence gathering, and environmental situations.

There is no standardized and globally accepted definition for what Artificial Intelligence stands for. Available definitions for AI can be classified into two groups. There are the "essentialist" definitions that define AI based on the end-goal expected from systems and there are the "analytical" definitions basing their definition on a list of essential abilities needed to create AI [22]. But all those definitions have one thing in common. AI-enabled systems should be "systems capable of performing tasks that would require a human factor". Artificial Intelligence tasks can range from simple tasks such as speech recognition, language translation or computer vision to more complex ones such as theorem proving, fault diagnosis, scientific analysis, or medication diagnosis. However, drone's ability to perform those various tasks depends on the level of Artificial intelligence they have been embedded with. They are three main class of Artificial Intelligence.

4.2 Type of Artificial Intelligence (AI)

- **Weak Artificial Intelligence/ Artificial Narrow Intelligence (ANI)**

This type of AI is the most common type of AI. Drones equipped with Weak AI can only perform tasks they were programmed to do. Hence have "narrow" capabilities. Searle explains that "Weak artificial intelligence refers to a scientific theory that recognises that computers imitate certain mental abilities and do not claim that computers can understand or that are intelligent" [23]. In this case, drones embedded with weak AI are capable of facial recognition, voice recognition, language processing (language generation and translation), reasoning, planning, mapping, data analyses for inconsistencies detection and much more. In France, Weak AI embedded drones were used to monitor compliance with lockdown measures in public spaces such as parks and beaches [24].

- **Artificial General Intelligence (AGI)/Strong Artificial Intelligence**

Strong AI refers to machines' ability to "learn, perceive, understand and function completely like a human being" [25]. Strong AI's should enable drones to perform intelligent actions and specific problem-solving.

To achieve that, drones equipped with AI technology use the advantage provided by four elements. Machine perception and Computer Vision (CV) for optimal data capturing/analysis and Machine Learning (ML) and Deep Learning (DL) to enable learning and decision-making ability.

- Machine Perception allows drones to better observe and capture their surrounding environment using electro-optical, stereo-optical and LiDAR [26].
- Computer Vision (CV) enables drones to perform automatic extraction and analysis to obtain substantial material from raw data previously gathered [26].
- Machine Learning (ML) "algorithms are designed to learn and improve when exposed to new data" [26]. Its implementation gives drones the ability to learn and improve from data previously analysed.
- Deep Learning (DL) is used as an alternative to Computer Vision (CV) and Machine Learning (ML) due to its price to performance ratio and hardware requirements. However, it is hard to train Deep learning algorithms since they require a huge quantity of data especially when it comes to training for image recognition. Despite that, with appropriate data for training and processing power, Deep Learning algorithms prove to be more capable than ML and CV and improve drones' decision-making process [26].

Strong AI's ability to process data captured was critical in some countries during the COVID-19 pandemic. With Strong AI, some drones were capable of using image-processing algorithms and other various algorithms to extract human's heart rate from video taken by drones and to detect actions such as sneezing and coughing [27]. In China, India, Italy, Oman and Colombia Strong AI was used on drones with the help of thermal cameras to identify potentially infected citizens based on their body temperature [28].

- **Artificial General Intelligence (AGI)**

The development of ASI is still in progress and if achieved, will change the way machines functions, and greatly improve human lives. The main goal of ASI is to create machines capable of higher cognitive function than a human [25].

4.3   AI Techniques and Applications in Drones

Artificial intelligence uses various techniques to enable machines to perform the way they do. Each technique is used depending on its functionalities and domain of application. With time, those techniques have evolved and as a result, hybrid AI methods have been developed.

The most noticeable and commonly used are Artificial Neural Networks, Fuzzy-Cluster Analysis, Evolutionary Algorithms, Genetic Optimization, Fuzzy Inference Analysis, Particle Swarm Optimization, Expert Systems.

Some of those techniques are used in drones to improve control in dynamic environmental conditions like wind, mapping, motion and path planning, stabilization system, enhancement of landing on moving targets and much more. All those factors have proven for drones while performing activities such as Telemedicine, Medical Supply contact-free delivery, surveillance and enforcement and hygiene applications. Here are some applications of those techniques to improve drones' functions:

- **Path Planning for Quadrotor UAV Using Genetic Algorithm**

Genetic algorithms can be classified as a searching type of algorithm, they were designed to study adaptation as it occurs in nature and devise a way to bring this process into computer systems [29]. They search for the best path to achieve a maximum problem-solving rate.

Genetic algorithms are made of chromosomes with different genes and each chromosome represents a solution. To come to the best solution, the genetic algorithm generates a population of solutions and lets it evolve [30]. Population, in this case, represents possible solutions considered by the algorithm. The focus of this process is to identify the fittest during the process of the algorithm meaning to find the best possible solutions over many generations. This is done by recombination and mutation in the population [31]. One of the problems during the use of a genetic algorithm is to define a constant that can be used to determines which chromosomes (solutions) are the fittest. Figure 7 illustrates the process used to obtain the best path for the UAV.

The process begins with an Initial Point (IP) and a Target Point (TP) in 3-dimensional coordinates. All possible routes are inserted and processed by the Generic Algorithm (GA). The final output (the fittest) will be the shortest path to travel by the drone [30].

Fig. 7 Genetic algorithm system process

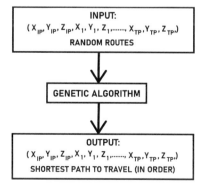

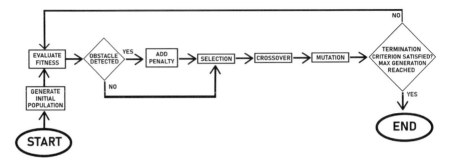

Fig. 8 Generic algorithm flowchart

Figure 8 shows the GA flowchart. The new element brought by the flow chart is obstacle detection. If an obstacle is present in the fitness chromosome, the penalty option is implemented, and the travel distance is augmented [30]. Next, at the selection part, the first half of the high fitness chromosome is copied and used to generate a set of new generations. They are then crossed and mutated to bring out a list of the fittest chromosomes. The fittest is the one with the lowest distance to travel for the drone [30].

In summary, the genetic algorithm considers various solutions available to the drone, combines them and through an elaborated process, determines the fitness chromosome (solution). With the obstacle detection process and genetic algorithm, the risk of damage to the drone navigating is reduced.

- **Path Planning for Quadrotor UAV Using Genetic Algorithm**

Artificial Neural Network (ANN) is a "parallel distributed processor consisting of simple neurons that enabled memorizing the knowledge of a system by mimicking its nonlinear model after training" [32]. ANN control and model are based on the activity and rules of the human brain neurons. Adopting those model and rules improve ANN algorithms ability to perform practical problem-solving. Its implementation in drone's architecture increases UAV's capabilities and range of abilities.

One of the ways ANN is used in drones is for Fault Diagnosis in UAV blades. Failure of propellers in a flying UAV can cause damage to both the drone and the people around the flying area. To be able to detect such failure, the method proposed uses noise measurement and ANN algorithms (AI) for the detection of unbalanced blades. For this method to be implemented, the noise made by the drone in various states (balance, unbalanced 1, unbalanced 2) is measured and data collected is processed and used to train and test a classifier based on ANN [33]. Figure 9 shows the process used to achieve the desired outcome.

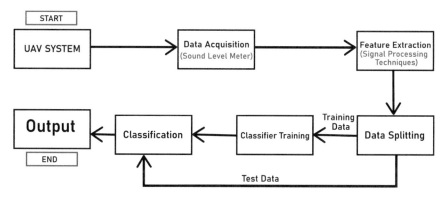

Fig. 9 ANN process

Fig. 10 Position of the sound level meter

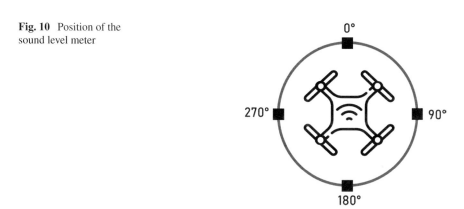

- **Data Acquisition (Sound Level Meter)**

As previously mentioned, data acquisition is made in three states [33]. The "balanced blade" refers to a state where no modification is made to the drone, the "unbalanced blade 1" refers to a state where one strip of paper tape is applied to the upper surface of the drone (at middle) and the "unbalanced blade 2" refers to the state where two strips of paper tape are applied to the upper surface of the drone (at the right). The paper strips are used to create a modification in the aerodynamics of the blade similar to those of a faulty propeller [33].

The measurements are then performed in an anechoic chamber covered by absorptive fibreglass wedges to reduce signal reflection as much as possible and the drone is positioned at the centre of the chamber on a tripod to reduce positioning errors. To ensure the data acquired is viable, the equipment used to measure noise must be compliant with the UNI EN ISO 3745:2012 standard [34].

In summary, the measurements are made over 3 different sections using the states mentioned earlier. In each section, the measurement equipment assumes different positions to achieve maximal quality [33]. Figure 10 gives an insight into the positions adopted during the measurements.

- **Feature Extraction (Signal Processing Techniques)**

At this stage, the measurements are not made of pure tone but a wide range of frequencies with various levels. To refine the measurements, it undergoes a frequency analysis process using the "bank" of band-pass filters method. The aim is to perform a sound analysis of the signals from their waveform to determine their spectrum (amplitude of the sound). By making the stationary signals go through a series of devices (filtering method) that only allow a certain range of frequencies, by excluding sound components of higher and lower frequencies and by measuring the output of each filter with a sound meter level, it is possible to obtain a certain level of measurement within a specific frequency range [33].

- **Data Splitting**

While working with ANN one important step is the "validation of model" problem [33]. It refers to the network's ability or inability to maintain its classification process in the event of new data being inserted. In case the network is unable to maintain its classification abilities, it overfits. Data Splitting suggests solving the overfitting problem by dividing the whole dataset collected from the drones into two subsets. The training set is used for the actual training and building of the model and the testing test is not used immediately. The model will eventually encounter the testing set during future training [33].

- **Classifier Training**

As previously mentioned, Neural Networks mimic the way the brain operates during the learning process. Artificial Neuron is used to refer to units in the Neural Network in charge of signal transfer. It is made of an input termination, a central body, and an output termination [33].

A neuronal network contains Neutrons in a structure with parallelism meaning they do not work together despite being connected. These connections are characterised by a parameter called weight and the neuronal network displays sign of intelligence when all the various unit interact.

The way a Neuron operates is simple. The neuron activates if the signal received exceeds a predetermined amount. Once active, it emits a signal transmitted over the communication channels to other neurons it relates to.

This weight can then be adjusted based on the experience gained during the training phase and an update rule is applied for the update to be done repeatedly. This process is repeated until the error rates drop to a certain level.

During the training phase, a multi-layer neural network is used with 1 hidden layer and 3 output classes representing the three conditions mentioned earlier and 70% of the input data and known outputs are selected and submitted to the network for weighing and error minimization. A backpropagation algorithm through various steps further reduces the error occurrence rate [33].

- **Classification**

Once the training stage is done, the testing phase is done to make sure that the network reacts appropriately when it faces new data. That is done by adding the 30% of unused data to the testing set and submitting it to the model to make sure it works as intended.

Using this process, the drone's ANN trains with the data collected and can apply its learning experience in live situations.

5 Susceptible Attack Methods Against Autonomous Drones

Attack methods refer to techniques or means used to gain unauthorized access into systems or network servers. In most cases, attack methods attempt to exploit vulnerabilities in the communication medium or other features specific to their target. The possibility of drones being exploited during the pandemic is, in our opinion, severe, since the drones were rushed into employment without security in-mind. To grasp the way attack methods, exploit drones' vulnerabilities and impact their operation, it is then critical to review the different communication methods used by drones.

Drone control is usually handled by a remote controller. The process is simple: using its radio transmitter the controller sends a PPM signal containing information such as throttle, yaw, pitch or roll to the drone's receiver [35]. That signal (PPM) is then translated into control signals for the electronic speed controller of each motor of the drone.

To reach the drone, the PPM signal is first transformed by a transmitter module into an electromagnetic oscillation and is then sent by a transmission standard in the air to the drone's receiver [35].

In this case, the transmission standard refers to the mean used to transfer data and signal between the drone and the controller. In most cases of drones' attack, the transmission standard is exploited, used or tampered with to gain unauthorised access to drones [35].

They are several transmission standards (communication methods) used in drone communication but the most common are:

- *GPS*: Global Positioning System (GPS) is a key component in drone communication. It uses satellite signals to provide accurate navigation and is usually combined with Inertial Navigation Systems (INS) to offer a more thorough UAV navigation solution. With the use of GPS, it is easy to determine a drones' position and plan the movements of autonomous drones in heavily populated UAV areas.
- *Bluetooth*: Bluetooth is a wireless short-range communication technology standard that uses radio waves to transmit data to and from the controller (mobile phone). It is usually used for data transmission in tiny drones and the outdoor

range is 100 m. In some drones, Bluetooth is used to broadcast information about the location and direction of the drone using a line of sight method [35].

- **Wi-Fi**: Most drones nowadays are Wi-Fi enabled. Wi-Fi is used in various ways by drones. It can be used to transmit a huge amount of data from drones to storage mediums such as cloud servers, web servers etc.., can be used to broadcast live video feeds to various devices (smartphones, computers, tablets) or can be used for remote control purposes. Its range is usually between 300 and 600 m; however, some remote controllers using a remote controller extender can reach a range of 2 km [35].

5.1 Man-in-the-Middle Attack

Man-in-The-Middle Attack (MiTM) is "the act of unauthorized individuals or parties placing themselves in the path of communication to eavesdrop, intercept and possibly modify legitimate communications" [36]. Figure 11 gives an insight into the structure of a MITM attack.

During a standard exchange between a drone and its remote controller, the drone exchanges data with the controller through a Wi-Fi connection established between the two. However, during a MiTM attack, the attacker aims at intercepting, sending, and receiving data meant for other parties on the network. To achieve that, a commonly used method is ARP Cache Poisoning. Using this method, the attacker sends false or spoofed ARP replies to corrupt ARP tables and can impersonate devices on the network. Simply put, attackers try to be considered as the remote controller by the drone and to be considered as the drone by the remote controller. Hence making all communications pass through their device. Table 1 presents the popular tools used to perform the MITM attack [36]. The MiTM Attack can have a disastrous

Fig. 11 MiTM attack structure

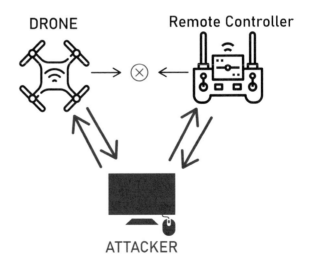

Table 1 MITM attack tools

Man-in-the-middle attack tools	
Cain & Abel	dnsniff
Ettercap	Karma
AirJack	wsniff
Ucsniff	ARPPoison
Wireshark	The Middler
SSLStrip	webmitm

impact during drone activities during COVID-19 where Telemedicine is done by medical professionals to communicate and emit diagnosis of patients. Information communicated by patients can be captured, tampered with or use for unintended purposes.

5.2 Denial of Service/Distributed Denial of Service (DDoS)

"The purpose of a DoS attack is to render a network inaccessible by generating a type or amount of network traffic that crashes the servers, overwhelms the routers, or otherwise prevents the network's devices from functioning properly." [37]. A DoS attack is a one-to-one availability attack meaning that it can be performed from a single machine. In contrast, a DDOS attack is a many-to-one availability attack. On a technical level, DDoS attacks require the use of other systems (computers, drones etc.) previously compromised and running programs called zombies. Once the attacker wants to perform a DDoS attack, he activates the zombies. They then send enormous network traffic toward the target to create a denial of service. There is various type of DoS attacks. Table 2 presents some examples.

A DoS/DDoS attack can be performed on drones by targeting either the drone or the remote controller. In the case of DoS attacks directed toward the Remote Controller, an efficient attack is the Deauthentication Flood attack.

In this attack, the attacker sends Deauthentication packets continuously to the remote controller causing it to disconnect from the drone. To be able to do that, the first step is to gain control over communication between the drone and the Remote Controller. It is done by performing a spoofing attack that gives the ability to send signals to the drone while pretending to be the Remote Controller [38]. Once Deauthentication packets are sent and the connection is lost with the Remote Controller, the drone either crashes or initiates the RTH mode to return to the last recorded Home Point automatically. Before that happens, the attacker can connect to the drone, change network settings, use FTP privilege to gain root access, install a backdoor to main access or even manipulate and steal the drone. This attack is described in the case where the drones' Wi-Fi is not password protected. In case it is protected, attackers can use Aircrack to capture packets, crack the Wi-Fi password using a dictionary attack and then initiate the Deauthentication Flood attack.

Table 2 Type of DoS attack

DoS name	Description	Type
SYN flood	Is executed by initiating many connections from a machine or several machines to the victim without replying to the victim's SYN/ACK packets. Resulting in accumulation and inability to process any new connections on the victim's side	Resource exhaustion
DNS reflection	Is executed by redirecting a high number of requests spoofed from several victims to an available DNS server	
Ping of death	Is executed by sending fragments of ICMP Echo Requests. Those fragments reunite and cause systems to crash due to their size being larger than the IP packet's maximum size	Malformed packet
Smurf	Is executed by using the spoofed source address of the victim to send ICMP Echo Request messages to a direct broadcast address of the network known as Smurf Amplifier. The aim is to amplify single requests into multiple ones. The Smurf Amplifier then sends large numbers of responses from the previously received request to hosts in the network	ICMP flooding
Fraggle	Is a variation of the Smurf attack. Instead of using ICMP, Fraggle uses UDP request and ICMP port unreachable message instead of ICMP Echo Response	Resource exhaustion

In case of DoS attack directed toward the Drone, an efficient attack is the JSON attack. This attack is performed by sending a request to the drone while pretending to be the Remote Controller. The first step in this attack is to perform an ARP Cache Poising or Spoofing attack to appear to be the Remote Controller. The next step is to identify open ports on the drone using a simple Nmap scanning command. Once it is confirmed that port 32/tcp used by telnet is open, the request is sent as a JSON record wrongly formatted with 1000 characters in the first field [39]. This wrong format causes the CPU and memory usage of the drone to drop gradually. This method creates an overflow in the drone's memory and causes the drone to stop functioning [39].

In both cases, the damages can be life-threatening especially if they are performed on drones used to deliver medical supply, test samples, medication or other critical supplies during the COVID-19 pandemic. By performing a DoS attack, supply can be lost or stolen due to the drone losing and crashing or the drone being hijacked after DoS is executed. The damages can be irreversible.

5.3 GPS Spoofing

Global Positioning System (GPS) is a satellite-based radio navigation system that offers geolocation and time information to GPS receivers. Military GPS signals are designed with strong encryption protocol and are therefore more resistant to most attacks. In contrast, Civil GPS signals are designed for public access and are freely accessible by anyone. Making it vulnerable to spoofing, signal blockage and jamming attacks [40].

Given this, many devices rely heavily on Civil GPS, particularly drones whose navigation, sensors and other features operate almost exclusively with GPS signals. (Global Positioning System Directorate, 2013) claims that the Civil GPS is "unencrypted, unauthenticated and openly specified in publicly available documents" [40]. This aspect makes devices relying on Civil GPS easy target to spoofing attacks.

When performing a GPS spoofing attack, overt and covert strategies are the main spoofing strategies used. Overt strategy refers to a method where the spoofer does not attempt to conceal the detection of the attack while in the Covert strategy the spoofer tries to avoid detection from both the GPS receiver and the drone's navigation system [40].

They are various GPS Spoofing attacks; this section reviews the "Aligned Attack". For the execution of the "Aligned Attack", the spoofer needs a "receive" and "transmit antenna" since the attack is performed away from the drone. The "receive antenna" is used to receive authentic signals from available GPS satellites and the "transmit antenna" emits a counterfeit signal toward the drone's antenna. The counterfeit signal must be sent to the drone's antenna in such a way that each counterfeit signal aligns within a few meters to the authentic signal. The counterfeit signal must be spreading-code-phase aligned with the authentic signal [40].

After the capture of the drone's carrier– and code-phase tracking loops, the spoofer can adjust the code-phase of the spoofing signals to influence the drone's receiver to report a simulated (false) location compared to its true location. The drone's time can also be influenced by making a little adjustment in the counterfeit signal code-phase [40].

It is important to remember that the "aligned attack" can only be performed if the spoofer is capable of measuring the system delays to the nanosecond and make up for it by emitting signals based on the delay predicted [40].

5.4   De-authentication Attack

De-authentication is an IEEE 802.11 protocols' feature part of the management packets system. It allows the access point or station to de-authenticate users by sending De-authentication packets. This is done by the access point to save resources in case of very resource-demanding periods or in case they are inactive users over the network [41].

With most civil UAVs not using a security protocol on their network or having a default password set by the manufacturer, it is easy for attackers to send de-authentication packets to the drones to de-authenticate the remote controller. Figure 12 gives an overview of the De-authentication process [41].

Fig. 12 De-authentication
attack

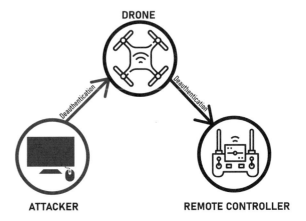

This attack can be performed in two scenarios. In the first scenario, the drones' access point is set on "No Security" and connected to its remote controller. In the second scenario, the access point is set on "WPA/WPA2" encryption. Each of these scenarios requires a different approach.

In the "No Security" scenario, Deauthentication packets are sent to the drone resulting in a loss of connection between the drone and the access point. The attacker can then connect to the drone without the need for a password and control it. The attack is performed using the following process. Figure 13 shows the steps resulting in a successful De-Authentication attack with no security.

Fig. 13 Steps to perform
de-authentication attack (No
Security)

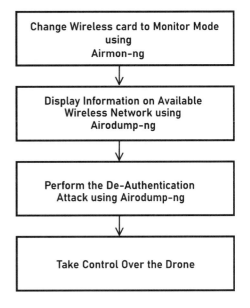

- First, the attacker sets his machine on monitor mode using the following command "*sudo airmon-ng start wlan0mon*". This mode allows the attacker to "read all the packets of data, even if they are not sent through this mode and control the traffic received on wireless-only networks" [42].
- Next, it is critical to determine the MAC address of the drone and its remote controller since it will be used to send the de-authentication packets. By using Airodump-ng, through this command "*sudo airodump-ng wlan0*", the attacker can get critical information such as the drones' "*MAC Address*", "*Channel*" and "**Encryption Type (WEP or WPA/WPA2)**" that will next be used to get the remote controllers' "*MAC Address*".
- Using the previous information, the following command is used to determine the "**MAC address**" of the remote controller. Using airodump-ng, the following command is used "*sudo airodump-ng –bssid droneBSSID –channel droneChannel wlan0mon*" to capture incoming traffic to the drone. Since the remote controller is connected to the drone, it is certain to capture traffic incoming traffic with "*MAC Address*" of the remote controller.
- With the MAC address of both the remote controller and the drone known, the de-authentication attack can be performed. De-authentication packets are sent by the attacker to the drone's MAC address, which in turn sends de-authentication packets to the remote controller, causing a loss of connection. The following command is used to perform the attack "*sudo aireplay-ng -0 0 -a droneBSSID -c remotecontrolBSSID wlan0mon*" [38]. The attributes used in this command are "*-0*" for the number of De-authentication packets to be sent, "*-a*" for the drone MAC Address and "*-c*" for the remote controllers' MAC address.
- Once the connection is lost between the drone and the remote controller, the drone either crashes, remains in position until the connection is re-established or initiates its RTH to return to the last saved point. During that time, the attacker simply connects to the drone since there is "No Security".

In the "WPA/WPA2" scenario, Deauthentication packets are sent to the drone resulting in a loss of connection between the drone and the access point. However, in this case, de-authentication is done to make the remote controller re-authenticate. The goal is to capture four-way WPA handshake and perform a **dictionary attack** to get the password of the access point. Figure 14 shows the steps resulting in a successful De-Authentication attack with WPA/WPA2.

- In the first part of this attack, the **steps** followed in the "No Security" scenario are reproduced. The difference is that the attacker cannot connect to the access point since it has a password. So, a dictionary attack is performed to determine the password and connect to the drone.
- Using Airodump-ng, the attacker collects four-way WPA2 handshake packets sent to the access point using the following command "*airodump-ng –bssid droneB-SSID –channel droneChannel –write nameoffile wlan0mon*". The attributes used in this command are "*–bssid*" for the MAC Address of the drone, "*–channel*" for the channel of the access point and "*–write*" for the name of the file where

Fig. 14 Steps to perform
de-authentication attack
(WPA/WPA2)

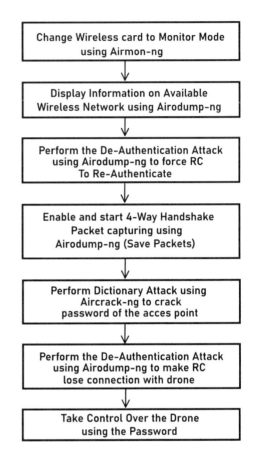

captured packets are to be saved. To be sure to capture enough packets, it is best
to perform the Deauthentication and packet collection process more than once.

- Once the packets are captured, the dictionary attack can be performed by using the
  following command "***aircrack-ng nameoffile -w pathtothepassworddictionary***".
  The attributes used in this command are "***nameoffile***" for the name of the file where
  the captured four-way handshakes are saved, "***-w***" for the path to the file where
  the word list (dictionary) of possible passwords is saved. Aircrack-ng duplicates
  the four-way handshakes and uses the word list (dictionary) to verify if any of the
  words in the word list matches results of the four-way handshake. If it does the
  key to the access point is identified. Word lists of possible passwords for drones'
  access point is available for download on various platforms for free.
- Once the password is identified; the attacker can launch another ***Deauthentication***
  to de-authenticate the remote controller from the drone and use the ***password*** to
  connect to the drone and take control of it.

6 Threat Actors Targeting Autonomous Drones and Their Incentive

The deployment of drones during the COVID-19 pandemic has brought convenience and efficiency in the way the delivery of goods and security surveillance is been carried out and many more. During the pandemic, the employment of drones has been utilised in diverse ways. The threat actor's incentive is to target these drones for financial gain, at a time where nations are vulnerable to attacks due to other ongoing concerns. The autonomous performance of drones faces numerous threats and fallbacks in its operation due to sophisticated attacks from major threat actors and vulnerabilities in the security hierarchy. This section will emphasise on threat actors and frequently used attack vectors used to exploit and compromise drone security and operations. There are various types of threat actors that could pose a threat to the deployment of drone amid COVID-19. Among these can be distinguished individuals, cyber-criminals, organizations, hacktivists, and organized corporate espionage [43]. Most of these threat actors aim at either intercepting the communication medium, gaining unauthorized access to on-board information of a drone, or disturb operation.

To delve into this, it is paramount to reflect briefly on the building blocks of a drone. Drone as a UAV consist of three fundamental building blocks which are unmanned aircraft (UMA), Ground control station (GCS), and Communication data link (CDL) [44]. In most attacks' scenarios, the communication datalink is highly exploited by threat actors due to the adaptation of wireless communication either through Bluetooth, Wi-Fi, radio waves, and satellite transmission. Each communication and data storage medium present a level of security vulnerabilities which are constantly exploited by threat vectors within several attack vector medium. When the security chain is compromised, it is an indication that a vulnerability is been exploited by a threat actor.

6.1 CIA

As initially highlighted, threat actors aim at intercepting the flow of communication of a drone as well as it is on-board information. As drone usage during this pandemic is increasing in popularity, the threshold of threat actors could increase in a sophisticated manner due to the physical exposure drones are subject to within the aerospace as well as the increasing demand for drones during this period [45]. This makes sensors, ground control, imagery, and sensor data vulnerable to attacks. These threats breach the privacy, safety, and security of data of a deployed drone. The attack execution from these threat vectors aims at compromising the integrity of payload, gaining access to confidential payloads, and depriving the access of services, these break the security chain of the system [43]. The security chain within the cyberspace focuses on three fundamental building blocks, which are data confidentiality, integrity, and

Fig. 15 CIA diagram

availability. Confidentiality ensures data in and out of a system can only be accessed by authorized personnel while integrity ensures originality and authenticity of data [46]. Finally, availability ensures consistency and accessibility of service or data within a system setup. These elements are illustrated in Fig. 15.

Threat actors depend on attack vectors as a means of carrying out attacks against drones during the pandemic. The attack vectors could be technical or intellectual procedures use in executing attacks. Some of these attacks' vectors may include GPS spoofing, Denial of service, backdoor access, and many more. By deploying these attack vectors, a threat actor can compromise the security system of a drone and gain unauthorized access. This is illustrated in Fig. 16.

6.2   Threat Actors and Security Integrity

Considering the integrity of data, threat actors aim at compromising or altering flight history, schedules, destination profiles, and major security protocols of a drone designated for operation during the COVID-19 pandemic. In some instances, hacktivists might alter the information of a scheduled drone by deploying spoofing, a methodology used in falsifying data of a system [47]. In this case, the threat actor has the privilege to tamper with the originality of data to suit the objective of the attack either by altering GPS coordinates or redirect its communication link. When the coordinates of a drone are altered by a threat actor, its mission also gets diverted. A drone scheduled for security surveillance or delivery of personal protective kits and medicine could be redirected to a different geographical area which it is not intended for. As reported in 2016 by **the US Department of Homeland Security (DHS) and the US Customs and Border Protection (CBP) agency**, a US border patrol drone was diverted from its flight path by drug traffickers to transport drugs along

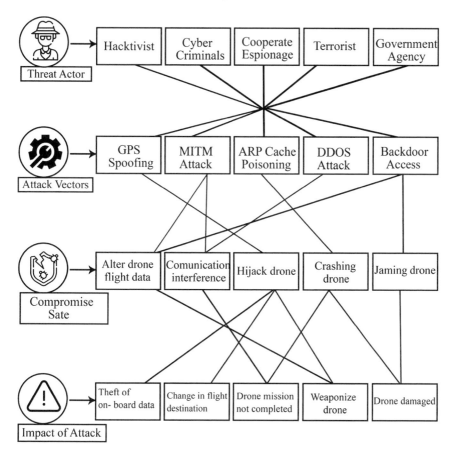

Fig. 16 Threat actors attack tree

the U.S-Mexico border with impunity [48]. In this case, the surveillance drone was redirected to create a blind spot for a major crime to be carried by a threat actor. The threat actors in this scenario did not tamper with the availability of service and data confidentiality. However, the integrity of data was compromised when the drone was diverted from its original task. Another incident occurred in 2011 when the Iranian military made progress in hijacking a U.S operated drone and redirected its destination to Iranian territory [49]. This was done by jamming the drone's frequency which it depended on in navigating its GPS coordinates. New frequency coordinates were then communicated to the drone to assume it was near its intended destination when the drone was far from home. In this scenario, the threat actor was able to decrypt the GPS coordinates of a US military-operated drone and transmitted false data to change its intended destination.

6.3 Threat Actors and Security Confidentiality

Confidentiality is a core security element that is broken by threat actors when access is gained through the data transmission of a drone or its data storage platform either cloud or web base storage. This is done by eavesdropping and sniffing on a drone transmission network or third-party software. An attacker sniffs on the network to collect packets of a drone then deploys sophisticated tools such as aircrack-ng to carry out the attack. Aircrack-ng is a software suite designed for detecting and sniffing on network packets and cracking of security encryption protocols of a wireless network interface [50]. Also, a threat actor could have access to the on-board camera of a deployed drone and have the privilege of getting video and image feed from the drone. When a threat actor gets to have a clue of ongoing activities between a deployed drone and a network interface, it provides the attacker with information helpful to carry out a major attack. This could result in a Man-in-the-Middle attack, where a threat actor intercepts all communication links from the GCS to the drone. An adversary with relevant information on drone deployment can execute a sophisticated attack base on the information at their disposal. When the confidentiality of the drone security system is compromised, it exposes the drone to potential attacks that might be executed instantly or after some time when data flow over the network is carefully analysed by a threat actor. A hacktivist might gain access to a military or government-operated drone and leak the captured information to a public domain as a means of creating chaos or fuelling a propaganda mission regarding health cases and the number of supplies sent to a region. This reveals the impact created by a breach of security (in this case confidentiality) in the case of drones, especially when sensitive information gets into the wrong hands.

6.4 Threat Actors and Service Availability

The reliability of a system is important especially during this pandemic period. One major security element threat actors target most is service availability of dispatched drones. There have been countless cases where drones have gone missing or jammed within the aerospace. A threat actor aims at taking control over a dispatched drone in flight and redirect its mission or crash it. A drone scheduled to deliver some packages could get hijacked in-mid-air without completing its intended mission. What a threat actor does is initiate a Denial of Service (DoS) attack by overwhelming the associated network interface. Denial of service (DoS) or Distributed Denial of Service (DDoS) occurs when a threat vector causes a drone service or network to be unable to provide its intended services either temporally or permanently [51]. In some instances, when a service is unavailable, it gives room for the attacker to plant malicious code into related drone systems before getting it back in full operation. In this case, there is a temporal Denial of Service (DoS) geared towards a major security breach that might be in operation for a long period especially when a back door is created. A drone being hijacked by a threat actor could be weaponized rather than providing the aid intended amid the pandemic and released back to its original operators without the operators knowing the hidden damages created. A drone scheduled to deliver medical supplies in a remote area could be jammed by attackers, hence denying access to those supplies when needed. Each of these scenarios provides an insight into the impact Denial of Service (DoS) attack can have on a larger scale in the case of drones used during the COVID-19 pandemic. Commercials drones designed for high profile missions such as humanitarian aid and supplies face a lot of threat of such nature.

The privileges attackers have after compromising a dispatched drone seem to be limitless. It reveals the threat hackers pose to drones and their related transmission mediums. The threat level varies from one drone mission to another as well as the estimated impact. Terrorists for instance target drones designated for special missions with relevant payloads onboard. This is directed towards government and high profiles institutional drones to fuel massive online propaganda and cooperate espionage. In 2016 it was reported that a Palestinian terror group for two years could see what the Israeli military surveillance drone saw [52]. Having access to such privilege always allows threat actors to monitor surveillance drones and determine their course of action. This is a typical example of having the confidentiality of the drone security system been compromised. Unauthorized personnel could gain access to information which they can capitalize on to exploit other venues. As demonstrated by Samy Kankar in 2016, a threat actor can hijack a drone about a mile away by making use of standard radio, reprogram the drone's onboard software to hijack the system and take control over it [53]. By reprogramming the onboard software, threat actors can impersonate legitimate drone pilots and be granted full control. This was done by jamming the drone's frequency which it depended on in navigating its GPS coordinates. New frequency coordinates were then communicated to the drone to assume it was near its intended destination when the drone was far from home. In this scenario, the threat actor was able to decrypt the GPS coordinates of a US military-operated drone and transmitted false data to change its intended destination.

The common pattern in most of these attacks requires the threat actor to disrupt the signal between a drone and its GCS by deploying malware or altering the drone's central data. Once this is successful, the threat actor creates a sync between the drone and its programmed script designed to initiate the attack. The attack patterns discussed reveal that, in some instances, service availability of the drone was first affected by jammed the system then proceeded to alter data records by compromising integrity. This shows how one security element is compromised to initiate another attack and to pave way for another security element to also be compromised. It can be concluded that the three-security element, confidentiality, integrity, and availability are all interdependent elements of drones' systems.

6.5 Threat Actors and Attack Motivation

Threat vectors targeting drones during this pandemic are motivated by a purpose that compels them to execute attacks against drones. The purpose varies from one threat actor to another. While some might be motivated by minor goals, the impact of the successful execution of the attacks could result in massive damages and breaches. Some of these purposes are but not limited to:

- Threat actors such as terrorists execute drone attacks to weaponize them for attacks or suicide mission.
- Curiosity is one major purpose that motivates most threat actors, especially teenagers who have access to tools and means to exploit drones.
- Governments motivated by rivalry with other nations can often launch drone attacks to gather intelligence especially when it comes to surveillance data.
- Hacktivists are mostly motivated by online propaganda and political interference.
- Cybercriminals are motivated by money. Most cybercriminals execute ransomware attacks against drones to demand ransom from owners.
- A threat actor could launch a drone attack to carry out a prank.
- Theft is one purpose that motivates threat actors to launch attacks against drones scheduled to deliver goods.

7 Countermeasures

As already established, most drones used amid this pandemic are subjected to attacks that are geared towards their communication channel. It is needful to set up frameworks, security measures, and policies that safeguard drones from attacks launched against them and limits the impact of those attacks on drone's effective operations. These countermeasures will be addressed in connection with the already discussed attacks in the previous section. The countermeasures are in-line with the hasty, yet appropriate deployment of drones during the pandemic. It is paramount to note that, no system is completely secured, however, the measures addressed in this section is geared towards narrowing down the possibility at which an attack may occur as well as managing the impact of an attack.

7.1 Countermeasures Against GPS Spoofing

A successful GPS Spoofing attack leads to the drone's receiver antenna being unable to provide accurate information to the drone, sometimes causing irreversible damages. It is necessary to have in place a mechanism to ensure that GPS remains fully functional in the case of GPS Spoofing.

An encoded binary code is embedded in the legacy GPS signal known as the Y-code (encrypted precision code) transmitting on the frequencies L1 and L2 [54], mostly designated for military use. There is a dynamic sequence that causes the encrypted binary code to change 10.23 million times per second, causing the Y-code to change its combination key uniquely without regular repetition. In the absence of the encrypted key, attackers cannot generate the Y-code, hence basically impossible to spoof a GPS set in the direction of a Y-code. Coarse Acquisition (C/A) code is included in the GPS legacy signal which was initially intended to acquire a Y-code [55]. The change in the sequence of the C/A code is less compared to that of the Y-code as its 1 and 0 s changes within 1.023 million times per second with a potential repetition every millisecond [56]. C/A code can be recreated by attackers since is made available in a public signal-in-space interface specification, hence subjected to potential spoofing of GPS signal [57]. Direct tracking of the encrypted Y-code by using a GPS receiver with a Selective Availability Anti-Spoofing Module (SAASM), will provide high-level protection against GPS spoofing.

An additional security mechanism can be provided to the dispatched drone by enabling the navigation system with an inertial measurement unit (IMU) since an attacker cannot manipulate or spoof the gravitational field or vehicle dynamics of the Earth to persuade the inertial unit to assume it has shifted in a way it has not.

7.2 Countermeasures Against DoS/DDoS Attack

Drone communications depend greatly on the Global Positioning System (GPS) for the drone's accurate positioning, navigation, and timing (PNT) [58]. However, operating on civil GPS (an insecure and unauthenticated system), drones are vulnerable to GPS Jamming which is a major contributor to a DOS attack.

In the case of GPS Jamming, the implementation of an adaptive antenna (ADA) in the drone's architecture ensures an uninterrupted workflow and assured PNT functionalities by overcoming GPS jamming and making sure there is consistency in operation with related GPS. With ADA, an advanced digital signal processing method is deployed to immune the drone from single or multi-jammer occurrences or interference [59]. This is possible due to the multi-band jamming immunity that ADA provides.

Another way of avoiding interference is by filtering in the receivers. In this line of defence, it is necessary to filter the majority of the interference as it reaches the receiver [60]. This is to eliminate signals that are not in the direction of the GPS frequencies intended to be received. However, unexpected signals that fall in-band could overwhelm the receiver.

7.3 Countermeasures Against Man-in-the-Middle Attack

MiTM attack is established on the concept of unauthorized monitoring of data flow, interception of data flow, and altering data variables [61]. In setting up security systems to counter such attacks on the communication datalink of a drone, it is needful to consider these elements. As initially discussed, the drone architecture is designed to have communication with the Ground Control Station (GCS) through wireless transmission. One security infrastructure recommended in the process of securing the communication datalink is the use of a virtual private network (VPN). A VPN is a designed private network developed within a public network infrastructure to serve as a secured and anonymous means of transmitting data [62]. The inclusion of VPN into a drone network infrastructure should be done by first doing a close examination of potential threat level, followed by a review of the dynamics involved in the implementation process, as well as the maintenance and troubleshooting of a specific VPN [63]. By introducing a VPN service into a drone network infrastructure, the network layer gets encrypted hence providing a level of anonymity with regards to geographical location, data in transit, and user credentials. With such security measures in place, data confidentiality is assured within the security chain of the drone's system. This will ensure a narrow success rate of MITM attack by depriving attackers of the privilege of accessing the communication data link between a drone and the GCS.

7.4 Countermeasures Against De-authentication Attack

De-authentication breaks the confidentiality of every security system and goes further to affect the integrity state of the security chain. As a means of preventing or ensuring constraints towards de-authentication against 802.11 networks, there are several security measures and controls that could be considered.

Setting up an external firewall that is made up of Egress, Ingress, and Address filters. This helps to monitor traffic flow within the drone's network and examine the data flow leaving the network related to already issued policies by the network administrator [64]. An external firewall can be directly located on the drone. The drone might be in direct connection to servers, routers as well as network switches. With the aid of an external firewall, there is direct control of the inflow of data from outside sources as well as the type of data that flows from the workstation. This reduces the success rate of attackers sending de-authentication packets to the drone or its controller as an initial means of initiating the attack. The address filter sees to it that any address not recognized by the network is filtered out. It is also recommended that a strong password is used in securing data across the network. Though a strong password made up of alphanumeric elements and symbols might reduce the success rate of a dictionary attack; however, multi-factor authentication must be set up to enforce a level of constraints. With multi-factor authentication, the system requires not just a username and password. It goes on to request One Time Password (OTP) either through SMS or mail notification. OTP is a two-factor authentication that provides mobile devices with temporal tokens which are bound to expire after a given period [65]. In some instances, a multi-factor authentication might require a fingerprint or facial recognition to complete the process of de-authentication [66]. All this increases the security level of the drone system. This means when one security layer is compromised, another layer makes up for it. This will ensure the safe flight of drones dispatched in the fight against COVID-19.

8 Proposed Framework for Drone Operation

Developers are now paying attention to privacy-protection and security technologies for drones, due to the social and legal implications on privacy intrusions and safety standards [67]. With the recent deployment of AI and drones, the proposed framework provides a robust and secure mechanism against malicious actors. The pandemic has resulted in a significant number of drones being deployed, and it is critical for these systems to be protected appropriately. This section proposes a framework that encompasses the already discussed countermeasures such as ADA, IMU, Y-code encryption protocol, legislative measures and controls needed to ensure the privacy and safe operation of commercial drones. The framework considers major industry playmakers when it comes to drone usage. These playmakers have been categorized as operators, service providers, and manufacturers. It is critical for activities carried out by these playmakers to be guided by laid down policies, legislations, security standards, and mitigation plan. In certain jurisdictions, there are laid down legislation that needs to be adhered to in operating or manufacturing a drone either for private use or commercial use. The framework considers policies on geographical movement such as restricted flight zone and limited frequency bandwidth. The framework will be more applicable to drones and operated in a jurisdiction that recognizes these factors as vital to adhere to. Figure 17 gives an understanding of the various processes and elements of the framework.

   The application of this framework is well discussed by incorporating the framework into an applicable scenario such as the deployment of drones during health crises such as COVID-19. First, the signal between the drone and antenna is encrypted with the Y-code binary to safeguard signals from being corrupted or jammed. SAASM is recommended for GPS receiver as a counter mechanism against GPS spoofing. Access to credentials such as login is secured with multi-factor authentication. This could be a two factor or three-factor authentication procedure. This requires users to undergo layers of verifying their identity before been granted access. The framework sees to it that, access to drone data storage is granted to legitimate persons with the help of multi-factor authentication. The framework takes into consideration a means of identifying anomalies within the network connectivity by embedding an intrusion detection system. An intrusion detection system monitors the network for any anomalies or intruders [68]. This is embedded in the network transmission

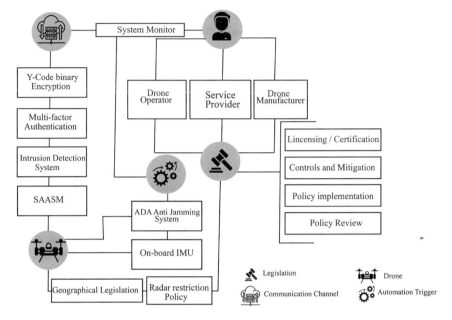

Fig. 17 Proposed framework

channel between the drone and its data source. In this case, service operators and providers are alerted on any unauthorized intrusion within the network channel. The ADA anti-jamming system will perform as an automated mechanism that counters any jamming attack against the drone. Both the IMU and ADA are automated and have a connection with the monitoring system which provides relevant operators and service providers with information on its performance and status. By automating their performance, the drone is enabled with security measures that are autonomous and triggers instantly against jamming or spoofing.

Despite all these technical features, the framework acknowledges the essence of legislative and security standards that might be relevant to the drone base on jurisdiction and purpose of drone usage. This has been categorized as licensing and certification, controls and mitigation, policy implementation and review, geographical legislation, and radar restriction which are solely directed to the drone and its flight. Figure 18 shows the framework and its impact on various attacks.

9 Conclusion

This paper critically analyses the unique combination of drones with artificial intelligence. The paper initially reviews drone's architecture, along with their vulnerabilities to better understand the security issues it faces. It then analyses drone's current deployment, especially during the recent global pandemic (COVID-19) where drones

Fig. 18 Impact of framework on attacks

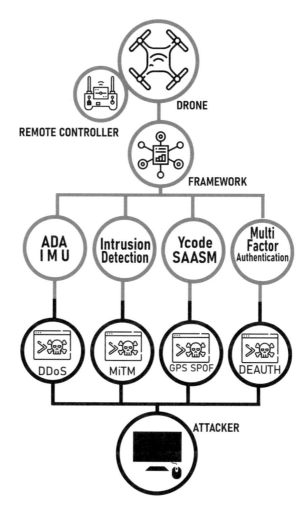

have been used for logistical purposes, for crowd dispersal, for urban disinfection, and even for facial recognition. The paper also critically evaluates artificial intelligence and indicates how it enhances drone design. Futhermore, the paper takes a look at the way AI was embedded in drones and gives an overview on how drones without an efficient security mechanism in place are vulnerable to cyber attacks.The paper then reviews threat actors, their incentives and possible attack methods they could potentially use against commercial drones. Finally, a robust framework is proposed for the drone industry in order to improve on drone's security and enable drones to face some of the aforementioned attacks. The framework would bridge the gap between commercial and government grade security, and improve trust with the public in order for trust, transparency, and privacy to be formed. The framework is designed to act as a robust mechanism against malicious threat actor who target commercial drones for financial and malicious gain.

References

1. ECDPC (2020) COVID-19 situation update worldwide, as of 27 July 2020, 27 July 2020. [Online]. https://www.ecdc.europa.eu/en/geographical-distribution-2019-ncov-cases
2. Foster P, Neville S (2020) How poor planning left the UK without enough PPE | Free to read. Financial Times [Online]. https://www.ft.com/content/9680c20f-7b71-4f65-9bec-0e9554a8e0a7. Accessed 27 July 2020
3. Tavakoli T, Carriere J, Torabi A (2020) Robotics, smart wearable technologies, and autonomous intelligent systems for healthcare during the COVID-19 pandemic: an analysis of the state of the art and future vision. Advanced intelligent systems
4. Kharpal A (2020) Use of surveillance to fight coronavirus raises concerns about government power after pandemic ends. CNBC [Online]. https://www.cnbc.com/2020/03/27/coronavirus-surveillance-used-by-governments-to-fight-pandemic-privacy-concerns.html. Accessed 27 July 2020
5. Reuters T (2020) Google's wing aviation gets OK for drone deliveries in U.S. CBC [Online]. https://www.cbc.ca/news/technology/google-wing-aviation-drone-deliveries-1.5108679. Accessed 27 July 2020
6. Taylor P (2020) Could 'Pandemic Drones' help slow coronavirus? Probably not—but COVID-19 is a boom for business. Forbes [Online]. https://www.forbes.com/sites/petertaylor/2020/04/25/could-pandemic-drones-help-slow-coronavirus-probably-not-but-covid-19-is-a-boom-for-business/#234caaf562a4. Accessed 27 July 2020
7. Wood L (2020) 5 ways drones are being used in efforts to fight COVID-19: virus detection, sprayer, food delivery, surveillance and emergency medical delivery—ResearchAndMarkets.com. GlobeNewswire [Online]. https://www.globenewswire.com/news-release/2020/04/07/2012735/0/en/5-Ways-Drones-are-Being-Used-in-Efforts-to-Fight-COVID-19-Virus-Detection-Sprayer-Food-Delivery-Surveillance-and-Emergency-Medical-Delivery-ResearchAndMarkets-com.html. Accessed 27 July 2020
8. Moguel E, Conejero JM, Sánchez-Figueroa F, Hernández J, Rodríguez-Echeverría R (2018) Towards the use of unmanned aerial systems for providing sustainable services in smart cities. Sensors (Switzerland) [Online]
9. Morley T (2020) The UK drone users report introduction [Online]. Accessed 16 Aug 2020
10. Wagner I (2020) Projected commercial drone revenue worldwide 2016–2025. Statista [Online]. https://www.statista.com/statistics/607922/commercial-drone-market-revenue-worldwide-projection/
11. Mirzaeinia A, Hassanalian K, Lee K, Mirzaeinia M (2019) Energy conservation of V-shaped swarming fixed-wing drones through position reconfiguration. Aerosp Sci Technol [Online]
12. Gunarathna JK, Munasinghe R (2018) Development of a quad-rotor fixed-wing hybrid unmanned aerial vehicle. In: MERCon 2018—4th international multidisciplinary Moratuwa engineering research conference [Online]
13. Yaacoub J-P, Salman O, Chehab A, Noura H (2020) Security analysis of drones systems: attacks, limitations, and recommendations. Internet of Things 11 [Online]
14. UST (2019) UAV ground control stations. Unmanned Syst Technol [Online]. https://www.unmannedsystemstechnology.com/category/supplier-directory/ground-control-systems/ground-control-stations-gcs/. Accessed 27 July 2020
15. Çuhadar I, Dursun M (2016) Unmanned air vehicle system's data links. J Autom Control Eng [Online]
16. Altawy R, Youssef AM (2017) Security, privacy, and safety aspects of civilian drones: a survey. ACM Trans Cyber-Phys Syst [Online]
17. Corrigan F (2020) How do drones work and what is drone technology—DroneZon [Online]. https://www.dronezon.com/learn-about-drones-quadcopters/what-is-drone-technology-or-how-does-drone-technology-work/. Accessed 05 Aug 2020
18. Statista (2020) Global IoT enterprise drones shipment by use case 2019–2023. Statista [Online]. https://www.statista.com/statistics/1079495/global-drone-enterprise-use-cases-iot/. Accessed 16 Aug 2020

19. Shakarian P, Shakarian J, Reuf A (2013) Introduction to cyber-warfare. Syngress [Online]
20. Dey V, Pudhi V, Chattopadhyay A, Elovici Y (2018) Security vulnerabilities of unmanned aerial vehicles and countermeasures: an experimental study. In: Proceedings of the IEEE international conference on VLSI design [Online]
21. Astaburuaga I, Lombardi A, La Torre B, Hughes C, Sengupta S (2019) Vulnerability analysis of AR.drone 2.0, an embedded linux system. In: 2019 IEEE 9th annual computing and communication workshop and conference, CCWC 2019 [Online]
22. Miailhe N, Hodes C (2017) The third age of artificial intelligence. Institut Veolia
23. Searle JR (1990) Is the brain's mind a computer program? Sci Am
24. Mogg T (2020) French police the latest to use speaker drones to enforce coronavirus lockdown. DigitalTrends, 23 Mar 2020 [Online]. https://www.digitaltrends.com/cool-tech/french-cops-use-speaker-drones-to-enforce-coronavirus-lockdown/. Accessed 2 Sept 2020
25. Hassani H, Silva ES, Unger S, TajMazinani M, Mac Feely S (2020) Artificial intelligence (AI) or intelligence augmentation (IA): what is the future? MDPI Open Access J
26. Schroth L (2018) Drones and artificial intelligence, 28 Aug 2018 [Online]. https://www.droneii.com/drones-and-artificial-intelligence.
27. Pennic F (2020) 'Pandemic Drone' could detect virus symptoms like COVID-19 in crowds. HIT Consultant, 27 Mar 2020 [Online]. https://hitconsultant.net/2020/03/27/pandemic-drone-could-detect-virus-symptoms-like-covid-19-in-crowds/#.X1DnznlKiHs. Accessed 2 Sept 2020
28. Greenwood F (2020) Why fever-detecting drones won't help spot potential COVID-19 patients. Science The Wire, 23 May 2020 [Online]. https://science.thewire.in/health/fever-detecting-drones-dont-work/. Accessed 2 Sept 2020
29. Mitchell M (1998) An introduction to genetic algorithms. First MIT Press Paperback Edition
30. Galvez RL, Dadios EP, Bandala AA (2014) Path planning for quadrotor UAV using genetic algorithm. IEEE, Palawan
31. Thede MS (2004) An introduction to genetic algorithms. J Comput Sci Colleges
32. Muliadi J, Kusumoputro B (2018) Neural network control system of UAV altitude dynamics and its comparison with the PID control system. J Adv Transp
33. Iannace G, Ciaburro G, Trematerra A (2019) Fault diagnosis for UAV blades using artificial, robotics
34. ISO (2020) Acoustics. Determination of sound power levels and sound energy levels of noise sources using sound pressure. Precision methods for anechoic rooms and hemi-anechoic rooms
35. Kuss M (2020) Can drones fly without Wi-Fi? [Online]. https://tipsfordrones.com/can-drones-fly-without-wi-fi/. Accessed 26 Aug 2020
36. Prowell S, Kraus R, Borkin M (2010) Seven deadliest network attacks. Elsevier
37. Shinder L, Cross M (2008) Scene of the cybercrime. Elsevier
38. Gudla C, Rana S, Sung AH (2018) Defense techniques against cyber attacks. In: International conference on embedded systems, cyber-physical systems, & applications, Mississipi
39. Bonilla CATB, Parra OJS, Forero JHD (2018) Common security attacks on drones. Int J Appl Eng Res 13:4982–4988
40. Shepard DP, Bhatti JA, Humphreys TE (2014) Unmanned aircraft capture and control. Austin. J Field Robot
41. Bertoli GdC, Osamu S (2019) IEEE 802.11 de-authentication attack detection using machine learning on unmanned aerial system: emerging trends and challenges in technology. In: Proceedings of the 3rd Brazilian technology symposium. ResearchGate, pp 23–30
42. Said Y (2020) Using monitor mode in Kali Linux 2020, 11 Aug 2020 [Online]. https://linuxhint.com/monitor_mode_kali_linux_2020/. Accessed 25 Aug 2020
43. Lillian A (2018) Data thieves: the motivations of cyber threat actors and their use and monetization of stolen data. RAND Corporation, Santa Monica, Calif.
44. Rencüzoğulları EA, Kılıç A, Kapucu S (2018) Design consideration: UAV system architecture for smart applications. In: International Eurasian conference on science, engineering and technology (EurasianSciEnTech 2018), Ankara

45. Jackman A (2019) Consumer drone evolutions: trends, spaces, temporalities, threats. Taylor & Francis Group, Routledge
46. Nweke LO (2017) Using the CIA and AAA models to explain cybersecurity activities. PM World J
47. Jokar P, Arianpoo N, Leung VC (2012) Spoofing detection in IEEE 802.15.4 networks based on received signal strength. Elsevier
48. Cimpanu C (2016) Border patrol drones vulnerable to GPS spoofing attacks. Softpedia News
49. Peterson S, Faramarzi P (2011) Exclusive: Iran hijacked US drone, says Iranian engineer. The Christian Science Monitor, Turkey
50. Čisar P, Maravić Čisar S (2018) Ethical hacking of wireless networks in kali linux environment. Ann Fac Eng Hunedoara Int J Eng
51. Virupakshar KB, Asundi M, Channal K, Shettar P, Patil S, Narayan DG (2020) Distributed denial of service (DDoS) attacks detection system for openstack-based private cloud. Elsevier
52. Axe D (2016) How Islamic Jihad hacked Israel's drones. The Daily Beast
53. Sanders G (2019) The very real dangers of hacked drones. Tractica Omdia
54. O'Hanlon BW (2017) Signal processing and the global positioning system: three applications. Cornell University
55. Wei Z, Heejong S, Ke Z (2009) Simulation and analysis of coarse acquisition code generation algorithm in GPS system. In: The 1st international conference on information science and engineering (ICISE)
56. Manfredini EG (2017) Signal processing techniques for GNSS anti-spoofing algorithms. Politecnico di TorinoPorto Institutional Repository
57. Zeng Q, Qiu W, Zhang P, Zhu X, Pei L (2018) A fast acquisition algorithm based on division of GNSS signals. J Navig
58. Fisher KA, Raquet JF (2011) Precision position, navigation, and timing without the global positioning system. Air Force Research Institute (AFRI)
59. Vaseghi SV (2006) Advanced digital signal processing and noise reduction. Wiley
60. Nichols S, Dygert R (2013) Filtering interference for antenna measurements. MI Technologies
61. Aliyua F, Sheltamia T, Shakshuki EM (2018) A detection and prevention technique for man in the middle attack in fog computing. In: The 9th international conference on emerging ubiquitous systems and pervasive networks
62. Jeff T (2013) How virtual private networks work. Cisco System
63. Fraser M (2020) Understanding virtual private networks (VPN). Global information assurance certification paper
64. Dickey C, Kornegay K (2015) Assessing the vulnerability of wireless networks. Morgan State School Department of Electrical and Computer
65. Dmitrienko A, Liebchen C, Rossow C, Sadeghi A-R (2014) On the (In)Security of mobile two-factor authentication. Vrije Universiteit Amsterdam, Amsterdam
66. Aldwairi M, Aldhanhani S (2017) Multi-factor authentication system. In: 2017 international conference on research and innovation in computer engineering and computer sciences (RICCES), Langkawi
67. Ottavio M (2015) Privacy and data protection implications of the civil use of drones. European Parliament, Brussels
68. Ayodejia A, Liua Y-k, Chaoa N, Yang L-q (2020) A new perspective towards the development of robust data-driven intrusion detection for industrial control systems. Elsevier, p 4

Effective Splicing Localization Based on Image Local Statistics

P. N. R. L. Chandra Sekhar and T. N. Shankar

Abstract In the digital era, people freely share pictures with their loved ones and others using smartphones or social networking sites. The news industry and the court of law use the pictures as evidence for their investigation. Simultaneously, user-friendly photo editing tools make the validity of pictures on the internet are questionable to trust. Intense research work is going on in image forensics over the last two decades to bring out such a picture's trustworthiness. In this paper, an efficient statistical method based on Block Artificial Grids in double compressed JPEG images is proposed to identify areas attacked by image manipulation. In contrast to existing approaches, the proposed approach extracts the local characteristics from individual objects of the manipulated image instead of the entire image, and pair-wise dissimilarity is obtained between those objects and exploits the manipulated region, which has the highest variance among other objects. The experimental results reveal the proposed method's superiority over other current methods.

Keywords Splicing localization · Object segmentation · Block artificial grids · Cosine dissimilarity

1 Introduction

Nowadays, in digitization, people strongly connect with social networking sites and freely share their ideas, pictures, and comments. In present-day society, images are used extensively in many ways, such as evidence in the court of law, journalism, science, and forensics discovery [1]. The Government is also taking positive steps towards digitizing all fields to reach the public. Simultaneously, the rapid growth of technology in developing powerful image editing tools induced an interest to make

P. N. R. L. C. Sekhar (✉) · T. N. Shankar
Department of Computer Science and Engineering, Koneru Lakshmaiah Education Foundation, Vaddeswaram, Guntur, AP, India
e-mail: cpnrl@gitam.edu

T. N. Shankar
e-mail: tnshankar2004@kluniversity.in

the images or videos manipulate with ease and cannot be traceable with human vision. Copy-move, splicing, re-sampling, cloning few manipulation attacks which are frequently to tamper digital images. If an image has undergone these attacks and uses for evidence, it significantly impacts the trustworthiness of such evidence [2]. It is a challenge to distinguish the original with manipulated and establish the integrity and authenticity of digital images [3].

Image Forensics, a branch of Multimedia Forensics, aims at developing powerful techniques and tools towards detecting manipulation attacks on images [4]. In traditional methods like watermarking, authentication is considered an active method where authentic code is embedded in the original image to verify authenticity. Whereas blind or passive methods do not require any external clue to assess the authenticity of the image. Many image tampering techniques work on the assumption that pictures taken from different cameras or different processing operations introduce different inherent patterns into tampered image [5]. Furthermore, these underlying patterns consistent throughout the original image, and when any manipulation attacks it, there will be inconsistency in those patterns of tampered image. These intrinsic inconsistency statistics can thus be used as forensic features to identify image tampering [6].

In image splicing, a part of the source image copied and pasted into the donor image. The post-processing techniques applied to the tampered image made human vision challenging to find such attacks [7]. This challenge attracted many researchers to find methods for image splicing detection. These techniques extract image features and use classification techniques to reveal the forgery, and achieve even high success rates [8]. However, it is worth locating the tampered region in many real-time purposes to gain confidence [9]. Splicing localization brings many more challenges as it requires pixel-level analysis rather than image-level analysis [4, 10].

The images captured by digital cameras are stored in Joint Photographic Experts Group (JPEG) format. The JPEG format uses lossy compression and responsible for the proliferation of images on the internet and social networking sites. In JPEG compression, the digital image divides into 8×8 non-overlapping blocks, and for every block, the discrete cosine transform (DCT) is evaluated and then quantized using a standard quantization matrix. When a splicing attack manipulates the image, it introduces discontinuities, and these statistical traces, such as JPEG quantization artefacts and JPEG grid alignment discontinuities, are used to exploit tampering attacks [11, 12].

The tampered region blocks will undergo single compression in a splicing attack while the remaining blocks will have double compression. For double compression (DQ) artefacts, a model of periodic DCT patterns is created in [13] and evaluated each block of the image concerning its conformance of the model. Any block whose probability distribution distinguishes from the original classifies as blocks manipulated by a tampering attack. A similar approach found in [14] where the authors assume that the distribution of JPEG coefficients changes with the number of recompressions and proposes a training a set of support vector machines (SVM) for the first digit artefacts and estimated the probability distribution of each block as single or double compressed thereby exposed the splicing attack.

In [15] proposed an alternative method to exploit DQ artefacts. They compare the discontinuities using the quality factor adopted in the tampered region with the principle that a JPEG ghosts—a local spatial minimum- will correspond to the tampering attack. The limitation of the method is; it works only if the tampered region has a lower quality factor than the rest of the image. An alternative to the DQ discontinuities, in [16], the authors created a model on the entire image DCT coefficient distributions using the degree of quantization. The inconsistencies become indicative of the tampering attack. The difference between this method and DCT-based is that the output is not probabilistic, making the technique relatively difficult to interpret although efficient.

Other techniques use JPEG grid discontinuities as they occur during compression by placing spliced objects that misaligned the 8×8 block grid. The 8×8 block creates the grid even when the image compress with high quality and these discontinuities are invisible to human eyes but can be exposed using filtering. The absence or misalignment of the 8×8 grid with the rest of the image can become a fingerprint to exploit a tampering attack in [17]. In [18], the authors extracted local features from the intensity of the blocking pattern. Any variations in those features indicate the block grid's absence or misalignment to detect splicing attacks. In continuation the authors of [12] expose tampering detection and localization by the probability distribution of its DCT coefficients. They used three features that can truly distinguish tampered regions from original ones and obtain accurate localization results. The drawback of their method is the refining of the probability map in post-processing, and it is a remediate strategy that influences localization results. To eliminate the drawback, [19] used a mixture model based on normalized grey level co-occurrence matrix (NGLCM) and obtained more accurate localization with the prior knowledge of both tampered and original regions. To get this, they used conditional probabilities of tampered regions and original regions of DCT blocks in first, second, and third-order statistics. Still, their method is time consuming and the rate of false alarm is high.

In this paper, we move towards proposing a forensic technique that can localize the tampered region from a single JPEG image with double compression. Unlike other techniques that produce probability maps from 8×8 DCT coefficients, we proposed an efficient statistical model that uses block artificial grids (BAG) [18] and expose localization of spliced object.

1.1 Our Contribution

Over the years, various splicing localization techniques have proposed that there is still scope for robustness and effectiveness, as splicing is complex in nature. In this regard, we are offering the following contributions to our proposed work. (i) Instead of extracting features on the image level, we segment the image into individual objects and obtain features from each object. (ii) For each object, we estimate the variance of the BAG noise (iii) Instead of probability maps, we used pair-wise dissimilarity to classify the suspicious objects from original ones to expose tampered object.

The paper is organized as follows: Sect. 1 describes JPEG fingerprints from block artificial grids to speed up computation time. In Sect. 2 the proposed statistical method to expose splicing attack in JPEG image is described. The experimental and evaluation results present in Sect. 3, and finally, the paper concluded in Sect. 4.

2 Proposed Method

In this work, our primary goal is to localize the tampered region of the spliced JPEG image. The proposed work is framed into three levels: object-level image segmentation to extract individual objects in the spliced image, estimate the variance of each object using block artificial grids, and pair-wise dissimilarity among objects to localize tampered region as shown in Fig. 1.

2.1 Object Segmentation

Object detection is a challenging computer vision problem that solves object detection and classification. Among several object detection techniques, Mask R-CNN [20] is a widely used framework for object detection developed by Facebook research. It outperforms COCO suite challenge consisting of instance segmentation, bounding-box object detection, and person key point detection. It is a simple extension of Faster R-CNN with predicting the object's mask and easy to estimate human poses.

Using the Mask R-CNN framework, as shown in Fig. 2, we performed object detection and segmentation [21] for the given spliced image and extracted individual masks of all objects. Then for each mask, find its object from the input image along with the bounding box area. We split each object into foreground object consisting of the object mask region and the background object consisting of the remaining part in the bounding box from each bounding box object.

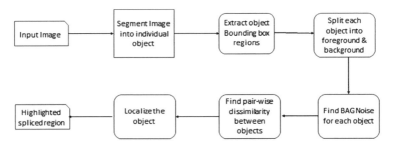

Fig. 1 The Proposed Frame Work

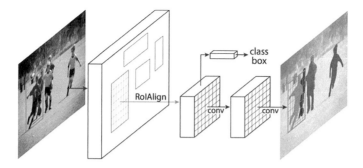

Fig. 2 Mask R-CNN Frame Work adopted from [20]

2.2 Block Artificial Grids

When the image is compressed with lossy JPEG, it leaves horizontal and vertical breaks in the image and is commonly refers as Block Artificial Grids (BAG). The BAGs of the entire image are roughly at the border an 8×8 block with a periodicity of 8 at both horizontal and vertical edges. When any attack alters the image, then the BAGs appear within the block instead of at borders. Thus this JPEG fingerprint is used in image forensics [12]. While compress the image using a digital camera, it introduces noise such as natural noise, BAG noise due to the JPEG compression factor. The artificial grid lines in an 8×8 block are feeble than the border edges. In [18], the authors extracted weak horizontal and vertical lines with a periodicity of 8 separately to enhance these weak lines, and then combined them is referred to as BAGs.

In this paper, we focus on extracting BAGs in colour images. Since the luminance component in the JPEG standard is 8×8 block, we used only the luminance component rather than Cb and Cr of components of YCbCr image. The second-order difference of an image regards as weak horizontal edges of an image. For the given image $I(m, n)$, the absolute second-order difference $d(m, n)$ is obtained by

$$d(m, n) = |2I(m, n) - I(m + 1, n) - I(m - 1, n)| \qquad (1)$$

To enhance the weak edges and remove the interference coming from strong image edges, a median filter is applied. To further reduce the edge influence as in [18] ignored differentials greater than an experimental threshold. Then the enlarged horizontal edges are accumulated for every two subsequent blocks as:

$$e(m, n) = \sum_{i=n-16}^{16} d(m, i) \qquad (2)$$

Then to equalize the amplitudes throughout the resultant image, a local median reduces from each element.

$$e_r(m, n) = e(m, n) - median[\{e(i, n)|m - 16 \leq i \leq m + 16\}] \qquad (3)$$

Thus, the weak horizontal edge image w_h is obtained by applying periodical median filter as:

$$w_h(m, n) = median[\{e_r(i, n)|i = m - 16, m - 8, m, m + 8, m + 16\}] \quad (4)$$

where $w_h(m, n)$ are elements of extracted horizontal BAG lines. The five elements in Eq. 4 with spacing 8 used in the median filter, makes the strong BAGs and weak BAGs smooth, and others can be removed. As more elements used in the median filter, BAGs can be extracted in a better way.

The vertical BAGs w_v are extracted similarly.

$$w_v(m, n) = median[\{e_r(m, i)|i = n - 16, n - 8, n, n + 8, n + 16\}] \qquad (5)$$

The final BAG is obtained by combining Eqs. 4 and 5 as

$$w_b(m, n) = w_h(m, n) + w_v(m, n) \qquad (6)$$

Equation 6 gives BAGs for the original image. When an image is attacked by tampering, the BAGs appear at some abnormal position such as centre of the block. So, for a fixed 8×8 block w_{mn} these abnormal BAGs can be obtained as [5]:

$$\begin{aligned} w_{mn} = \; & \mathrm{Max}\{\sum_{i=2}^{7} w_b(i, n)|2 \leq n \leq 7\} \\ & - \mathrm{Min}\{\sum_{i=2}^{7} w_b(i, n)|n = 1, 8\} \\ & + \mathrm{Max}\{\sum_{i=2}^{7} w_b(m, i)|2 \leq m \leq 7\} \\ & + Min\{\sum_{i=2}^{7} w_b(m, i)|m = 1, 8\} \end{aligned} \qquad (7)$$

2.3   Localization of Splicing Region

Using mark R-CNN object detection framework [21] as discussed in Sect. 2.1 the individual objects split into foreground object and background object-which is assumed to have similar characteristics of the whole image. Then for object extract the BAGs as discussed in Sect. 2.2. To expose discrepancies in BAGs of individual objects, we find BAG noise as:

$$\mu = \frac{1}{R} \sum w_{mn}(i, j) \tag{8}$$

$$\sigma = \frac{1}{R} \sum (w_{mn}(i, j) - \mu)^2 \tag{9}$$

μ is mean, σ is variance, and R represents the no of BAG features in w_{mn}.

After that, we used pair-wise dissimilarity between foreground and background objects to detect the tampered object. For each pair of distinct objects, let the BAG noise is estimated be S_1 and S_2. Then the cosine dissimilarity between the objects defined as:

$$L_D = 1 - \frac{C(S_1, S_2) + 1.0}{2} \tag{10}$$

where

$$C(S_1, S_2) = \frac{S_1^T . S_2}{\|S_1\| . \|S_2\|.} \tag{11}$$

$C(S_1, S_2)$ is the cosine angle between two BAG noises. This metric L_D gives values in the range [0, 1]. Where the values near to 0 represent similar BAG noise levels of both objects, and near to 1 represents different levels.

From the dissimilarity matrix, we find the pair that gives maximum dissimilarity, and for each object in the pair, identify the object which has maximum dissimilarity with other objects and expose as tampered object.

In summary, the method described above includes three aspects. (i) We extract features from individual segments rather than whole image. (ii) For each individual object, extract the BAG noise. (iii) Using pair-wise dissimilarity between objects localize the tampered object which has maximum dissimilarity. As a result, localizing accuracy, as well as computational complexity are improved.

3   Experimental and Performance Analysis

This section evaluates the proposed method on two datasets and compares its performance with other recent technique.

Typically, CASIA dataset [22] is a widely used evaluation dataset for JPEG image splicing forgery detection, and it consists of 7491 authentic and 5123 spliced images with JPEG, TIFF, and BMP types of images. We used the Mask R-CNN framework for object segmentation, so we randomly selected 1000 tampered images of animals, persons, birds, vehicles with the size 384×256. The proposed method test on those chosen tampered images of the CASIA dataset for localizing spliced regions.

The qualitative evaluation of splicing images on the CASIA dataset shows in Fig. 3. The first row consisting of randomly chosen four images, and the ground truth masks are in the second row. The proposed method results are in the third row, except the spliced region masked as white. From the results, our method's superiority is very clearly evident to localize the spliced region. The advantage of object segmentation is clear evidence in our results.

To increase the robustness of the proposed method, we evaluated our approach on the Image Manipulation Dataset (IMD) [23]. The dataset contains 48 high resolution JPEG compressed images with size 3264 × 2448 and different quality factors ranging from 20 to 100%. The images were cropped to 2048 × 1536 to reduce the computational complexity and spliced each other and obtained 600 spliced images. The proposed method applies to those images.

The evaluation results on the customized IMD spliced dataset from [23] are shown in Fig. 4. The first row contains four sample images from the dataset. The ground truth masks are in the second row, and the proposed method results are in the third row. From the results, the proposed method works well on the high-resolution images.

                    (a)   (b)   (c)   (d)

Fig. 3 Visual evaluation of proposed method on CASIA dataset

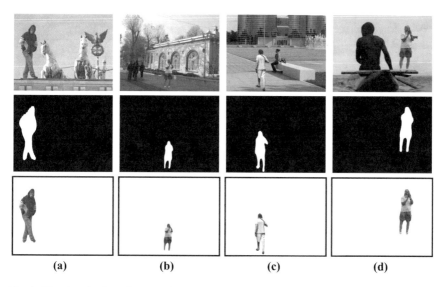

Fig. 4 Visual evaluation of proposed method on the high-resolution images from [23]

3.1 Localization Accuracy

The accuracy of splicing localization evaluates based on pixel-level F-measure. To evaluate, we used two metrics True Positive Rate (TPR), a measure of the rate of pixels that are truly detected as spliced, and False Positive Rate (FPR), a measure of the rate of pixels that are falsely detected as spliced.

$$TPR = \frac{TP}{TP + FN} * 100 \tag{12}$$

$$FPR = \frac{FP}{FP + N} * 100 \tag{13}$$

where TP is True Positive, FP is False Positive, TN is True Negative, and FN is False Negative. It expects to have high TPR and low FPR in the results. From these metrics, the F-measure defines as follows:

$$F = 2 * \frac{TPR * FPR}{TPR + FPR} \tag{14}$$

We evaluated average TPR and FPR and F-measure for all the selected images from the CASIA dataset and compared them with a recent method to analyse the proposed method. [19]. The method of [19] is based on a normalized grey level co-occurrence matrix on 8×8 DCT coefficients and using the Bayesian posterior

Table 1 Comparative results on CASIA and IMD datasets using average F-measure

Method	CASIA 2.0	IMD
NGLCM	0.6524	0.5572
Proposed	0.7852	0.0692

probability map, localized the tampering objects. To evaluate the superiority of the proposed method, we compared our results with recent methods.

Table 1 contains the Comparative results of the proposed method with [19] method on both datasets based on average F-measure. From the results, it is evident that BAG noise on individual objects in the proposed method enables us to have much superior performance than [19].

The method is robust when it has stable performance even after applying some post-processing operations on the spliced image. To evaluate the proposed method robustness, we applied JPEG compression with different quality factors, Gaussian blur, and added Gaussian noise to all the spliced images and tested.

For JPEG compression, 8 different quality factors ranging from 20 to 90 are considered. For Gaussian blur, Gaussian smoothing kernel with standard deviation $\sigma = 1.0$ is considered and for Gaussian noise, variance of 0.03 and 0.05 are considered.

The evaluation results on IM Dataset are shows in Table 2. As the quality factor (QF) in JPEG compression decreases and additional post-processing operations included, the NGLCM method decrease in its average F-measure values. In contrast, the proposed method has superior as well as stable performance even in such situations.

The IM dataset images are very high-resolution, and we try to downscale the quality factor to the lowest level 20. Figure 5 is a graph showing the performance of the proposed method with other existing method. NGLCM method decreases its average F-measure as the JPEG compression quality factory is reduced to 20. The proposed method outperforms and gives stable performance even when the quality factor reduces because the BAGs are affected only in those objects than the rest of the image.

Table 2 Comparative results for robustness on IM dataset using average F-measure

Method	(JPEG compression)		(Gaussian blur)	(Gaussian noise)	
	QF = 50	QF = 70	$\sigma = 1.0$	Variance = 0.03	Variance = 0.05
NGLCM	0.3934	0.4323	0.5412	0.5389	0.5395
Proposed	0.6418	0.6596	0.7520	0.7514	0.7520

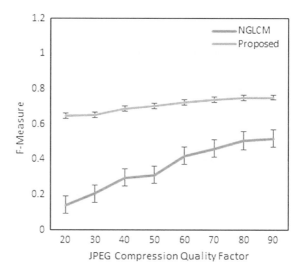

Fig. 5 Comparative results of JPEG quality factory with F-measure

3.2 Computational Complexity

The effectiveness of any method depends on its average computation time spent is minimal to get the desired result. In the proposed method, after segmenting the individual objects, we obtain BAG features from each object instead of the whole image by saving a lot of computation time. For localization, also we used a simple statistical method instead of unsupervised learning techniques. Table 3 gives the average running time spent by each method. Among the two methods, the proposed method takes less time than other existing methods.

Table 3 Average running time

Method	Proposed	NGLCM
(Average running time in secs)	16.8	78.9

4 Conclusion

This paper proposed an effective splicing localization method using local statistics of the image. When the JPEG image is splicing with another image's object, the block artificial grids in the tampered area move from 8×8 grid lines to its centre. Taking this clue as a feature descriptor, we exposed splicing forgery through object segmentation. The method is straightforward, effective than other conventional methods that use JPEG fingerprints. The proposed method also robust even when the quality factor is low in high-resolution JPEG compression. The method fails on low-resolution images, and we considered it as our future work.

References

1. Ali Qureshi M, Deriche M (2014) A review on copy move image forgery detection techniques. In: IEEE 11th international multi-conference on systems, pp 1–5
2. Redi JA, Taktak W, Dugelay J (2011) Digital image forensics: a booklet for beginners. Multimed Tools Appl 51:133–162
3. Farid H (2009) Image forgery detection a survey. IEEE Signal Process Mag 26(2):16–25
4. Birajdar GK, Mankar VH (2013) Digital image forgery detection using passive techniques. Digit Invest Int J Digit Forensics Incident Response 10(3):226–245
5. Liu B, Pun CM, Yuan XC (2014) Digital image forgery detection using JPEG features and local noise discrepancies. Sci World J 1–12
6. Chandra Sekhar PNRL, Shankar TN (2016) Review on image splicing forgery detection. Int J Comput Sci Inf Secur 14(11):471–475
7. Bahrami K, Kot AC, Li L (2015) Blurred image splicing localization by exposing blur type inconsistency. IEEE Trans Inf Forensics Secur 10(5):999–1009
8. Zhang Y, Zhao C, Pi Y, Li S (2012) Revealing Image splicing forgery using local binary patterns of DCT coefficients. In: Liang Q et al (eds) Communications, signal processing, and systems. Lecture notes in electrical engineering, vol 202, pp 181–189
9. Chandra Sekhar PNRL, Shankar TN (2019) Splicing localization based on noise level inconsistencies in residuals of color channel differences. IJRTE 8(3):764–769
10. He Z, Lu W, Sun W, Huang J (2012) Digital image splicing detection based on Markov features in DCT and DWT domain. IEEE Trans Pattern Recognit 45(12):4292–4299
11. Zampoglou M, Papadopoulos S, Kompatsiaris Y (2017) Large-scale evaluation of splicing localization algorithms for web images. Multimed Tools Appl 76(4):4801–4834
12. Bianchi T, Piva A (2012) Image forgery localization via block-grained analysis of JPEG artifacts. IEEE Trans Inf Forensics Secur 7(3):1003–1017
13. Lin Z, He J, Tang X, Tang CK (2009) Fast, automatic and fine-grained tampered JPEG image detection via DCT coefficient analysis. Pattern Recognit 2492–2501
14. Amerini I, Becarelli R, Caldelli R, Del Mastio A (2014) Splicing forgeries localization through the use of first digit features. In: IEEE international workshop on information forensics and security (WIFS), pp 143–148
15. Farid H (2009) Exposing digital forgeries from JPEG ghosts. IEEE Trans Inf Forensics Secur 4(1):154–160
16. Bianchi T, De Rosa A, Piva A (2011) Improved DCT coefficient analysis for forgery localization in JPEG images. In: IEEE international conference on acoustics, speech and signal processing (ICASSP), pp 2444–2447
17. Luo W, Qu Z, Huang J, Qiu G (2007) A novel method for detecting cropped and recompressed image block. In: International conference on acoustics speech and signal processing, vol 2, pp 117–220

18. Li W, Yuan Y, Yu N (2009) Passive detection of doctored JPEG image via block artifact grid extraction. Signal Process 89:1821–1829
19. Xue F, Wei Lu, Ye Z, Liu H (2019) JPEG image tampering localization based on normalized gray level co-occurrence matrix. Multimed Tools Appl 78:9895–9918
20. He K, Gkioxari G, Dollar P, Girshick R (2017) 'Mask R-CNN'. In: EEE international conference on computer vision (ICCV), pp 2980–2988
21. Abdulla W (2017) Mask r-cnn for object detection and instance segmentation on keras and tensorflow. https://github.com/matterport/Mask RCNN
22. Dong J, Wang W, Tan T (2013) CASIA image tampering detection evaluation database. In: IEEE China summit and international conference on signal and information processing, Beijing, pp 422–426
23. Christlein V, Riess C, Jordan J, Riess C, Angelopoulou E (2012) An evaluation of popular copy move forgery detection approaches. IEEE Trans Inf Forensics Secur 7(6):1841–1854

Applying Big Data Analytics in DDos Forensics: Challenges and Opportunities

Augusto Gonzaga Sarmento, Kheng Cher Yeo, Sami Azam⬤, Asif Karim⬤, Abdullah Al Mamun⬤, and Bharanidharan Shanmugam⬤

Abstract DDoS (Distributed Denial-of-Service) attacks greatly affect the internet users, but mostly it's a catastrophe for the organization in terms of business productivity and financial cost. During the DDoS attack, the network log file rapidly increases and using forensics traditional framework make it almost impossible for DDoS forensics investigation to succeed. This paper mainly focuses on finding the most suitable techniques, tools, and frameworks in big data analytics that help forensics investigation to successfully identify DDoS attacks. This paper reviewed numbers of previous research that related to the topic to find and understand general terms, challenges and opportunities of using big data in forensics investigation. The data mining tools used in this paper for simulation was RapidMiner because of its ability to prepare the data before the analysis and optimizes it for quicker subsequent processing, and the dataset used was taken from University of New Brunswick's website. Algorithms that were used to evaluate the DDoS attack training dataset are Naïve Bayes, Decision Tree, Gradient Boost and Random Forest. The evaluation results projected that the majority of algorithms has above 90% of accuracy, precision and recall respectively. Using the data mining tools and recommended

A. G. Sarmento · K. C. Yeo · S. Azam · A. Karim (✉) · B. Shanmugam
College of Engineering, IT and Environment, Charles Darwin University, Casuarina 0810, NT, Australia
e-mail: asif.karim@cdu.edu.au

A. G. Sarmento
e-mail: amgsarmento@yahoo.com

K. C. Yeo
e-mail: charles.yeo@cdu.edu.au

S. Azam
e-mail: sami.azam@cdu.edu.au

B. Shanmugam
e-mail: bharanidharan.shanmugam@cdu.edu.au

A. Al Mamun
Institute of Information Technology, Jahangirnagar University, Dhaka, Bangladesh
e-mail: abdullah.iiuceee@gmail.com

© The Author(s), under exclusive license to Springer Nature Switzerland AG 2021
H. Jahankhani et al. (eds.), *Cybersecurity, Privacy and Freedom Protection in the Connected World*, Advanced Sciences and Technologies for Security Applications, https://doi.org/10.1007/978-3-030-68534-8_15

algorithms will help reduce processing time associated with data analysis, reduce cost and improve the quality of information. Future research is recommended to install in an actual network environment for different DDoS detection models and compare the efficiency and accuracy in real attacks.

Keywords DDoS attacks · DDoS forensics · Big data analytics · Bid data forensics · Forensic investigation

1 Introduction

All information about network, protocols, application, and web are stored in a log file and this log file usually saves indiscriminately everything [1]. As we all aware that the network traffic is continuously increasing, which means that the size of logs files also increasing. Since all the information regarding DDoS attacks also stores in log files, investigation to find some meaningful insight regarding the attackers' details has become extremely difficult due to the big amount of data. Furthermore, using the current conventional forensic investigation method is time-consuming, costly and sometimes impossible to succeed.

Distributed Denial-of-Services (DDoS) is a type of cyber-attacks to an organization network where multiple systems flood the resources or bandwidth of the organization's systems. Malicious people use multiples zombie's computers to overwhelm the network's available resources which could be application or service with the request so that legitimate users not able to access the system [2]. This greatly affects the internet users in a computer network but mostly it's a catastrophe for the organization in terms of business productivity and financial cost. This is an ongoing issue for government agencies, financial institutions or any organization that need to be prevented and solved with a watchful approach [3].

This paper aims to identify the most suitable techniques, tools and framework in big data analytics that help forensics investigation to successfully identify DDoS attacks. This paper will deliver suitable data mining tools that will facilitate the forensic investigation in DDoS attack using big data, good forensics investigation methods that will be more suitable for big data investigation and also a report of experiment's result.

The research is focusing more on three things such as DDoS attacks itself (why and how it happens and also how to prevent it), big data analytics in forensic investigation and investigates DDoS attacks using big data analytics in forensics investigation. In addition, different data mining tools were evaluated and the chosen one was used in this paper. Moreover, current DDoS forensics methods were explored, and also numbers of algorithms that are used in DDoS forensics was explored as well. It is assumed that the evaluated data mining and algorithms will help to reduce cost and time spending on forensics investigation as well get an insightful pattern.

The rest of the paper is organized as follows: Sect. 2 provides the literature review of background knowledge regarding DDoS attacks, big data and forensics investigations. Section 3 discusses the methodology used in this paper. Section 4 presents the result of dataset evaluation using data mining tools and algorithms. Section 5 talks about the recommendation and discussion. Finally, Sect. 6 concluded the paper.

2 Literature Review

Several researches that have been focused on how difficult and challenging it is to do forensics investigation on big data. There are a number of solutions that have been proposed as well to overcome those difficulties and challenges and those solutions will be discussed more details in the later sections of the paper. In this section we elaborated on some related works that have been done prior to this paper alongside their statement and explanation. The evaluation outcome of related research attempts is arranged in 6 sections: information about digital forensics and its framework in Sect. 2.1; big data and its characteristics in Sect. 2.2; big data forensics and its challenge in Sect. 2.3; DDoS attacks and DDoS forensics in Sect. 2.4; algorithm used in related work in Sect. 2.5; and data mining tools comparison in Sect. 4.1.

2.1 Digital Forensic

Digital forensics is part of forensic science that responsible to identify an incident along with collection, examination, and analysis of evidence data. It is also responsible for investigating the cyber-crime and cyber-incidents, find the possible evidence and present it to the court for further judgment. Digital forensic has four main frameworks process [4–7]:

Identification

In this step, the investigator identifies the evidence of the crime or incidents and prosecute litigation. This step usually considered as the stage of preparation and preservation as well. The preparation includes preparing the tool, resources alongside with the necessary authorization or approval to collect data. Preservation involves securing the crime or incidents and possible evidence.

Collection

In this step, the investigator team starting to collect physical and digital evidence at the crime scene. Everything will be recorded in this stage and all the evidence is collected using standardised techniques. In this stage, while collecting the data, the investigator needs to make sure preserving the confidentiality and integrity of the data.

Organization

In this step, the investigation team efficiently collects the evidence which can lead to finding information regarding the criminal incidents. First, the investigators examine the collected data to find the potential pattern that can lead to the crime and the suspect. After that, the investigators analyse the correlation between found patterns and suspect to determine the fact.

Presentation

The investigator prepares the report of the result to present it in the court to prosecute litigation. The investigators have to make sure that the result they present must be easy to understand without requiring any specific knowledge.

2.2 Big Data

Nowadays, people define big data as a dataset that is too big, too fast and too difficult for traditional tools and frameworks to process. Big data is characterised by followings [4, 5]:

Variety

It describes different data that exist. Since big data comes from multiple sources like network or process logs, web pages, social media, emails, and any other various sensors, the data can be categorised as structured, semi-structured and unstructured.

Volume

It refers to the large amount of data that can be generated and stored. For example, in this era, many organizations like Google and Woolworths deal with terabytes or petabytes of data.

Velocity

Velocity refers to how big data getting bigger due to the new different systems that come every day. The velocity can be categorised as a real-time, batch, stream, etc. It is not only referring to the speed of incoming data but also about the speed of data flow inside the system.

Veracity

Veracity refers to the integrity and confidentiality of the data. It also involves data governance, quality of data and metadata management alongside the legal concerns.

Value

It refers to how big data can be turned to something that valuable for economy and investigation. Bid data can reveal all the important pattern that is searched for which is previously unknown and those can lead to something that valuable.

2.3 Big Data Forensic and Its Challenge

Big data forensics is defined as a branch of digital forensics that deals with evidence identification, collection, organization, and presentation to establish the fact using a very large-scale of dataset. Big data forensics can be looked at from two perspectives: first, a shred of small evidence can be found in the big dataset and second, by analysing big data, a crucial piece of information can be revealed [4–7].

To enable high-velocity capture, discovery, and/or analysis and to efficiently extract patterns and value from large volume and a wide variety of data, big data requires a new design generation of technology and architectures. Unfortunately, digital forensics' traditional tools and technologies are incapable of handling big data. Following are the challenges that encounter is each step of digital forensics investigation when dealing with big data [4–7]:

- *Identification*: When the amount of possible evidence is very large, it can be difficult to identify the important pieces of evidence to determine the fact.
- *Collection*: If there is an error that occurs during the collection stage, it will affect the whole investigation process. Because the *Collection* is considered as the most crucial steps.
- *Organization*: Since the existing analysis techniques do not comply with the characteristic of the big data, it can be challenging to organize big data set and identify the facts about the incidents.
- *Presentation*: It will be hard for the jury to understand the technicalities behind filtering, analysing big data and identifying value. Because it is not as easy as traditional computer forensics.

2.4 DDoS Attacks and DDoS Forensic Methods

The following details are taken previous related works that were conducted by [8–10]. The authors explained about DDoS attacks and DDoS forensic very precise and understandable. Table 1 summarizes the DDoS attack architectures from previous related works, Table 2 summarizes launching steps. There is not a lot that can be done apart from disconnecting the victim system from the network and fix it manually when DDoS attacks occur. However, the defence mechanism can be used to detect the DDoS as soon as possible and prevent it immediately, showed in Table 3. Table 4 summarizes DDoS Detection strategies and Fig. 1 showed the classification of DDoS defense mechanism. Also, Table 5 summarizes different algorithms used in the works reviewed earlier.

Table 1 Summary of DDoS attack architectures from previous related work

Attacks architecture	Description
Agent-Handler architecture	It also considers as Botnet based architecture. The attackers use the Botnet to conduct an attack and the Botnet consist of masters, handlers and bots
IRC (Internet Relay Chat)-based architecture	Instead of doing an attack using the original address, the attack is launched through a public chat system. Because IRC allows users to communicate without requiring any authentication check and no security
Web-based architecture	The attackers launch the attack by hidden themselves within legitimate HTTP and HTTPS traffic

Table 2 Summary of DDoS attack launching steps from previous related work

Steps	Description
Discover vulnerable host and agents	Attackers using tools and resources to find any system of the network that does not run with the antivirus virus and weak security defence system
Compromise	After the attackers finding the vulnerable system, they exploit the vulnerable system and install the attack code
Communication	The attackers communicate with the agents to schedule attacks, to identify active agents or to upgrade agents. The communication can be done via TCP, UDP and ICMP
Launching an attack	The attackers select the victim system and launch the attack

2.5   Algorithm Used in Related Work

See Table 5.

3   Methodology

To measure the accuracy and compare the efficiency of different learning models in detecting the DDoS attack, this paper utilised simulation. The simulation has been used in the past researcher that related to the same topic as this paper. What the past researches did are calculating the percentage of true negative, true positive, false negative and false positive. The dataset was divided into two parts throughout the simulation such as training as testing. Using this approach will help to simplify the complexity of DDoS attacks. Instead of capturing the real net flow data of DDoS attack, simulation aids in simplifying the data gathering process. Off course that this

Table 3 Summary of DDoS defence architectures from previous related work

Defence architectures	Description
Source-end defence mechanism	To prevent network users from generating the DDoS attacks, the source-end defence mechanism is deployed at the source of the attack. In this approach, all the malicious packet is identified by a source device in outgoing traffic and filter the traffic
Victim-end defence mechanism	It filters, detects or rate malicious incoming traffic at the routers of victim networks for instance network providing Web services. In this detection system, an anomaly intrusion detection system can be used
Core-end or intermediate router defence mechanism	Any router in the network can try independently to identify the malicious traffic and filter the traffic. For example, it is a better place to filter the traffic because both attack and legitimate packets arrive at the router
Distributed end or hybrid defence mechanism	One of the best strategies against DDoS attacks could be attack detection and mitigation at the distributed end. The core-end is suitable to filter all kinds of traffic and the victim-end can detect traffic accurately

Table 4 Summary of DDoS detection strategies from previous related work

Strategies	Description
Statistical	Utilizing the statistical properties of normal attack patterns for DDoS attacks' detections. Calculate a general statistical model for normal traffic and used it to test the incoming traffic to determine if it is legitimate traffic or not
Soft computing based	Using learning paradigms such as ANN (Artificial Neural Networks) which has self-learning characteristics to identify unknown disturbance or attacks in a system
Knowledge based	The rules that already established in advance are used to test against network events or actions. All the known attacks are defined as attack signatures and use the signatures to identify the actual attack
Data mining and machine learning	Protecting network devices and applications using an effective defensive system called NetShield from becoming a victim of DDoS flood attacks. It eliminates vulnerabilities of the system on the target machine using preventive and filter and protecting IP-based public networks on the Internet

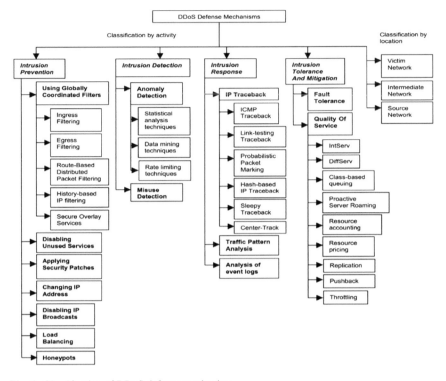

Fig. 1 Classification of DDoS defense mechanism

Table 5 Different types of algorithm using in previous related work

Algorithms	Paper and Authors
MapReduce by Hadoop	"DOFUR: DDoS Forensics Using MapReduce" by [1]
Hadoop Distributed File System (HDFS)	"Digital Forensics in the Age of Big Data: Challenges, Approaches, and Opportunities" by [4]
MapReduce, Decision Tree and Random Forest, Image Forensics, Neural Network and Neural Language Processing (NLP)	"Digital Forensics as a Big Data Challenge" by [7]
Decision Tree, Baysian, Neural Network, Nearest Neighbour, Genetic Algorithms, Case-based Reasoning, Rough Set and Fuzzy Logic	"Dealing with Terabyte Data Sets in Digital Investigation" by [11]
Gaussian Naïve Bayes	"A Novel DDoS Attack Detection Based on Gaussian Naïve Bayes" by [2]

approach comes with its drawback which is it may over-simplify the real situation of DDoS attack. Because the dataset that was used in this paper have been pre-processed before training and testing whereas in actual situation it is not. In addition, the constant changing actual threats environment may not reflect in the captured dataset.

The dataset that is used in this paper is taken from the University of New Brunswick (Canadian Institute for Cybersecurity)'s website. The dataset is divided into 7 different groups according to a different type of DDoS attack respectively as follows: Portman, UDPLags, LDAP, NetBIOS, UDP, MSSQL and Syn. The variables that is used to determine the DDoS attacks are time stamp, source and destination IP address. Since the datasets have been pre-processed and labeled, the data is ready to be evaluated using data learning algorithm and data mining tools.

RapidMiner is used in this study and it has been used extensively in data science. It is best in the area of future predictive analytics because it predicts future development based on collected data. The program can import Excel tables, SPSS files and data sets from many databases. In addition, it can be used for data mining, text mining, opinion mining and sentiment mining.

Power BI was used to confirm or visualize whether the dataset contains DDoS attack packet. The dataset was grouped by timestamp and then count the number of packets per timestamp. The DDoS attacks can be verified as shown on the spikes or the sudden increase in the number of packets (see Sect. 4 Part II). Power BI was chosen because it is more intuitive than the RapidMiner built-in visualization tools.

4 Results

The results are arranged in 3 sections, different data mining tools are compared in Sect. 4.1; finding the characteristics of DDoS attacks using Power BI in Sect. 4.2; and the result of dataset evaluation to find the accuracy, precision and recall of the algorithms in Sect. 4.3.

4.1 Software Comparison

Before deciding which data mining tools to be used for the evaluation, different data mining tools have been examined and explored such as RapidMiner, WEKA, Orange, KNIME and SAS. The characteristics and support of the data mining tools are summarised in Fig. 2. All these data mining tools have libraries that can be extended and used in the programming language.

After evaluating the performance of different data mining tools, RapidMiner was chosen for Analysis. RapidMiner can design modular operator concept even for very complex problems. To describe the operator modelling knowledge discovery (KD) processes, RapidMiner uses XML. It can also take input and output for and from

Tools	Characteristics	Programming Language	Operating System	Price/License
RapidMiner	• It predicts future developments based on collected data. • The program can import. Excel tables, SPSS files, and data sets from many databases. • It prepares the data before analysis and optimizes it for quicker subsequent processing.	Java	Windows, Mac, Linux	Free but also cost based on Versions
WEKA	• It has many classification methods such as artificial neural networks, ID3, decision trees and C4.5 algorithms. • Its machine learning capabilities support major data mining task like association, classification, clustering and regression • It is really useful for teaching and research purposes.	Java	Windows, Mac, Linux	Free Software
Orange	• Without extension of prior knowledge, it creates appealing and interesting data visualizations. • Its machine learning support data mining task such as clustering, regression, classification and much more. • It has the capabilities of learning about user's preference over time and reacts accordingly.	C++ Python (Extensions and query language)	Windows, Mac, Linux	Free Software
KNIME	• Helps to reveal hidden data structures. • Enables data mining and numbers of machine learning's methods to be integrated. • It is really effective when pre-processing data for example: loading data and extracting transforming.	Java	Windows, Mac, Linux	Free Software
SAS	• One of the best data mining tools for business analytics. • Good for large presentation in terms of prognostic sector and interactive data visualization. • It has high scalability so it can possible increase the performance proportionally by adding additional harder or other resources.	SAS language	Windows, Mac, Linux	Limited freeware through educational institutions.

Fig. 2 Data mining tools evaluation

any different form of dataset. RapidMiner has more than 100 learning schemes for clustering task, classification and regression.

4.2 Timestamp Visualization of Dataset Using Power BI

The graphs below shows the number of requests per protocol for each second. Using Power BI, the visualization is achieved by creating timestamp bins one second in duration. Then a measure is calculated as the count of records based on the column named "Flow ID". For each graph, the total of BENIGN (not harmful or safe) packets are shown to visualize what normal series of packet looks like before or after DDoS attacks. For the dataset used, this attack occurred in 3rd November 2018.

Figure 3 shows the flow of Portmap packets suddenly increases at around 10:01 am reaching 8.2 thousand request per second.

Figure 4 shows the flow of UDPLags packets suddenly increase at around 11:29–11:31 am reaching 13.5 thousand request per second. There is also an increase in UDPLags although the increase on other protocol is more significant. Overall, algorithm is able to identify BENIGN from malicious packets with high accuracy, precision and recall.

Figure 5 shows the flow of LDAP packets suddenly increase at around 10:21–10:27 am reaching 19.7 thousand request per second.

Number of Portmap Requests by Timestamp (bins) and Label

Label ●BENIGN ●Portmap

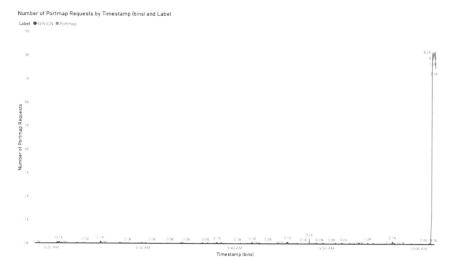

Fig. 3 Portmap DDoS attacks and timestamp

Number of UDPLag by Timestamp (bins) and Label

Label ●BENIGN ●Syn ●UDP ●UDPLag

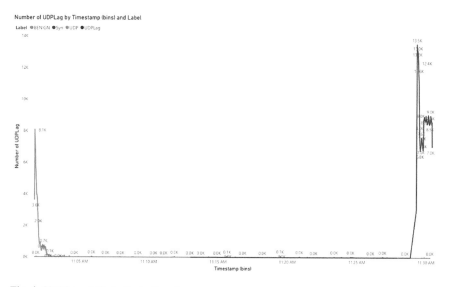

Fig. 4 UDPLags DDoS attacks timestamp

Additionally, Fig. 6 shows the flow of NetBIOS packets suddenly increase at around 10:02–10:09 am reaching 8.1 thousand requests per second.

Figure 7 shows the flow of UDP packets suddenly increase at around 10:53–11:01 am reaching between 9.3 and 11.7 thousand requests per second.

Figure 8 shows the flow of SYN packets suddenly increase at around 11:35–11:37 am reaching 13.6 to 27.2 thousand request per second.

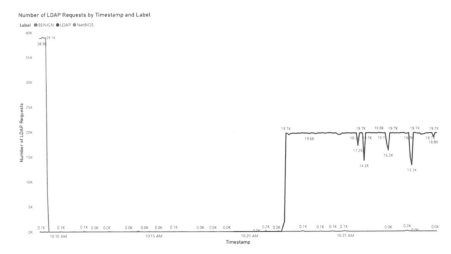

Fig. 5 LDAP DDoS attacks and timestamp

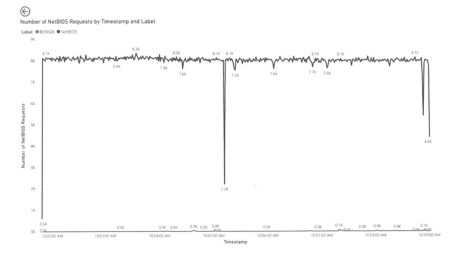

Fig. 6 NetBIOS DDoS attacks and timestamp

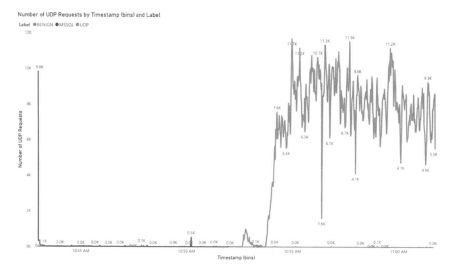

Fig. 7 UDP DDoS attacks and timestamp

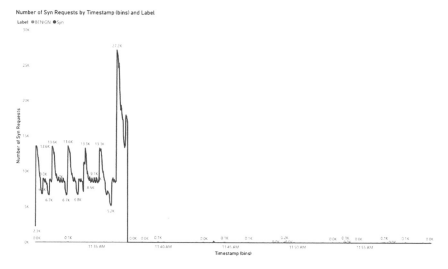

Fig. 8 SYN DDoS attacks and timestamp

Figure 9 demonstrates the flow of MSSQL packets suddenly increase at around 10:34–10:42 am reaching between 11.9 and 12.5 thousand requests per second.

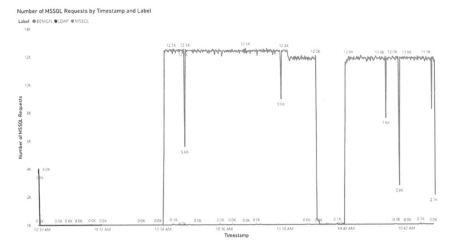

Fig. 9 MSSQL DDoS attacks and timestamp

4.3 Comparison of Machine Learning Algorithms

Distributed Random Forest.

Random Forests are based on "classification trees" which trains a 'forest' of decision trees and performs binomial classification predictions by introducing a training input to the individual trained trees in the 'forest' and promoting the dominant classification for each tree as the prediction result. In the distributed implementation of Random Forests, each cluster node is reassigned an identical division of the whole training dataset. Each computing cluster then trains an individual Random Forest cluster and majority classification for each cluster is identified as the prediction result [12].

Decision Tree.

Decision trees categorize the training data by sorting them from the root of the tree down to some leaf node, with the leaf node as the prediction result. Each leaf node in the tree serves as a test case for the highlighted attribute, and each path to the root is the possible answer to the test case. This algorithm is naturally recursive and is iterated for every subtree starting at the branch node. Decision trees use a variety algorithm to strategically decide where and how many splits to make. Each split increases the homogeneity of consequent splits. The integrity of the node increases depending on the target variable. The decision tree splits the nodes on all available variables and then selects the split which results in most consistent splits [13].

Gradient Boosting Machines.

Gradient Boosting trains many models in a steady increasing pattern. Gradient boosting performs by using gradients in the loss function $y = ax + b + e$ where e is the error variable. The loss function is a measure indicating how good the model's coefficients are at fitting the underlying data. A logical understanding of loss function would depend on what we are trying to optimise. One of the biggest motivations of using gradient boosting is that it allows one to optimise a user- specified cost function, instead of a loss function [14].

Naive Bayes.

Naive Bayes (or Idiot Bayes) is a classification algorithm for binomial and polynomial classification problems. The calculation of the probabilities for each hypothesis is simplified to make their calculation tractable. Rather than attempting to calculate the values of each attribute value P (d1, d2, d3|h), they are assumed to be conditionally independent given the target value and calculated as P(d1|h) * P(d2|H). The approach executes well on data where the assumption that the attributes do not interact is disregarded [13].

The precision is reliable but still depends on the split between training data and testing data. For instance, for MSSQL requires 50/50 split in order to get result for Decision Tree and Naïve Bayes and NetBIOS requires 90/10 split to get result for Gradient Boosting Machine and Naïve Bayes. The split is needed to change because for above dataset, there is not enough information to accurately train the model. For the rest of the dataset, only requires 30% training and 70% testing to get an accurate and precise result.

Since recall is close to 100%, as shown in Table 6, most of the true positive was found therefore proving that the training covers almost all the dataset. Since the accuracy is also almost close to 100% for the average, it simply means that the models predict most of the data correctly. In most test except for the two (Naïve Bayes' precision for NetBIOS and SYN dataset), precision close to 100% means how useful the generated model is. Generally, the algorithms that are used in this simulation paper can be used to examine DDoS attacks.

5 Discussion and Recommendation

After evaluating the different machine learning algorithms, for those algorithms that resulted in high precision, accuracy and recall, it can be recommended to use the algorithm model for DDoS forensics investigations. For lower values, further modelling is required to generate a model that is accurate and precise enough for it to be used in DDoS forensics investigation. It is also recommended for further research to use different dataset and up to date tools to confirm the findings of this research.

After evaluating previous research papers and related work, it can be said the traditional forensic framework is not suitable for big data investigation or DDoS forensics. Khattak et al. [1] and Zawoad and Hasab [4] proposed to use Hadoop's MapReduce for the forensic investigation of DDoS attacks. This method will help to find out whether the system is under attack, who attacks the system and which

Table 6 Algorithmic performance

System	Algorithm	Accuracy (%)	Precision (%)	Recall (%)
Portmap	Distributed Random Forest	99.97	100.00	99.97
	Decision Tree	99.73	99.76	100.00
	Gradient Boosting Machine	96.30	100.00	95.73
	Naïve Bayes	99.91	99.97	97.90
UDPLags	Distributed Random Forest	99.93	100.00	99.93
	Decision Tree	99.93	99.97	94.10
	Gradient Boosting Machine	100.00	99.92	99.43
	Naïve Bayes	99.93	100.00	99.93
LDAP	Distributed Random Forest	99.99	98.21	99.74
	Decision Tree	99.99	100.00	99.99
	Gradient Boosting Machine	100.00	99.93	98.31
	Naïve Bayes	99.92	100.00	99.92
NetBIOS	Distributed Random Forest	–	–	–
	Decision Tree	100.00	100.00	99.49
	Gradient Boosting Machine (90/10)	100.00	100.00	97.54
	Decision Tree	100.00	100.00	99.49
UDP	Distributed Random Forest	–	–	–
	Decision Tree	100.00	94.56	99.79
	Gradient Boosting Machine	99.92	100.00	100.00
	Naïve Bayes	99.96	100.00	99.96
SYN	Distributed Random Forest	–	–	–
	Decision Tree	–	–	–
	Gradient Boosting Machine	100.00	100.00	99.40
	Naïve Bayes	99.36	56.37	100.00
MSSQL (50/50)	Distributed Random Forest	–	–	–
	Decision Tree	100.00	100.00	100.00
	Gradient Boosting Machine	–	–	–
	Naïve Bayes	99.94	100.00	99.94

incoming traffic is part of the attack. Hadoop provides MapReduce to use for parallel processing of distributed data. Adedayo [5] reassessed the digital forensic examination stages and proposed additional techniques and algorithms that help to handle big data issues in the investigation Fig. 10. The author continues stating that the proposed solution is not intended to stand alone rather than to support the existing framework and to solve the challenge facing by existing methods.

Another study conducted [2] talks about the DDoS attacks and the impacts. The authors proposed a new approach based on network traffic to analyse and detect

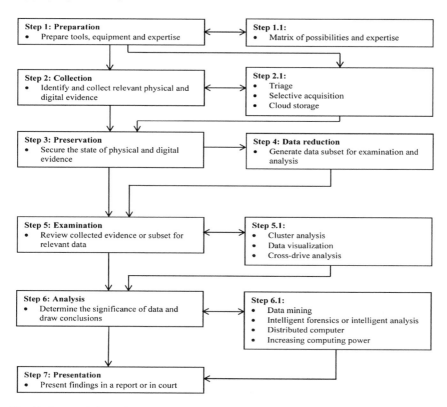

Fig. 10 Proposed digital forensics framework

DDoS attacks using Gaussian Naïve Bayes method. Hoon et al. [15] stated supervised learning algorithms such as Deep Learning, Gradient Boosting, Distributed Random Forest and Naïve Bayes performed better than unsupervised learning algorithms such as farthest first, canopy, make a density-based cluster and filtered cluster. To tackle the big data challenge, Guarino [7] suggested using decision trees and random forest to find anomalous behaviour or anomalous transaction and neural network to recognise application normal behaviours (it is suitable for network forensics to recognise complex patter). Beebe and Clark [11] said implementing data mining tools and research to the forensic investigation will help to reduce processing time associated with data analysis, reduce cost and improve the quality of information.

6 Conclusions

This paper proposes a more practical solution and framework to facilitate DDoS forensics investigation as illustrated in Fig. 10. This paper also carried out simulation using RapidMiner and compare different accuracy, precision and recall of the algorithms in detecting DDoS attacks. This paper evaluates 4 different machines learning algorithm and compares its usefulness and effectiveness. This research initiative used RapidMiner, unlike the previous studies which is majority used WEKA because RapidMiner accept any data format and it prepares the data prior to analysis and optimizes it for faster subsequent.

References

1. Khattak R, Bano S, Hussain S, Anwar Z (2011) DOFUR: DDoS using mapReduce. Front Inf Technol 117–120
2. Fadil A, Riadi I, Aji S (2017) A novel DDoS attack detection based on Gaussian Naïve Bayes. Bull Electr Eng Inform 6(2):140–148
3. Kupreev O, Badovskaya E, Gutnikov A (2019) DDoS attacks in Q3 2019. SECURELIST [post-print]. https://securelist.com/ddos-report-q3-2019/94958/
4. Zawoad, S, Hasan R (2015) Digital forensics in the age of big data: challenges, approaches, and opportunities. In: 17th international conference on high performance computing and communication (HPCC). 7th international symposium on cyberspace safety and security (CSS), pp 1320–1325
5. Adedayo OM (2016) Big data and digital forensics. In: International conference on cybercrime and computer forensics (ICCCF), pp 1–7
6. Guo H, Jin B, Shang T (2012) Forensic investigations in cloud environment. In: International conference on computer science and information processing (CSIP)
7. Guarino A (2013) Digital forensics as a big data challenge. ISSE
8. Peng T, Leckie C, Ramamohanarao K (2007) Survey of network-based defense mechanism countering the DoS and DDoS problems. ACM Comput Surv 39(1):3-es
9. Prasad KM, Redy ARM, Rao KV (2014) 'DoS and DDoS attacks: defence, detection and traceback mechanisms—a survey. Glob J Comput Sci Technol: E Netw Web Secur 14
10. Douligeris C, Mitrokotsa A (2004) DDoS attacks and defense mechanism: classification and state-of-the-art. Comput Netw 44(5):643–666
11. Beebe N, Clark J (2005) Dealing with terabyte data sets in digital investigations. In: IFIP international conference on digital forensics, vol 194, pp 3–16
12. Breiman L (2001) Random forest. https://libguides.ioe.ac.uk/c.php?g=482485&p=3299839
13. Karim A, Azam S, Shanmugam B, Kannoorpatti K, Alazab M (2019) A comprehensive survey for intelligent spam email detection. IEEE Access 7:168261–168295
14. Towards Data Science (2018) Understanding gradient boosting machines. https://towardsdatascience.com/understanding-gradient-boosting-machines-9be756fe76ab
15. Hoon KS, Yeo KC, Azam S, Shanmugam B, Boer FD (2018) Critical review of machine learning approaches to apply big data analytics in DDoS forensics. In: International conference on computer communication and informatics (ICCCI-2018)

Cyber Security in the Global Village and Challenges for Bangladesh: An Overview on Legal Context

Kudrat-E-Khuda (Babu)

Abstract With the rapid penetration of the Internet and other information and communication technology worldwide, cyber-crime is emerging as a threat to personal data stored in computers and likely to affect the entire data systems. Even the United States, one of the most technologically advanced countries, is also subjected to such crimes. Bangladesh, being a less developed country, is also at the risk of cyber-crimes that might jeopardize the country's national security. As the incumbent government eyes to ensure internet connectivity at all government institutions by 2021 upholding the motto of 'Digital Bangladesh', more and more national and multinational companies are offering online services to their services through the internet following the government's agenda. From shopping to Banking, all are just a click away with the higher rate of internet penetration. However, criminals are also using the online platform where they are committing various sorts of criminal activities including phishing, hacking, and stealing personal data. Hence, the state-owned, as well as private organizations, might fall prey to cyber-attacks which might affect the lives of the entire population. Moreover, the country's 90% of software is unlicensed that also intensifies the risk of cybercrimes thanks to their compromised security issue. In addition, the recent tug of wars between Bangladeshi and Indian hackers impacted the diplomatic relations between the two nations. More importantly, there have been scores of media reports saying that terror groups use online platforms for financing and maintaining intra-group communications. In this context, the existing laws and government moves against cyber-crimes are apparently very scanty to combat the burgeoning threat. The study attempts to shed light upon the threat posed by cyber-crimes in the context of the global village with an emphasis on the perspective of Bangladesh.

Keywords Bangladesh · Cyber-crimes · Cyber-security · Globalization · The ICT Act, 2006 · The Digital Security Act, 2018

K.-E.-K. (Babu) (✉)
Daffodil International University, Dhaka 1207, Bangladesh
e-mail: kekbabu@gmail.com

© The Author(s), under exclusive license to Springer Nature Switzerland AG 2021   253
H. Jahankhani et al. (eds.), *Cybersecurity, Privacy and Freedom Protection in the Connected World*, Advanced Sciences and Technologies for Security Applications, https://doi.org/10.1007/978-3-030-68534-8_16

1 Introduction

The era of globalization is characterized by the rapid proliferation of information technology and communication. Secure cyberspace is the era of globalization and is a crucial element in maintaining national security. This plays an important role in achieving a country's economic stability and effective security [17]. Cyberspace is the world of computer networks (and the users behind them) where information is stored, exchanged and revealed [14]. With the rapid and dramatic growth of information and communication technology (ICT), cyber-crime has become a major security issue in the international arena. Both individual cybercriminals and state-sponsored cyber-attacks pose threats to states protecting their confidential data. Apart from having a profound impact on the economic progress and defence systems, these threats escalate diplomatic tensions leading to anarchy in the world order. Global peace, stability, and development might be affected by the abuse of information communication technology. Bangladesh with its less sophisticated cyber-surveillance system and cybersecurity tools may easily become a safe haven for cybercriminals committing phishing, hacking and stealing personal data. Digital services extended to people by the government and non-government sectors and personal and organizational data are targeted by criminals. Proper security measures are not often ensured while providing services through digital platforms. In addition, the Information Communication Technology Act, 2006 [8] might do little to secure cyberspace. This study endeavours to explore the major challenges for Bangladesh with its disarrayed cybersecurity and countermeasures in the context of the globalized world. In this respect, the study re-examined the efficacy of the existing information and telecommunication laws and proffers restitutive measures to ensure cybersecurity in Bangladesh. The study concludes with the utterance that the time is ripe for Bangladesh to enhance its cybersecurity and secure its cyberspace.

2 Cyber Security in the Global Village

As the Internet connects virtually every human being living on the planet, a new coinage terming the global citizens as netizens. Cyber threats are no longer being seen as national security concerns, they are indeed global phenomena. Cybercrimes pose harm not only to individuals or specific target groups even to the states. Cyber-criminals tend to exploit any potential loopholes at networks, systems, data, and operators to garner money. According to B. Williams, there are four Cybercrimes groups. First of all, cyber-criminals just after the money. Such an example came in April 2013 when the U.S. stock market suffered $130 billion in minutes only because of a hacked Twitter news stream propagating a false story of an explosion at the White House [17]. Second, the competing organizations pursuing sensitive knowledge or intellectual property that could exploit them over others. In both the civil and security industries, this is worrying. A Russian crime organization recently cumulated the largest documented set of stolen internet data, consisting of 1.2 billion usernames and combinations of passwords, more than 500 million email addresses [13]. Thirdly, an insider de facto might pose a threat from within. Recent breaches of IT systems ranging from Iran's nuclear facilities to thousands of American diplomatic cables have underscored the importance of ensuring cyber-security in the Information Age. Cybercrimes because of their transnational nature and anonymity of the criminals are more exacerbating and their potential damage is disproportionate.

While a striking issue in its very own right, cyber-crime forecasts the inescapable clashes that will emerge from the close contact facilitated by the Internet between diverse cultural practices. The emergence of the Information Age has created an unparalleled network between people all over the world and also established connectivity at organizational scales. Intra-organizational and governmental communications have never been so rapid, cheap, and specific as the internet has taken the whole process of connectivity to an unprecedented dimension. Even information transmission to non-networked is also facilitated by common software platforms. Such connectivity, while helpful to all, comes at a potential cost. Globally governments and associations are while encountering umpteen cybersecurity occurrences, focusing on the management of cybersecurity threats and dealing with their fallout. For some associations, the most common cybersecurity threat is the danger of classified data being gotten to and possibly abused by an outside or potentially antagonistic party i.e. data breaches. One of the key difficulties in reacting to data breaches is that information, ruptured from one or more jurisdictions, can be passed instantly to other jurisdictions. The transboundary nature of occurrences can make investigating a data breach, distinguishing our alternatives for managing the breach, a mind-boggling and overwhelming procedure. This is particularly so on the grounds that speed is quite often a basic factor in exacting a compelling response. In the Asia Pacific region, there has been a rush of new digital security enactment in recent years, Governments establishing bodies to regulate or monitor digital security, and governments and controllers regularly issue guidelines/reports on this. For example, Indonesia and Singapore both launched cyber agencies in 2015, Japan approved the Cyber Security Basic Act, and a report on cyber resilience was issued by the Australian Securities and Investments Commission. Laws or guidelines on these matters are being formulated out of the blue for different nations in the Asia Pacific. Also countries, for instance, the United States, where the Justice Department released in April 2015 its "Best Practices for Victim Response and Reporting of Cyber Incidents", are adding to already existing frameworks of cybersecurity guidelines. Despite the intensive and exhaustive administrative action, there is, sadly, no combined approach to cybersecurity regulation or potential legal recourse with regards to data breaches in the Asia Pacific. Subject to change under varying jurisdictions, data breaches may include Responsibilities under data protection laws, employment/labour laws, equal rights and obligations, equity rules, corporate governance, fiduciary duties, and business or sector-specific legislation, in addition to cybersecurity laws. When data is believed to have been moved out of a jurisdiction, in some jurisdictions, state laws on national secrets can come into force. Similarly, local knowledge of responsibilities in each nation and how each applicable regulator or court works by and by is crucial for reacting to an episode of the data breach and knowing the legal remedies could be accessible and which would be better. Utilizing this learning, can help the clients to examine data ruptures, to distinguish obligations, to devise plans to limit the further revelation of the data and moderation of impact or harm, and to recognize, where accessible, lawful solutions for recouping the information or loss related with the information rupture. Many of the Government's websites use international servers and foreign vendors. As a result, these are always in a vulnerable position and at risk of being sabotaged by the system's insiders [1]. Potentially the fourth strategy is the biggest threat to our national security. This relates to a state-sponsored cyberattack aimed at undermining a national security framework such as critical infrastructure or important national economic components to some degree in order to achieve strategic advantages over that specific country [17]. In this context, the instance of China can be cited. Some of the powerful countries in the world such as the U.S., U.K., France, Germany, and India always consider China as a potential threat to cybersecurity and charged the country in connection with espionage for gaining strategic advantages. In 2007, it is confirmed that China launched a series of network-based cyberattacks on the above-mentioned countries. In addition, these countries do have greater military ambitions to boost the capacity of the country to engage in the information or cyber warfare, if necessary in the near future [7].

3   Cyber Violence Against Women in Bangladesh

In Bangladesh, women are lopsidedly subjected to violence and harassment; cyberbullying to pornography are mentionable phenomena that are facilitated by the internet and other electronic devices. While the extension of Information and Communication Technology (ICT) and burgeoning Internet infiltration are considered as positive markers of development in the country, yet their association with certain existing socio physiological settings and insufficient legal protections have paved the way for extensive cyber brutalities against women. By and large, the type of this glaring infringement of human rights ranges from cyberstalking, vengeance pornography, cyberbullying, and trolling. Women are the primary targets of hostile and frequently forceful lewd gestures and disparaging messages on the internet from unidentified and counterfeit sources. Doctored nude pictures of women alongside spam, sex-act recordings, rape threats, and obscene proposition have turned into the new standard of social media. Mobile telephony has taken internet penetration by storm with the number of active internet connections in Bangladesh hit 90.5 million in August 2018, of them, 80.47 million are connected with mobile internet [5]. The ever-increasing internet penetration and mobile phone devices have seen an upsurge in Facebook use. Of the 29 million registered Facebook users, 86% use mobile phones to access the social media networking site. Women population constitutes 1% of cell phone and internet subscribers. Young women in Bangladesh are more likely to be victims of sexualized and abusive online violence in nature. Though legal framework and organizational protection is feeble, a sizeable number of women lodge formal complaints in connection with badgering, abuse, and violence emanating from cyberspace. Cybercrime has been reported by 73% of women internet users [18]. The Cyber-Help Desk of the government's Information and Communication Technology Division has received more than 17,000 complaints as of December 2017, 70% of complainants being women. Exposure to pornographic content among the youths, whether intentional or unintentional, aggravates the other associated risks, for example, picture-based maltreatment of users where women are exceedingly victimized. In the digital world, with around 78% of cases of doctored photos containing pornographic contents, women are found to be the victims. It can be noted that nearly 77% of teenagers in the country regularly watch pornography [12]. In June 2019, the Bangladesh National Women Lawyers' Association reported that badgering remained an issue and inadequate preventive and counteracting laws caused some young women to drop out of their classes or works because of trauma and stigma. The establishment of complaint committees and the installation of complaint boxes at educational institutions and workplaces mandated by the directive of the court have rarely been implemented [16]. Very often social media accounts are hacked with malicious intent. The criminals usually upload manufactured indecent photos of the victims, send provocative messages to the victims' contacts (i.e. Facebook friends) in order to disparage and humiliate them. Some of the key motives of these perpetrators include smearing the victims, taking revenge, coercing them to establish physical relationships, pressing for hush money, physically torment the victims, and so on. A pattern is seen while reviewing the lawsuits, investigations, and media reports of cyber violence against women in Bangladesh. Most commonly the perpetrators establish consensual physical contact with the victims earning their trust. They film the intimate moments with hidden cameras installed in the scenes, it is obvious if the pattern of these heinous crimes is observed closely. Unfortunately, it doesn't stop here, the criminals then go on to blackmail the victims and coerce them to gain their ill motives. Those clips are used later in order to force the victims to submit themselves to the will of the criminals demanding continuation of physical relationships and hush money. Meanwhile, criminals often record the nefarious acts of rape and film the incidents. Those video recordings are used later by the perpetrators to silence the victims to abuse them furthermore. Those recordings are most commonly released on the internet despite submissions of the victims making them traumatized and stigmatized in society. There are reports of deaths by suicide as the victims feel utterly helpless and do not find any headway. Another pattern is also noticed where vindictive ex-husbands and lovers post intimate videos or photographs on the internet to satisfy their grudges. As young women are less experienced with the internet they are most vulnerable to falling prey to the traps of cybercriminals.

3.1 Effects of Cyber Violence

The effects of cyber violence against women in a somewhat conservative society like Bangladesh are pervasive. The families of the victims are also greatly affected by such incidents. The series of events that follow are a double blow equally for the victims and their families often leaving them ostracized. People tend to believe whatever stuff they come across on social media. Such an indiscriminate belief system is the result of hollow public psychology stemming from a lack of awareness, ignorance, and education. Consequently, when a purportedly released photo of a girl surfaces on the internet mixed up with raunchy gossip, general internet users bother little to verify whether the photo is real or fabricated. Merrily they go on to ogle at the contents and make those viral. This tendency to spread sex-related gossip amplifies the victims' sufferings manifolds. Not to mention the misery of the victim's family members who face social exclusion, humiliation, and public resentment [10]. The consequences of these cyber violence are disastrous at individual levels leading to severe depression, a sense of guilt, paranoia, and fear of harm to self and family members. Victim's careers, education, and social life are jeopardized by these issues with some of them taking the path of drug addiction some choosing to end their lives. Very few of the victims recover from the trauma in a handful of exceptional cases. Bangladesh National Women Lawyers' Association tallied 65 reported suicide attempts by female victims subjected to such violence from 2010 to 2014. According to the findings of the association, on average there are 11 cases of suicide attempts by women due to cyber violence. Whereas the number of such cases was 8 in 2008, the data reveals an upward trend. However, the official statistics are nascent in comparison to the actual number of such incidents. The number of unreported cases far outweighs the reported ones [4].

4   Challenges to Bangladesh

The major concern for Bangladesh is that most of the software used in the country is pirated. In such a situation, it a big challenge for the country to protect its cyberspace in the poor infrastructural system. In Bangladesh, around 90% of software is pirated [3]. Right now, it has become a common practice and culture among the country people of using the pirated software, leading its cyberspace to the most vulnerable position in the cybersecurity domain. This is the major challenge the country is facing right now, but its consequences and impact cannot be ignored.

Apart from the concern, there are some other serious challenges for cybersecurity in Bangladesh that cannot be ignored any longer. According to Bangladesh Telecommunication Regulatory Commission (BTRC), in August 2018, the number of active internet connections in Bangladesh reached 9, 05 crores, which a matter of thanks to the introduction of around 18 lakh new connections to the network in one month. Among these, 8.47 crore connects to mobile internet, 57.33 lakh connects to fixed broadband internet while the rest use WiMAX. The total number of active Internet connections exceeded the seven-crore plateau in April 2017, the six-crore mark in August 2016, five crore in August 2015, and four crores in September 2014, respectively (The Daily Star, 21 September 2018). Such rapid growth of internet users in Bangladesh has put the country's financial sector under persistent cyber threat. In such a situation, it is an urgent need for strong in-built cybersecurity in Bangladesh. A small group of experts who work regularly on cyber-threat intelligence, data security, and encryption is also in desperate need.

To understand the challenges, first of all, we need to be conscious of the dimension of the cyber-crimes we are facing in our daily life. This may break it up into four groups. First, Cyber-crimes against people, such as: hacking or cracking, unlawful/unauthorized entry, illegal surveillance, data intrusion, e-mail spoofing, spamming, cheating and fraud, abuse and cyber-slaughter, slander, drug trafficking, virus transmission. And worms, infringements of intellectual property, abuse of machine and network resources, Internet time and information theft, forgery, denial of services, dissemination of pornographic materials, etc. The second is property-related cybercrime, such as the robbery of credit card money, intellectual property violations, Internet time theft, etc. The third one is organized crime. Examples of these crimes include unauthorized control/download over network resources and websites, posting of indecent/obscene content on web pages, virus assault, e-mail bombing, logic the bombing, trojan horse, data dodging, download blocking, theft of valuable belongings, terrorism against government organizations, vandalizing the infrastructure of the network, etc. The fourth and last group of cyber-crimes are taking place against Bangladesh's society or social values. Such crimes include forgery, online gambling, prostitution, pornography (especially child pornography), financial crimes, youth pollution by indecent exposure, web jacking, etc. [11].

In Bangladesh, pornography is one of the major concerns in terms of the country's social culture and moral values. We can now communicate with anyone anywhere in the world and share or exchange our cultural values, thanks to the rapid digital expansion and globalization. From the cultural perspective, many harmful elements of different country's culture easily can intrude on our own culture due to the diffusion of culture. Pornography is a very untoward element for the country's culture where adult education is not welcomed even. Bangladesh police are receiving a huge number of complaints of demanding ransom by threatening with secret nude video footage and photoshopped pornographic photos, according to the lawmen. Most of the victims are teenage girls. Besides, women and children are also being targeted by criminals. When any crime is conducted from abroad, then it would be considered as 'dual criminality'. That means the crime is considered in both countries. But there is a complexity to deal with the crime like pornography as such crimes [in the context of Bangladesh] may not be considered as crimes in many countries like the U.S. in every case. In such a situation, the victims in Bangladesh will have to face difficulties to deal with such crimes. On the other hand, transnational crimes like child pornography, which are considered crimes in both countries and every country, can be dealing with international cooperation. Here an instance can be given of such an issue.

Several years back, Tipu Kibria, a well-known child litterateur in Bangladesh, was arrested by police red-handed for child pornography. He used street male kids in his home and lab to make pornographic videos and photo shooting for girls. He had already assaulted some 400–500 street kids at the time he was arrested by police for his filthy ambition. Throughout these illegal activities, he has two assistants to help him out, and police found 13 foreign buyer names from Tipu Kibria who regularly paid him for weekly supplies via foreign or online banking transactions. Bangladesh police also believe that there might also be several other manufacturers other than Tipu Kibria. We may therefore explicitly state that pornography is a major concern regarding cybersecurity in Bangladesh [1]. Cybersecurity threat, especially for the financial transaction including e-commerce and online banking is also a grave concern for Bangladesh. Besides, transnational crimes like drug smuggling, trafficking, terrorism, etc. are other big challenges to Bangladesh's cybersecurity. Due to the lack of proper cybersecurity measures, Bangladesh is also facing a serious concern of cyber threats in the banking sector.

In February 2016, Hackers stole $101 million from Bangladesh's central bank account with the Federal Reserve Bank of New York using the SWIFT payment network for fake orders. Cyber heist is one of the world's greatest cyber-crimes. If Bangladesh fails to take proper measures and adopt strong security policies, the country's banking sector may become the further victim of such cyber heists in the coming days. Widespread use of credit cards and that electronic payment methods often risk a large number of private customer details, such as bank account name, bank account number, cell phone, e-mail ID, etc. [3]. Law enforcement agencies are often receiving complaints or cases of direct or indirect cyber threats to financial transactions through online banking. On February 12, 2016, Eastern Bank, a private bank in Bangladesh 21 Suspicious card transactions found. A fraudster with a fake EBL card used one of United Commercial Bank Limited's ATM Booths, which set off the alarm in UCBL's network, causing the crime ring to unravel. On February 25, Dhaka Metropolitan Police said that the investigation of the ATM Card scam case has brought up names of various hotel travel agencies and some bank officials. The lawyers also detained a German citizen in connection with ATM fraud, and three official City Bank, a local private bank. Piotr was wanted in 3–4 countries on fraud charges, and according to police, we would be seeking information from those countries through Interpol (The Daily Star February 26, 2016).

Some Bangladesh-based foreigners have allegedly been involved in the skimming scam that robbed money from ATMs in signs of emerging financial crimes that terrified both banks and customers. In February 2016, Bangladesh Bank, three other commercial banks, and lawyers analysed video footage of four ATM Booths, which were skimmed off at least Taka 25 lakh. The spokesperson for the central bank said they mainly find the involvement of at least two foreign nationals in the crimes. There are similar concerns in different private banks like—Eastern Bank Limited, United Commercial Bank Limited, and City Bank—struck by ATM frauds (The Daily Star, February 16, 2016). All banks both private and government commercial banks are taking security measures to curb illegal transactions. But due to the lack of proper and adequate security measures and technological support and responsibility of the authorities concerned, the country's troubled banking sector is still struggling to face cybersecurity challenges. Besides, many people in Bangladesh are being victimized by phishing or fraudulent attempts to get confidential evidence like usernames, passwords, and credit card details through e-mails or attractive advertisements. In these cases, victims typically lose $100–500 per case and refuse to go to the police to complain, which makes the case in Bangladesh more difficult to deal with [1].

Hacking or unauthorized intrusion into a computer system without the owner or user's permission is also a concern for cybersecurity in Bangladesh [11]. Hackers most of the time targeted the financial websites of both the government and prominent privates organizations. Lack of adequate cybersecurity know-how, Bangladesh is in a more difficult position to tackle cyber-piracy by a weak cyberinfrastructure network such as reliance on international server system providers, etc. [1]. Data-stealing is another concern in Bangladesh. The leak of the Bangladesh War Crime Tribunal's verdict (partially) in 2014 is an example of the data-stealing. The data of the tribunal leaked through Skype's voice recording. It was a major backlash for the Bangladesh government and exposed the vulnerability of the Bangladesh cybersecurity arena [1]. Besides, the cybersecurity of social media platforms especially Facebook, Twitter, and Linkedin are in grave threat in Bangladesh. Though Bangladesh police, Bangladesh Telecommunicate Regulatory Commission have strengthened monitoring and established separate monitoring teams recently, such hacking of social media accounts are happening frequently till February 2019. Hackers are targeting mostly prominent personalities, celebrities, and females and taking money from the victims threatening of tarnishing their social image.

5 Existing Acts and Their Limitations

Right now, there is no debate about the level of cybersecurity risk in Bangladesh, but the major concern is whether the country would be able to address the risk properly and timely. It is sorry to say that the concerned authorities are still reluctant to take full-scale measures to combat the risk, thanks to the lack of understanding in different concerned stakeholders. To combat cyber-crime, the Bangladesh government has formulated a few laws including Information and Communication Technology (ICT) Act, and Digital Security Act. But literally, these laws are seen largely to be used in curbing the freedom of speech and expression. Some contradictory articles and sections of the laws are being used by the government and law enforcement agencies to gag the news media and social media. Bangladesh government passed the ICT Act on October 8, 2006. Seven years after its formulation, the parliament of the country amended the act keeping some controversial provisions on 6th October 2013. However, a cyber-crime victim can sue someone under the law for cyber-crime regardless of his or her place and location in the world. Victims may at least use this ICT Act as a starting point, but after that, they certainly need strong cooperation to make progress from first, regional law enforcers in Bangladesh with expertise in cybersecurity such as CID (Criminal Investigation Department) and from foreign law enforcers such as Interpol [1]. Human rights advocates, representatives of civil society, and media critics urgently demand that section 57 of the Information and Communication Technology Act be repealed as such clause of law provides room for widespread misuse. The maximum penalty for offenses under the section before its amendment was 10 years imprisonment and a fine of Taka 1 crore. However, law enforcers had to take permission from the appropriate authorities to file a case however arrest any person under the rule. After the 2013 amendment, the maximum jail term was raised to 14 years. In addition, legislators were granted the right to detain someone without a warrant. A rough translation of section 57(1) says, "If any person deliberately publishes or transmits or causes to be published or transmitted in the website or in any other electronic form any material which is false and obscene and if anyone sees, hears or reads it having regard to all relevant circumstances, its effect is such as to influence the reader to become dishonest or corrupt, or causes to deteriorate or creates the possibility to deteriorate law and order, prejudice the image of the state or person or causes to hurt or may hurt religious belief or instigate against any person or organization, then this activity will be regarded as an offense." Despite the reforms with a few significant changes, the 2006 key Act remains unchanged with all its inconsistencies and imposes unnecessarily harsh punishments [2] However, the 2013 ICT Act (amended) has become the Bangladesh government's tool for violating fundamental human rights, such as freedom of opinion and expression. The act includes a range of ambiguous imprecise and overboard clauses [9] that could help to further instigate rather than contain cyber-criminal activities. According to the ICJ, section 57 of the original ICT Act is 'incompatible with the obligations of Bangladesh under Article 19 of the ICCPR: the offenses imposed are ambiguous and excessive, the limitations on freedom of speech and opinion go beyond what is allowed under Article 19 (3) of the ICCPR' [9]. J. Barua said, "Section 57 is not specific and covers a wide area of offenses, there will be little chance to get an acquittal from any charge" [2]. After reviewing the ICT Act 2006 with its amendments, we may conclude that there should be legislation to cover cyber space-related crimes, but the current act is ambiguous and needs to be structured on a permanent basis as a modernist legal structure, not only based on the ad hoc system [6].

From the very beginning, rights activists and journalists were critical of Section 57, and the debate on the provision and demand for its abolition escalated after the arrest of journalist Probir Sikdar in 2015. In addition, under section 57 of the Information and Communication Technology Act, at least 21 journalists were sued in four months to July 2017 in the face of the growing demand for the abolition of the provision that is widely open to misuse (The Daily Star, 7 July 2017). Amid widespread criticism of the ICT Act, on May 2, 2017, Bangladesh Law Minister Anisul Huq said that section 57 would be withdrawn and a new "Information Technology Act in the pipeline" will be implemented. On 19 September 2018, Bangladesh's Parliament passed the 2018 Digital Security Act with a tough clause authorizing police officers to search or arrest someone without warrant Rights activists and journalists expressed concern that the act goes against the constitutional spirit and would restrict freedom of speech, freedom of expression, freedom of thought and hinder independent journalism. Section 43 of the new law specifies that when a police officer believes that an offense has been committed or is being committed at a given location, where there is the risk of committing offenses if the evidence is lost, the official can search the location or any person there. *Sampadak Parishad* (The Editors' Council), a daily editorial forum in Bangladesh expressed surprise, frustration, and shock in a statement on 16 September last year., it said the sections 8, 21, 25, 28, 29, 31, 32, and 43 of the act pose serious threats to freedom of expression and media operation. Section 3 of the Digital Protection Act incorporates a clause of the Access to Information Act 2009 which will extend to information-related matters. Where a person commits any crime or assists others in committing crimes under the Official Secrets Act, 1923, as provided for in section 32 of the law, through a computer, digital device, computer network, wireless network, or any other electronic medium, he or she may face a maximum of 14 years in prison or a fine of Tk 25 lakh or both. The law also includes a definition of the "Spirit of the Liberation War" in section 21, which says, "The high ideals of nationalism, socialism, democracy, and secularism, which inspired our heroic people to dedicate themselves to, and our brave martyrs to sacrifice their lives in, the national liberation struggle." Under section 29 of the law, a person can face up to three years' imprisonment or a fine of Tk 5 lakh or both if he or she commits the offenses provided for in section 499 of the Penal Code via a website or electronically. Section 31 of the Act states that a person may face up to seven years in prison or Taka 5 lakh in fine or both if he or she is found to have intentionally published or broadcast something on a website or in electronic form that may spread hate and build enmity between different groups and communities, and may cause deterioration in law and order (The Daily Star, 20 September 2018).

6 Policy Opinions

We can provide several remedial policy options in the above scenario regarding cybersecurity, cyberspace safety, and reducing cyber-crime rates in Bangladesh. We suggest policy options for the government in Bangladesh, but it also includes the individual security domain. These options could be as such:

6.1 Reform of Legal Structure

We resound with the ICJ's legal recommendations about the ICT Act 2006 and its amendments to both the Bangladesh Parliament and the Government of Bangladesh. The ICJ refers to the Bangladesh Parliament for all reasons, 'Repeal the Information and Communication Technology Act [8], as amended in 2013, as amended in 2013, or amend the ICT Act to bring it into line with international laws and standards including the legal obligations of Bangladesh under the ICCPR. At a minimum, this will require it (1) to amend section 57 of the ICT Act in order to ensure that any envisaged limitations on freedom of expression and opinion are in accordance with international law and standards, (2) to amend section 57 of the ICT Act to ensure that forbidden speech is clearly defined; (3) Amend the ICT Act to ensure that any restriction on freedom of speech and information, including any penalty provided for, is necessary for a valid purpose and proportionate to the harm caused by that speech "[9]. In this regard, the ICJ also proposed policy alternatives to the government of Bangladesh. Such policy options are: (i) 'Take action to ensure that the provisions of the ICT Act are not used to infringe the right to freedom of speech, including restricting the legitimate exercise of public opinion on matters which may include criticism of the Government, (ii) drop charges against bloggers for the legitimate exercise of their freedom of expression; (iii) Guide government agencies to refrain from filing unfairly limiting the freedom of speech in politically motivated cases and to pursue penalties disproportionate to the severity of the alleged offence [9].

6.2 Maintaining Rules of Cyber Security

In 2011, in his article 'Ten Rules of Cyber Security,' Eneken Tikk, the legal counsel at the NATO Cooperative Cyber Defense Center of Excellence, Tallinn, Estonia, provided a measured framework to preserve cyber security tenets considering national security issues as well as individual security concerns. Eneken Tikk's propositions are agreeable in many cases. He talked about 'the territorial rule' protecting cyber security as such, "Information infrastructure located within a state's the territory is subject to that state's territorial sovereignty" [15]. Tikk also spoke of 'the law of duty' where he proposed that the States behave responsibly to secure their own territories. He also spoke of the 'early warning statute' as such, "There is an obligation to notify potential victims about known, upcoming cyber-attacks" [15].

By examining Tikk's above cybersecurity rules, we can prescribe a solid national digitally insightful agency for Bangladesh to battle present and forthcoming potential cyber threats from anyplace of the world as they say counteractive action is superior to fix. Secondly, in Tikk's opinion, a state should adopt 'the data protection rule' protecting its vital national data. In his words, "Information infrastructure monitoring data are perceived as personal unless provided for otherwise." In this context, another rule of Tikk can be cited here, 'the duty to care rule'. He is suggesting everyone take a minimum level of responsibility to secure any kind of information infrastructure [15]. By resounding his idea, we can propose that the Bangladesh Government exploit her own resources, skills, and implement trend-setting innovation to secure the internet and national interests. Imparting training to our cyber experts, building up our own server frameworks and systems utilizing our own assets and labor, investing a sizeable amount of time to build up our cyber safety net, recruiting potential national programmers and so on can be beneficial to Bangladesh over the long haul as opposed to depending on foreign specialists. Thirdly, we can agree with 'the cooperation rule' of Tikk. He stated that "…cyber-attack has been conducted via information systems located in a state's territory creates a duty to cooperate with the victim state" [15]. Thirdly, we can agree with 'the cooperation rule' of Tikk. He stated that, "…cyber-attack has been conducted via information systems located in a state's territory creates a duty to cooperate with the victim state" [15]. We need solid worldwide participation to fight any sort of cybersecurity risk as to the majority of the cases, these threats have been included with transnational criminal exercises where the affected individuals might be victimized in one country and culprits may flee by taking advantage of international border boundaries. Bangladesh Government and Bangladesh Police have joined hands with international law enforcing agencies, for example, Interpol in such manner yet Bangladesh needs more collaboration particularly from the tech giants, for example, Microsoft, Google, Facebook, Yahoo, and others. In conclusion, Tikk's other two rules–' self-defense and the access to information rule–can be referred to. He said that "everyone has the right to self-defense" and "the public has a right to be informed about threats to their life, security, and well-being" [15]. As we quote him, we recommend that the Bangladesh government takes preemptive and precautionary measures to ensure cybersecurity at the individual and national levels.

6.3 Individual Awareness

The consequences and reality of globalization are undeniable. Awareness of personal data protection and safety must be developed at the individual level apart from government, initiatives to create secure cyberspace. Professionals irrespective of their hierarchy and varying organizational structures must gain a minimum level of expertise in handling cyber technologies and building awareness on cybersecurity threats does not seem to have any alternative. Only proper education and awareness can rescue Bangladesh from falling into the deep pitfall of cybersecurity threats (Alam, Md. Shah, personal communication, July 27, 2014). Basic precautionary measures should be exercised while using the internet. Here are some preemptive measures that can be taken:

(i) Keep trustworthy and restricted personal details (ii) Keep your privacy settings on (iii) Secure browsing (iv) Make sure your internet connection is safe (v) Be careful what you access (vi) Use good passwords (vii) Make online transactions from protected sites (viii) Be careful what you post (ix) Keep your antivirus software up to date. If we think if we have been a victim of cybercrime, we should go to our local police station, in some scenarios, contacts the FBI and Federal Trade Commission. Even if the crimes seem trivial, it is important to report such incidents. Our promptness may prevent the recurrence of such crimes. If we suspect identity theft, contact the financial institutions and companies where the fraudulence occurred.

7 Conclusion

Taken everything into account, the issue of cybercrimes is emerging as a global phenomenon that poses potential threats to the national security of any country and Bangladesh is no exception to that rather the issue of cybercrimes is more worrying for the country in the context of globalization. The absence of advanced cybersecurity tools and people's ignorance in handling tech gadgets coupled with a lack of awareness of cybersecurity threats might have disastrous impacts on the country. In addition, the country's laws seem inadequate to safeguard the cyberspace of the country. The international collaboration, enhancing technical know-how, gaining expertise, and campaigning on people's preparedness on how to deal with cybersecurity threats are some of the remedial aspects the country may take into consideration to combat ever-looming cybersecurity threats. The sharp increase in cyber-crimes in Bangladesh and all over the world validates the proposition that the issue of cyber-crimes is undeniable though some argue may that cyber threats may not be the possible near-future scenario for Bangladesh. Finally, on a note of conclusion, it can be stated that the time is ripe for Bangladesh to take preemptive and counteracting measures to thwart the threats posed by cybercriminals. In this regard, the Bangladesh government and the general people can mull over the suggestions provided in this paper.

References

1. Alam, S (2019) Cyber crime: a new challenge for law enforcers. City Univ J 2(1):75–84. https://www.prp.org.bd/cybercrime_files/Cybercrime. Accessed 25 Apr 2020
2. Barua J (2019) Amendment Information Technology and Communication Act. The Daily Star. https://www.thedailystar.net/supplements/amended-information-technology-and-communication-act-4688. Accessed 25 May 2019
3. Bleyder K (2012) Cyber security: the emerging threat landscape. Bangladesh Institute of Peace and Security Studies, Dhaka
4. BNWLA (2014) Survey on psychological health of women. Bangladesh National Women Lawyers' Association, Dhaka
5. BTRC (2018) Internet subscribers. Bangladesh Telecommunication Regulatory Commission. https://www.btrc.gov.bd/content/internet-subscribers-bangladesh-april-2018. Accessed 12 May 2020
6. Editorial (2013) Draft ICT (Amendment) ordinance-2013: a black law further blackened. The Daily Star. https://archive.thedailystar.net/beta2/news/draft-ict-amendment-ordinance-2013. Accessed 25 Dec 2019
7. Greenemeier L (2019) China's cyber attacks signal new battlefield is online. https://www.scientificamerican.com/article/chinas-cyber-attacks-sign. Accessed 12 Aug 2019
8. The Information & Communication Technology Act (2006). https://www.prp.org.bd/downloads/ICTAct2006English.pdf. Accessed 11 Jul 2020
9. International Commission of Jurists: Briefing Paper on the Amendments to the Bangladesh Information Communication Technology Act (2013). https://icj.wpengine.netdna-cdn.com/wp-content/uploads/2013/11/ICT-Brief-Final-Draft-20-November-2013.pdf. Accessed 10 May 2020
10. Karaman S (2017) Women support each other in the face of harassment online, but policy reform is needed. The LSE Women, Peace and Security blog. The London School of Economics and Political Science, London. https://blogs.lse.ac.uk/wps/2017/11/29/women-support-each-other-in-the-fa. Accessed 1 Mar 2020
11. Maruf AM, Islam MR, Ahamed B (2014) Emerging cyber threats in Bangladesh: in quest of effective legal remedies. Northern Univ J Law 1:112–124. https://www.banglajol.info/index.php/NUJL/article/view/18529. Accessed 21 Aug 2020
12. Elahi SM (2014) Porn addicted teenagers of Bangladesh. Manusher Jonno Foundation, Dhaka
13. Perlroth N, Gellesaug D (2014) Russian hackers amass over a billion internet passwords. https://www.nytimes.com/2014/08/06/technology/russian-gang-said-to-amass-more-than-a-billion-stoleninternet. Accessed 28 Jan 2019

14. Singer PW, Freidman A (2014) Cybersecurity and cyberwar: what everyone needs to know. Oxford University Press, Oxford
15. Tikk E (2011) Ten rules for cyber security-survival: global politics and strategy. Routledge, London
16. USSD (2017) Country report on human rights practices for 2016. US Department of State, Washington DC. https://www.state.gov/j/drl/rls/hrrpt/2016humanrightsreport/index.htm?ye. Accessed 2 Aug 2020
17. Williams B (2014) Cyberspace: what is it, where is it and who cares?. https://www.armedforc esjournal.com/cyberspace-what-is-it-where-is-it-and-who-cares/. Accessed 15 Jul 2020
18. Zaman S, Gansheimer L, Rolim SB, Mridha T (2017) Legal action on cyber violence against women. Bangladesh Legal Aid Services Trust (BLAST), Dhaka

Reasons Behind Poor Cybersecurity Readiness of Singapore's Small Organizations: Reveal by Case Studies

Nam Chie Sia, Amin Hosseinian-Far⦿, and Teoh Teik Toe

Abstract Digitalization and cybersecurity are two important trends that are affecting the business world tremendously. Digitalization, which drives data analytics, provides opportunities for organizations to create new models to beat competition. On the other hand, cybersecurity is a threat to organizations' financials, operations, and reputation. COVID-19 has accelerated the adoption of digitalization, which has opened up more opportunities for hackers for cyberattacks. In another word, digitalization underlines the importance of cybersecurity. With the foresight of the government, Singapore has promoted cybersecurity as one of the pillars for the nation's total defence to signal the government's attention and resources committed to fighting against cyberattacks. Notwithstanding the effort from the government, losses due to cyberattacks continue to rise. Furthermore, the network of the biggest healthcare provider in the country was compromised and its data, including that of the Prime Minister, was stolen. For small organizations where resources may be limited, the risks are even higher, pointing to the urgent need to address the situations. Therefore, this article uses two small organizations in Singapore as case study, to draw insights on the obstacles to implement digitalization and cybersecurity. With the insights, actions that can be taken by the government, businesses, and academies, are proposed to improve the digitalization and cybersecurity of small organizations, in Singapore and elsewhere.

Keywords Cybersecurity · Data analytics · Digitalization · Small organization · Singapore

N. C. Sia · A. Hosseinian-Far (✉)
University of Northampton, Northampton NN1 5PH, UK
e-mail: Amin.Hosseinian-Far@Northampton.ac.uk

N. C. Sia
e-mail: sia_nam_chie@hotmail.com

T. T. Toe
Amity Singapore, 101 Penang Rd, Singapore 238466, Singapore
e-mail: Teohteiktoe@gmail.com

© The Author(s), under exclusive license to Springer Nature Switzerland AG 2021
H. Jahankhani et al. (eds.), *Cybersecurity, Privacy and Freedom Protection in the Connected World*, Advanced Sciences and Technologies for Security Applications, https://doi.org/10.1007/978-3-030-68534-8_17

1 Introduction: Digitalization and Cybersecurity Trends

1.1 Global

It is well-known that the current world is being disrupted by new business models driven by digitalization. As organizations digitalized, more activities are being captured electronically [23], which enable the organizations to use insights drawn from data analytics to provide better services and products to customers, thus outperform their competitors [27].

The COVID-19 pandemics have accelerated the adoption of digitalization [20, 24]. Therefore, it is foreseeable that more consumers' activities will be captured digitally for data analytics, supported by machine learning and artificial intelligence capability, to assist companies to gain competitive advantages.

Some of the popular examples are companies such as Amazon, Airbnb, Netflix, and Uber, who use data analytics to alter the competitive landscape [18] create competitive advantages over their competitors [19].

With digitalization and data analytics, the world has become more connected. However, on the flip side, the risks and costs of cyber threats, where hackers can steal data from anywhere in the world, have increased tremendously. The risks are compounded as many organizations are adopting the "work from home" protocols due to COVID-19 pandemics [10].

1.2 Singapore

Singapore is a small country, located in South-east Asia, who was granted self-governance by the United Kingdom in 1959 and gained full independence in 1965. Although it has little natural resources, the country progresses economically and socially, due largely to the government's leadership. The gross domestic products (GDP) per capita grew from US$400 in 1959 to US$22,000 in 1999 [26], and US$64,000 in 2018 [16].

Singapore has a pro-business government, who can consistently execute policies to garner confidence from multi-national companies [26]. Sensing that digitalization and the power of data analytics are going to impact businesses, the government has launched a "smart nation" initiative to drive the entire nation to embrace digital transformation, as the government believes this will improve the lives of its population, create more jobs and enhance engagement with communities [26].

Singapore has not been spared from cyberattacks. In 2017, Singaporeans have fallen victim to various schemes, such as phishing, malicious software, and ransomware, which caused losses amounting to approximately US$30 million [12]. Between May and July 2018, personal data of approximately 1.5 million patients of

SingHealth Group, the largest healthcare service provider in the country, was illegally accessed and copied [37]. Among the patients whose data was breached, was the country's prime minister, who was specifically and repeatedly targeted [10].

In response to the cybersecurity risks, the government has promoted digital defence as the sixth pillar to the nation's Total Defence Framework [9]. Specifically, the government has led initiatives to formulate the Operational Technology Cybersecurity Masterplan 2019 to build resilient infrastructure and raise awareness to create a safer cybersecurity environment.

1.3   Small Organizations in Singapore

With regards to digitalization and data analytics, despite the "smart nation" initiative launched by the government, studies have shown that a substantial proportion of organizations, especially small and medium-sized enterprises (SMEs), have not adopted digitalization and data analytics [36]. It was found that 43% of Singapore SMEs are not familiar with the term "digital transformation" [30] and 85% of Singapore workers are not confident to perform data analytics [38]. Microsoft and ASME [30] further revealed that key decision-makers in small enterprises have much lower awareness of digitalization than their counterparts in medium-sized organizations. In fact, other than staff who work in the information technology department, the majority of the staff in small organizations find it challenging to understand digital transformation [20].

As Singapore prides herself as the gateway to South-east Asia and is one of the world most digitally connected cities [11], the low awareness of digital transformation and data analytics are worrying, as it shows that a pocket of the nation has not been keeping up with the rest. The situations are even direr in term of cybersecurity, as pointed out by the Cyber Security Authority (CSA) of Singapore, the majority of the cyberattack victims in Singapore are SMEs [10].

2   Rationale, Aim, Objectives, and Methodology of the Study

2.1   Rationale

As small organizations in Singapore have a comparatively low adoption rate of digitalization, it is worrying trends that the small organizations account for the majority of cyberattack victims. There is an urgent need to help these small organizations to gain competitive advantages in the digital world in a safe and secure manner. This is because SMEs employed two-thirds of the nation's workforce [42] and contributed to approximately half of her GDP [30].

Being the collective employers for two-thirds of the workforce, SMEs cannot be left behind in the digit transformation era. They also cannot continue to be the main victims of cyberattacks.

In addition, although studies on cybersecurity for SMEs are not new, our literature review shows that there is little being done in the Singapore context.

2.2 Aim

This study aims to identify and understand the root-causes of the low adoption rate for digitalization, and yet being the main victim of cyberattacks. Understanding the causes of the problems can help to formulate effective solutions to address the issues, as the first step of a change journey is to understand the situations and identify the problems [31].

While this study is performed in Singapore, it is believed that the lessons learned can be references for small organizations in other countries.

2.3 Objectives

With the aim to identify and understand the root-causes, the objectives of this study include:

- Identify the root-causes of the low adoption rate of digitalization and cybersecurity for small organizations in Singapore.
- Inquire board members and senior management of small organizations to understand the challenges from their perspective.
- Propose practical solutions to raise the adoption rate of digitalization and cybersecurity in small organizations.

2.4 Methodology

There is little study on the adoption of digitalization and cybersecurity for small organizations in Singapore, hence, this exploratory study will adopt qualitative methodology using case study method.

A qualitative research methodology is suitable for explorative and descriptive research [2]. It is also the recommended methodology to develop understanding, especially when there is little prior knowledge or research [6, 22].

Case study method is also appropriate for exploratory, explanatory and descriptive researches [41, 44]. Besides, case study is recommended for research that is focusing on real-life issues [43], especially when limited knowledge exists [29], and when in-depth investigations are required [17]. In addition, learning can be achieved through practical reflection [35] and practitioners are encouraged to use reflection-in-action to discover new knowledge [14].

The case study for this paper uses two small organizations in Singapore, with different financial resources and organization culture, to compare and contrast their readiness to adopt digitalization and cybersecurity.

The main data collection techniques are interview and archival record. Interview is adopted because it is more aligned to qualitative research, which tends to be exploratory [4], and it is the most important data collection methods in case study [41]. However, other data collection techniques should be used to supplement those collected via interviews [34]. Therefore, the authors also collect evidence using archival records to authenticate and corroborate those obtained from interviews.

After understanding the challenges, this study proposes actions that could be taken to raise the digitalization adoption rate and the cybersecurity standards of small organizations in Singapore.

3 Literature Review

3.1 Digital Transformation in Singapore

As the world is embracing digitalization, the Singapore ministers have been encouraging its communities, including SMEs, to embark on their journey in digital transformation and data analytics [28], to gain competitive advantages [40] and as a result, be a new engine for growth in Singapore [33].

Despite the encouragement, and support provided, by the government to embrace digitalization, the paces of adopting digitalization and data analytics among the Singapore SMEs are slow [30, 36, 38]. This trend is alarming, as the Singapore SMEs are running the risk of being left behind and losing out to their competitors [15, 32]. This is on the back that digitalized organizations are more productive than those who do not, as well as customers' increasing expectations for more personalized services that can only be provided through insights from data analytics [3]. Therefore, SMEs who are not embarking on digitalization may not survive the competition in the near future.

Some of the factors attributable to the slow take-up rate include lack of financial resources, constrain in staff resources, and availability of committed sponsors [39].

3.2 Cybersecurity

As organizations are embracing digitalization and data analytics to gain competitive advantages, this transformation has permeated almost every industry. Along with the growing trend where business organizations embrace digitalization, cybersecurity becomes a significant business issue that impacts customers, profitability, and reputation [25]. Cybersecurity can include many aspects such as data protection, integrity, confidentiality, encryption, and fundamental security functions [7].

Cybersecurity affects all industries and organizations of all sizes, including small business [1]. This assertion is similar to the evidence shown in Singapore, where the small organizations accounted for the majority of cyberattack victims [10]. The threats of cybersecurity, which include disruptions to businesses, negative publicity, litigation, and long-lasting reputational damages [25], can be costly for small organizations as they have little resources at their disposal [1]. It is a vicious cycle that due to the limited resources to strengthen their cybersecurity, small organizations are increasingly being targeted [5].

There is an indication of the poor cybersecurity readiness in the small organization that can be attributable to the poor awareness because they are too immersed in their day-to-day operations and did not spend enough time to proactively study emerging risks [5]. The lack of awareness then leads to delay in the investment in security and give priority to other urgent tasks [25].

3.3   Common Types of Cyberattacks

Phishing

Phishing is the most common type of attack. It is a form of social engineering where the hackers pose as a trustworthy organization [25]. For example, phishing can be initiated via an email that appears to be coming from a bank or government agencies to trick the victims to click on dubious links or attachments [12]. Once the victims clicked on the links or opened the attachments, a "secret" program will move into the laptops or devices without alerting the victims. From there, hackers can control or steal data from the victims' laptops or devices. Alternatively, the hackers can persuade the victims to disclose their confidential information, which will be used to access the victims' bank accounts or other information stored online.

A common consequence suffered by victims of phishing is to surrender the control rights of their organizations' websites to the hackers, who show little hesitance to alter or deface the websites. Unauthorized access and intentional alteration of information without rights are considered cybercrime [21]. For individuals, after disclosing their confidential information such as their bank account passwords to the hackers, they may lose their hard-earned money in their bank accounts.

In 2019, CSA detected an increase of 200% of phishing over the number in 2018 [10]. The situation just got worse, as in the first half of 2020, the number of cyber scams has increased by 2,500% compared to the same period one year ago [8].

Malicious software

Malicious software, or commonly known as malware, are programs that allow the hackers to control the laptops or devices, by compromising the security of laptops or devices, without the victims' knowledge [12]. It was noted that some of the malware was first detected 10 years earlier continue to successfully attack the victims in 2017, indicating that the victims did not update their scanning software to clean up their systems [12].

The malware can also deny access by the genuine owners of the devices. They do so by using an algorithm to encrypt files that deny the owners' access unless they know the passwords [10]. The hackers normally demand a certain amount of money before the victims are provided with passwords to unlock their devices. Such a technique is also known as ransomware. In 2019, there was an increase of 40% ransomware cases being reported by Singapore organizations, compared to 2018.

4 Cybersecurity Readiness: Case Study of Two Small Organizations in Singapore

As the first objective of this paper is to gain an in-depth understanding of the causes for poor adoption of digitalization but the proportion of cyberattack victims among the small organizations in Singapore, the authors performed an in-depth review of the two small organizations, using case study method, to understand the root-causes behind the phenomenon in Singapore.

The knowledge gain from the study is expected to provide insights for a better action plan to address the issues and to help successfully bring the small organizations up to speed on digitalization and cybersecurity.

For confidentiality, the two organizations are named as Organization A and Organization B.

4.1 Organization A

Organization A employs less than 25 staff and its annual revenue is less than US$3 million. Its annual surplus is less than US$0.5 million on average. Its main sources of revenue are the training courses and conferences it organizes for professionals, mainly working in Singapore but there are a minority who are working in South-East Asia countries.

Due to the COVID-19 pandemics, the government has capped the number of participants attending any single training and conferences at 5. This has adversely impacted Organization A, and it is expecting to incur losses in 2020. This has added challenges for it to pull through the crisis as its financial position was weak, even before COVID-19.

The board members of Organization A are mainly professionals working in the audit and risk management fields across various industries. As many of the board members are chief auditors or head of risk management, they are at the forefront of assisting their respective organizations to strengthen cybersecurity.

Despite its weak financial resources, the staff members or Organization A constantly attend training sessions and seminars to keep abreast of the latest development, include trends in digitalization, data analytics, and cybersecurity. Therefore, the staff is aware of the trends and importance of digitalization and cybersecurity.

Although it is small, Organization A is led by professionals, who adopt a relatively open and consultative leadership style.

In the last few years, Organization A has embarked on automating its financial, human resource, and payroll systems. In 2020, it has upgraded its customer management system. These automation projects have instilled a change mindset among its staff. According to the most senior person in the organization, those automation projects have provided an excellent foundation for further change in the organization, he is confident that the staff is more ready to take on data analytics projects to better engage its customers. After automating its customer management system, Organization B is in the process of taking "baby steps" to embark on a data analytics journey.

During the interviews with the staff members, all of them have a certain understanding of data analytics, while the majority of them view it as a necessary change going forward.

In addition, Organization A has engaged an external professional firm to assess its standard of cybersecurity. The organization is in the process of rectifying the gaps identified. Based on the interviews with the board and top management, they view cybersecurity as an important initiative, such that they will "look for the fund to do it even if we do not have the money." Organization A is fairly confident that its cybersecurity capability can protect the organization's data to a large extend, although they are aware that no controls can be foolproof.

4.2   Organization B

Organization B employs about 140 staff and its annual revenue is over US$20 million. On average, it has a profit exceeding US$5 million per year, in the last 5 years. In 2020, despite the impact of COVID-19 pandemics, it is on target to make a profit of approximately US$2 million, according to its revised budget.

At the end of 2019, the majority of the board members were entrepreneurs in their 70s. Data analytics is a term they rarely understood. As they grew up before the birth of the personal computer, they have little training and were reluctant to attend training, in technology and cybersecurity. The reluctance was raised in one of the correspondences with a regulator, who requested the board's training plan. Two years after the request, the board had not provided the training plan to the regulator, which resulted in a regulatory penalty.

Being Chinese entrepreneurs, the board members, in particular the board Chairman, adopted a relatively authoritative style. As the Chairman has little training and knowledge about data analytics and cybersecurity, there was no voice from the top to strengthen the organization to chart into these territories. Based on the observation and reading of archival documents, the board chairman has little understanding of the digitalization trends and has shown little interest to learn.

The top management members also have limited knowledge of cybersecurity. As a result, in the past few years, they did not engage any professional firm to review and assess their cybersecurity capability. Consequently, the statutory auditors issued a management letter in early 2020 to urge the organization to assess the cybersecurity capability.

Under the leadership of directors, who have little knowledge and interest to learn cybersecurity, the organization has undertaken little change management projects.

Table 1 contrasts various factors for Organizations A and B.

4.3 Insight Drawn from the Case Studies

Based on the study of organizations A and B, it is interesting to note that financial resources are not the main driver behind the slow adoption of digitalization and cybersecurity readiness. While Organization A has much weaker financial positions as compared to Organization B, it is in the process of adopting data analytics to provide better customer engagement. In addition, it has engaged a professional firm to assess its cybersecurity readiness. In contrast, although Organization B has more superior financial resources, it has no plan to adopt data analytics. It also needs the statutory auditor to nudge its management to engage a professional firm to review its cybersecurity. This observation is somehow contrary to the findings in Sia [39], who listed financial resources as the top challenge for a small organization in Singapore to adopt a data analytics strategy.

The more advancement of Organization A to embrace data analytics and cyber-security is mainly attributed to the awareness by its board and staff members, who have constant exposures and training in the two topics. On the other hand, the board and staff members of Organization B has little such exposures.

The situations in Organization B are made worse by their organization culture, which is authoritative. In some situations, the management may act as a sounding board to the board members by educating them on data analytics and cybersecurity strategies. However, the management needs a conducive and safe environment to voice their opinions. An authoritative style does not provide the management with a conducive and safe environment to do so. Therefore, with a board chairman who has little exposure and has shown great reluctance to attend training, the "ignorance' is deeply rooted throughout the entire organization, leading it to the poor state of adopting best practices for digitalization and cybersecurity. This demonstrates the importance of the tone from the top, and cybersecurity threats and risks must be managed from the boardroom [25].

Table 1 Compare and contrast between Organization A and Organization B

	Organization A	Organization B
Annual revenue	<US$3 million	>US$20 million
Annual profit	<US$0.5 million	>US$5 million
Main revenue source	Training courses and seminar	Education services
Target market	Professionals	Secondary school students
Staff strength	<25	140
Board members' background	Professionals with strong exposures to governance, risk management, and controls	Chinese entrepreneurs, mostly above 70 years old with little exposure to governance, risk management, and controls. Also, they have shown little interest to learn new skills
Staff exposure	Constantly updated on latest development in governance, risk management, and controls	Limited training and exposure to the latest governance, risk management, and controls. Rely on board members to give directions
Board culture	A consultative style where all the board members have an equal voice	An authoritative style where the most senior guy (Board Chairman) has the loudest voice and the rest are expected to follow him
Exposure to change management	Went through big changes, in term of structure, systems, and processes, in the last 3 years	Limited exposures to change from 2017 to 2019. Change in board members in 2020. The Chinese entrepreneurs retired and replaced by younger professionals in their 40s and 50s
Awareness of data analytics	Most board and staff members are aware of data analytics, although most may not have the skills to execute data analytics, they know it is an important way to bring the organization forward	Some board members may aware of data analytics but the one with the loudest voice among them do not know about data analytics
Awareness of cyber-security	Fully aware of cyber threats and have engaged a professional firm to review its security readiness	Limited awareness of cybersecurity. The statutory auditor raised a management letter point on the lack of cybersecurity assessment

As one of the board members in Organization A has put it: "they (data analytics and cybersecurity) are important projects, we need to do to survive. If we do not have the money, let's go and look for the fund." The strong awareness in Organization A has led it to be the more advanced organization, between the two, to adopt digitalization and cybersecurity, despite having weaker financial resources.

5 Fight Against Cybersecurity Threats

With the insights drawn from the two organizations, the authors are proposing key initiatives to help small organizations to embrace digitalization and cybersecurity in their pursuit for excellence.

5.1 Raising Awareness at the Top

There is an urgent need to communicate the importance of digitalization and cybersecurity to people, especially those serving in the senior roles and as board members of small organizations. This can be done through publicity and training. The Singapore government has put in tremendous effort to encourage Singaporeans and organizations to embrace digitalization and cybersecurity. However, the government cannot do it alone.

The government is of the view that cybersecurity is a collective responsibility of government, enterprises, and individuals [10]. Collaboration among these communities as well as academia is essential for digitalization and cybersecurity to be successful [11].

Academia can play an important role in this aspect as it can design interesting courses to help people overcome the fear of the unknown and to step out of their comfort zone to attend the training. This training cannot be too technical but to demonstrate the "what" digitalization and cybersecurity can help. The objectives of the training are not to teach the board to ask technical questions, but to equip them with the knowledge to ask the right questions for the business and governance structure [25].

5.2 Incentive and Financial Supports

Although the success story of Organization A demonstrates that financial resources are not the key obstacles, it cannot be generalized. Other organizations would need that extra help to fund the data analytics and cybersecurity projects. The government can either help to fund the projects directly or to do so via certain self-help groups.

5.3 Peer Support to Keep Abreast

As both data analytics and cybersecurity are relatively new areas, there are many learning opportunities, and organizations need to learn through trial and error during implementations. A common and easily accessible platform where like-minded organizations can gather and exchange experiences would help to facilitate more organizations to launch data analytics and cybersecurity projects.

5.4   Training to Fill the Shortage of Talents

There is a severe shortage of talents in digitalization and cybersecurity. The issue is not unique to Singapore as it is estimated that the Asia Pacific region has a shortage of 2.15 million. The global shortage is estimated to be 3 million [13]. The Singapore government is working with academia and businesses to train its workforce to meet the demand. In this regard, academia can help to train students with the right aptitude and skills to meet the demand of the industries.

5.5   Reinforcement—Regulatory Inspection, Internal Audit, and External Audit

With all the incentives provided, such as training, financial assistance, and peer group supports, there will still be some board members who are not engaged to tap on the resources to lead their organizations in the right direction. Therefore, there is a need to reinforce the implementation through inspection or audit, especially for cybersecurity. As a start, organizations can use their internal auditors to review their strategies and highlight weaknesses they noted [25].

Without good cybersecurity, it is a matter of time that the organizations will be violating regulations, such as the Personal Data Protection Act in Singapore, or the General Data Protection Regulation if they are dealing with European customers. Therefore, auditors and inspectors must highlight emerging risks.

In the case of Organization B, the statutory auditor has rightly raised a management letter point to highlight the potential risk to the board members, This demonstrates that auditors, and regulatory inspectors, have a significant part to play in enforcing organizations to strengthen their cybersecurity capability.

For organizations that continuously ignore the auditors' recommendation to review their cybersecurity, there should be penalties to deter such behaviors.

6   Conclusion

This paper has started by sharing the global and Singapore trends in adopting data analytics and cybersecurity strategies. It then focuses on the situations in small organizations, where it has used two organizations based in Singapore as case study to draw insights. Based on the study, financial resources, while important, is not the most critical element for organizations to embrace digitalization and cybersecurity. Instead, the awareness and willingness at the top of the house to embrace changes are the keys to success. With this understanding, various stakeholders, including policymakers and academies, can play an important role to raise awareness, provide training, and enforce the implementations.

The effort to successfully increase the pace of digitalization and cybersecurity adoption among the small organizations requires a concerted effort of the entire communities, as nobody, including the government, can single-handedly do it successfully [13].

7 Limitations and Suggestions for Future Research

This paper is exploratory research for small organizations in Singapore. It is performed using two small organizations as case study. While the characteristics, such as financial resource levels and compositions of the board members, are different in the two organizations, future research can be performed more comprehensively using a bigger sample size or organizations with different characteristics. Studies can also be extended to include small organizations in other countries.

References

1. Al-Moshaigeh A, Dickins D, Higgs JL (2019) Cybersecurity risks and controls. Is the AICPA's SOC for cybersecurity a solution? The CPA J 2019
2. Al Zefeiti SM, Mohamad NA (2015) Methodological considerations in studying transformational leadership and its outcomes. Int J Eng Bus Manag 1–11
3. Alibaba Cloud (2018) Digital transformation for SMEs. Alibaba Cloud. https://file-intl.ali cdn.com/event/file/a38c87de-dac9-40d5-9f3c-fdeda2be8a15pdf?Expires=1557646750&OSS AccessKeyId=5JK9n2yWStiegAGj&Signature=Yb%2B3LL4C7M4tpGwUtnrBpK5cP% 2BU%3D. Accessed 11 May 2019
4. Azorin JF (2007) Mixed method in strategy research. Research methodology in strategy and management, vol 4
5. Bada M, Nurse JRC (2019) Developing cybersecurity education and awareness programmes for small- and medium-sized enterprises (SMEs). Inf Comput Secur 27(3):393–410
6. Basias NP (2018) Quantitative and qualitative research in business and technology: justifying a suitable research methodology. Rev Integr Bus Econ Res 7(1):91–105
7. Bhattacharjya A, Zhong X, Wang J, Li X (2019) Present scenario of IoT projects with security aspects focused. In: Farsi M, Daneshkhah A, Hosseinian-Far A, Jahankhani H (eds) Internet of things: digital twin technology, communications, computing, and smart cities. Cham, Switzerland, Springer Nature AG, pp 95–123
8. CNA (2020) Channel News Asia Banking related phishing scams spike more than 2,500% in the first half of 2020. https://www.channelnewsasia.com/news/singapore/online-scams-inc rease-police-crime-social-media-impersonation-13053822. Accessed 27 Aug 2020
9. CSA1 (2020) Cyber Security Authority of Singapore: Opening speech by Mr Heng Chee How Senior Minister of State for Defence at the second Cybersecurity Awards and Gala

dinner. https://www.csa.gov.sg/news/speeches/second-cybersecurity-awards-and-gala-dinner. Accessed 15 Aug 2020

10. CSA2 (2020) Cyber Security Authority of Singapore: Singapore Cyber Landscape 2019
11. CSA3 (2020) Remarks by Mr David Koh, Chief Executive, Cyber Security Agency of Singapore, at the United Nations Security Council Arria Formula Meeting 2020
12. CSA (2018) Cyber Security Authority of Singapore: Singapore Cyber Landscape 2017
13. CSA (2019) Opening speech by Mr Heng Chee How Senior Minister of State for Defence at the second Cybersecurity Awards and Gala Dinner, 8 Nov 2019. Accessed 16 Aug 2020
14. Costley C, Elliott G, Gibbs P (2010) Doing work-based research. SAGE Publication
15. DBS (2018) 5 ways data analytics can help SME. DBS Business Class, 11 Sept 2018. https://www.dbs.com.sg/sme/businessclass/articles/innovation-and-technology/data-analytics-can-help-SME. Accessed 30 Aug 2020
16. DOS, Department of Statistics, Singapore (2018) CEIC data. Retrieved from Singapore GDP per capita. https://www.ceicdata.com/en/indicator/singapore/gdp-per-capita
17. Dasgupta M (2015) Exploring the relevance of case study research. Vision 19(2):147–160
18. Gartner (2018) Winning in a world of digital dragons. Gartner Executive Programs, Stamford
19. Hair JF (2018) Marketing research in the 21st centrury: opportunities and challenges. Braz J Mark 17(5)
20. IBM (2020) Briefing paper: companies are expecting more from digital transformation—and gaining more, too. Harv Bus Rev
21. Jahankhani H, Al-Nemrat A, Hosseinian-Far A (2014) Cybercrime classification and characteristics. Cyber crime and cyber terrorism investigator's handbook. Syngress, New York, pp 149–164
22. Kerr C, Nixon A, Wild D (2010) Assessing and demonstrating data saturation in qualitative inquiry supporting patient reported outcomes research. Expert Rev Pharmacoeconomics Outcomes Res 10(3):269–281
23. LaValle S, Lesser E, Shockley R, Hopkins MS, Kruschwitz N (2011) Big data, analytics and the path from insights to value. MIT Sloan Manag Rev 52(2) Winter
24. Lambert Y (2020) Digital skills win in-house lawyers a seat at the table. Financ Times
25. Lanz J (2014) Cybersecurity governance: the role of the audit committee and the CPA. CPA J
26. Lee KY (2000) From the third world to first, the Singapore story 1959–1999. The Straits Times Press, Singapore
27. Lemieux VL, Gormly B, Rowledge L (2014) Meeting big data challenges with visual analytics. Rec Manag J 24(2)
28. Lung N (2018) Singapore launches digital government blueprint to support its Smart Nation vision, 6 June 2018. https://www.opengovasia.com/singapore-launches-digital-government-blueprint-to-support-its-smart-nation-vision/. Accessed 30 Aug 2020
29. Marrelli AF (2007) Collecting data through case studies. Perform Improv 46(7):39–44
30. Microsoft and ASME (2018) Singapore SMEs who embrace digital transformation expect to see average revenue gains of 26%. https://news.microsoft.com/en-sg/2018/10/23/singapore-smes-who-embrace-digital-transformation-expect-to-see-average-revenue-gains-of-26-asme-microsoft-study/. Accessed 13 Apr 2019
31. Moore C (2011) The path to business process transformation. KM World 20(5)
32. OECD (2017) Enhancing the contributions of SMEs in a global and digitalized economy. In: Meeting of the OECD Council at ministerial level, 7–8 June 2017. https://www.oecd.org/mcm/documents/C-MIN-2017-8-EN.pdf. Accessed 30 Aug 2020
33. Ong-Webb GA (2017) Commentary: to benefit Singaporeans, Smart Nation must leverage big data, overcome privacy issues, 11 Aug 2017. Channel New Asia. https://www.channelnewsasia.com/news/singapore/commentary-to-benefit-singaporeans-smart-nation-must-leverage-9114644. Accessed 13 Apr 2019
34. Oplatka I, Hemsley-Brown J (2004) The research on school marketing current issues and future directions. J Educ Adm 42(3):375–400
35. Raelin J (2015) Action modes of research. A guide to professional doctorates in business and management. SAGE publication

36. Ramchandani N (2017) Government making it easier for SMEs to adopt data analytics and AI. https://ie.enterprisesg.gov.sg/Media-Centre/News/2017/10/Govt-making-it-easier-for-SMEs-to-adopt-data-analytics-and-AI--Yaacob. Accessed 11 May 2019
37. ST (2018) SingHealth cyberattack: how it unfolded. https://graphics.straitstimes.com/STI/STI MEDIA/Interactives/2018/07/sg-cyber-breach/index.html. Assessed 23 Aug 2020
38. Shivkumar S (2019) All companies in Singapore must look to data to compete or risk becoming obselete. Singap Bus Rev. https://sbr.com.sg/information-technology/commentary/all-compan ies-in-singapore-must-look-data-compete-or-risk-becoming. Accessed 19 May 2019
39. Sia NC (2018) The challenges for a small not-for-profit organization to embark on data analytics strategy. Amity Bus J 5(1):18
40. Tan WK (2016) Big data: Singapore's new economic resource. https://www.enterpriseinnov ation.net/article/big-data-singapores-new-economic-resource-847989586. Accessed 13 Apr 2019
41. Tellis WM (1997) Introduction to case study. Qual Rep 3(2)
42. Teo SL (2013) Welcome address by Mr Teo Ser Luck, Minister of State for Trade and Industry at the SME Talent Programme. Partnership ceremony between institute of higher learning and trade associations and chambers: https://www.mti.gov.sg/Newsroom/Speeches/2013/06/Mr-Teo-Ser-Luck-at-the-SME-Talent-Programme--Partnership-Ceremony-between-IHLs-and-TACs. Accessed 30 Aug 2020
43. Yin R (1984) Case study research—design and methods. Sage Publications, Beverly Hills
44. Yin R (2003) Case study research: design and methods, 3rd edn. Sage, London

Cloud and Its Security Impacts on Managing a Workforce Remotely: A Reflection to Cover Remote Working Challenges

Usman Javed Butt, William Richardson, Athar Nouman, Haiiel-Marie Agbo, Caleb Eghan, and Faisal Hashmi

Abstract Attacks against remote workers who are working from home due to the global pandemic has significantly increased. Cyber criminals have realised this and are exploiting users for financial gain and for espionage motives. Criminals are aiming to exploit vulnerable smart homes through Internet of Things and leverage access into corporate networks. This means that home users need to be extra vigilant against this contemporary technique. This paper will address these challenges with robust protocols for organisations to absorb, train, and implement. Additionally, this paper will align organisations expectations with user vulnerabilities to increase organisation resilience.

Keywords Home networks · Covid-19 · Cyberspace · Cyber criminals · Remote workers · Cyberattacks

U. J. Butt (✉) · W. Richardson · A. Nouman · H.-M. Agbo · C. Eghan · F. Hashmi
The Department of Computer and Information Sciences, Northumbria University, London, UK
e-mail: usman.butt@northumbria.ac.uk

W. Richardson
e-mail: william.j.richardson@northumbria.ac.uk

A. Nouman
e-mail: athar.nouman@northumbria.ac.uk

H.-M. Agbo
e-mail: haiiel-marie.agbo@northumbria.ac.uk

C. Eghan
e-mail: caleb.eghan@northumbria.ac.uk

F. Hashmi
e-mail: Faisal.hashmi@northumbria.ac.uk

© The Author(s), under exclusive license to Springer Nature Switzerland AG 2021   285
H. Jahankhani et al. (eds.), *Cybersecurity, Privacy and Freedom Protection in the Connected World*, Advanced Sciences and Technologies for Security Applications, https://doi.org/10.1007/978-3-030-68534-8_18

1 Introduction

Since the start of 2020 there has been a significant increase in remote workers working from home (WFH). This is due to the unforeseen global pandemic which has impacted all lives and the detrimental economic effects it has rendered. The precautions governments have taken are to minimise movements as much as possible, this means that organisations have mitigated human contact by ordering employees to WFH. This has created major challenges for organisations when leading geographically dispersed teams, these teams are expected to operate as they would in their normal workplace; however, WFH is a different environment and in this relaxed environment generates complacency. The problem with complacency when WFH is that the impact is much more serious, cyber criminals realise this and perceive remote workers as an easy target to attack to then pivot into corporate networks. Cyber criminal's incentive is to attack remote workers to gain access into corporate networks for financial gain. Similarly, some criminals are targeting remote workers to gain access and gather information on how organisations, such as national health services and governments, are handling COVID-19. The motives for cyber criminals attacking remote workers are therefore financial gain and espionage.

This paper will critically appraise the current threat landscape with contemporary data to support concluding remarks. The paper will critically analyse countries, and in particular the United Kingdom's (UK) approach to remote working and how they are currently dealing with challenges. Furthermore, the paper will review specific cyber criminals along with their motives, tools, and methods they are using against remote workers. Finally, the paper will analyse the data and construct secure, concise protocols that industries and organisations should utilise to protect remote workers and corporate networks from cyberattacks.

2 Critical Review of a Home Area Network

A Home Area Network (HAN) is a group of digital devices such as smartphones, tablets, and multimedia systems that are connected by an internet connection within a household, the HAN is a network where other devices can communicate and share files with each other [1]. The HAN can then access the internet from one centralised point, this is through the default gateway of a router. This means, that, all Internet traffic using IPv4 protocol and IPv6 protocol goes through the router [2]. The issue with this is that the traffic is not encrypted unless a Virtual Private Network (VPN) is being used. Similarly, the router comes with default credentials which could be exposed on the internet or they could be known to other entities. Securing the router with a VPN and configuring new credentials is detrimental to defending the HAN.

The importance of securing a home network is becoming more significant due to the ongoing pandemic. COVID-19 has rendered more employees to work from home, this means that employees need to be more security-aware against the possible

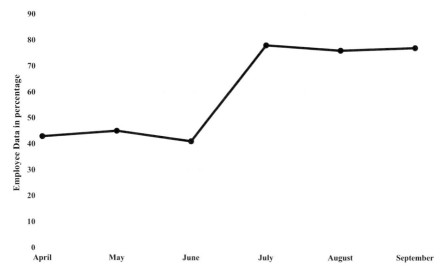

Fig. 1 Data from April to September 2020 showing UK employees working from home [4–9]

cyberattacks they could face. Figure 1 shows the profound increase in remote working employees in the United Kingdom (UK) from April to September 2020. The latest data shows that 77% of employees are working from home due to the pandemic. This is a colossal increase from 2019 when only 30% of employees remotely worked [3].

2.1 Types of Home Area Networks

There are two types of home networks, they are generally: wired and wireless, and they both use the same concept of connecting digital devices within the HAN and out to the Internet. The more common HAN is wireless since devices such as smartphones, and other wireless devices, come without an ethernet connection [10].

Wired Network: The wired network is a network with only wired connections using a networking cable such as an ethernet cable, and that is connected to the internet [1]. This connection gives users a faster internet connection and adds an element of security because of its wired infrastructure. The reason it offers more security is that a wired connection is not vulnerable to wireless attacks, a wired connection could only be accessed by a method called wiretapping, which would need physical contact with cables attached to the home network [11].

Wireless Network: The wireless network is what most digital devices can connect to the home router using radio frequencies as the communication medium. This connection is specified by IEEE 802.11 protocol standards, the connection typically uses either 2.4 gigahertz or 5 gigahertz [12]. A wireless connection is the most popular method of connecting to the internet since it is easier to connect, more flexible, cost-effective and most devices do not offer ethernet accessibility.

2.2 Vulnerabilities in Home Area Networks

The reason why a home HAN is vulnerable is because of the vast amount of Internet of Things (IoT) combined within a network. According to [10] the average family in the digital age uses ten IoT devices, this figure increases to fifteen with a family of five. The issue with this is that the more IoT connected to the HAN the more opportunities this renders to an attacker targeting the network. This then produces more access points to penetrate the HAN and thus maliciously access the content [13]. Once accessed, an attacker would have a gateway into the network and access information within. Subsequently, if the network were connected to a corporate network, they could then access this through the initial breach into the HAN. The National Cyber Security Centre (NCSC) [14] claims that remote workers attract significant amounts of risks, and these risks bring increased threats to the corporate network. Implementing the least privilege principle is imperative to restricting what employees can do and therefore minimising what attackers could do if their accounts were to be compromised.

Researcher [15] proved how vulnerable the home network really is, by hacking into his own network in twenty minutes. He was able to access his HAN through IoT devices such as smart televisions and network devices and elevate himself to administrative privileges. The reason he achieved this was his devices did not have automated software updates, therefore the software was vulnerable to attacks. He achieved this by targeting his network-attached storage (NAS) and used this as a backdoor into the network. He was then able to pivot through other vulnerable devices and escalate privileges throughout his network.

Similarly, [16] was able to circumvent the HAN firewall and release malware from a smartphone. They then scanned the HAN for vulnerable IoT devices to target before preparing an external attack. This supports previous evidence that IoT devices are largely vulnerable to cyberattacks and these two examples show how vulnerable a HAN can be.

Figure 2 shows data collected by [17] from sixteen million global households, the figures are shown in percentages. The most common vulnerability was 69.2% of household's routers had weak credentials. This means that these homes were especially vulnerable to credential guessing, dictionary attacks, and brute-force attacks. Further, the data shows that 59.7% of households had vulnerable routers, with attackers then being able to access the router and change the credentials. This would then present the opportunity into the corporate network from the initial foothold into the HAN. 59.1% of users had never accessed their routers or updated the software, this would leave the households vulnerable to attacks because they would be without

Fig. 2 Data from sixteen million homes showing security vulnerabilities [17]

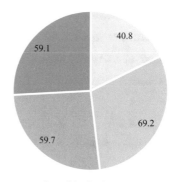

At least one vulnerable device to cyberattacks

⬜ Access by weak credentials

⬜ Vulnerable access through routers

⬜ Routers that have never been updated or accessed by user

the latest patch. Finally, 40.8% of households were vulnerable to at least one cyber-attacks. These high statistics show how vulnerable the HAN is, and emphasising this to remote workers is imperative to mitigating the problem.

A HAN can be accessed by any device that can access its perimeter through the Internet [18]. The main access point for internal and external traffic is the router; these stores imperative information such as Internet Protocol (IP) addresses, routing information, and Media Access Control (MAC) addresses [1]. Most internet providers now use Wi-Fi Protected Access 2 (WPA2) protocol in their routers, which HAS superseded the WPA and Wired Equivalent Privacy (WEP) protocol [19]. WPA2 offers more security by using the Counter Mode Cipher Block Chaining Message Authentication Code Protocol (CCMP) and Advance Encryption Standard (AES). As a result of CCMP and AES this ensures confidentiality and integrity; however, [20] reports of a vulnerability to Denial of Service (DoS) attacks and thus compromising the availability. A further problem with home routers is that reports have shown the software embedded is often five to six years older than the current models [21]. Users who are complacent or generally not security-aware could render their home networks vulnerable because of this issue. The annual threat report by [22] supports this with their research on routers: D-Link, Linksys, Netcore, and Shenzhen. These types of routers were significantly targeted between March and June 2020 using malware against remote workers. Weekly attacks of 160,000 suggest the increased activity was against remote workers intending to pivot into the corporate networks or use them for mass botnets.

Figure 3 illustrates data collected by [17] on the most popular devices used in smart homes and the most vulnerable to cyberattacks. The graph shows that the most used and vulnerable electronic device is network nodes. These include personal computers, laptops, routers, smartphone, and servers. All devices on the graph were shown to be vulnerable to cyberattacks apart from audio equipment and voice assistant devices. The graph shows the data in percentage of millions of people across western Europe.

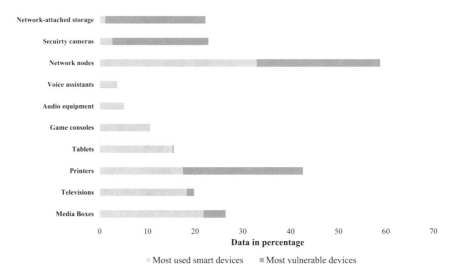

Fig. 3 Data showing the most used and vulnerable IoT devices in western European households

The problem with IoT is that they do not come with secure credentials and most users disregard implementing them. This causes an access point to be insecure and thus an avenue into the home network. Employees must consider this when connected to the internet and especially when connecting to their corporate network.

It is not only hardware, which is vulnerable within a home network, the user, if unaware of the threat, is also vulnerable. This means that users are vulnerable to social engineering through phishing and malicious hardware such as Universal Serial Bus (USB).

Social engineering is a concept of attackers tricking users into maliciously granting them access into the users' account. This then leverages access into the remote workers' corporate network, where the attacker could pivot from home to corporate within cyberspace [23] have predicted that home networks will be targeted by sophisticated attacks using artificial intelligence (AI). The attacks are called DeepAttacks initiated using polymorphic and metamorphic malware which evades firewalls by changing their code and appearance [23] further report that AI will use fake URL and HTML phishing sites orchestrated by vast amounts of bots to maliciously target home networks. The incentive will most defiantly be to gain a foothold into corporate networks and escalate privileges.

3 Cyber Threat and Their Motives Against Remote Workforce

Cyber threat actors are increasingly targeting remote workers because of the significant number of individuals who are working from home, and that are vulnerable to attacks. A report by [24] cites McAfee that a large number of users are being targeted through the Remote Desktop Protocol (RDP). RDP is a network protocol that is exposed to the Internet [25] and this is vital when connecting to corporate networks; however, criminals have realised this and are now targeting RDP to leverage access into the corporate network themselves. In mid-2020, the National Cyber Security Centre (NCSC) identified an increase of 127% in exposed endpoints. This means that an attacker could use RDP and execute remote code so that they could leverage access into that machine, or worse, the corporate network. This is how the worm 'BlueKeep' was used to compromise RDP and takeover Microsoft machines using a full graphical user interface [26].

Advanced Persistent Threats (APT) and Criminals. APT and criminals hold the same incentive, to steal data for financial gain or steal industry secrets for future production purposes. Since the start of 2020, these two types of attackers have been at the heart of most attacks and they continue to exploit remote workers using the pandemic as a malicious deception plan.

In the UK, the NCSC [27] thwarted 2000 online scams related to COVID-19, 555 malware sites and 200 phishing sites. This was at the height of the pandemic when high numbers of employees were working from home. Furthermore, the NCSC and the United States of America's (USA) Cybersecurity and Infrastructure Agency warned of increasing attacks from advanced persistent threat (APT) groups using malware and ransomware attacks against home workers. [28] cited sources on a rise in cloud-based attacks which rose by 630% between January to April 2020. Further, phishing scams rose by 600%, and ransomware rose by 148%. It is clear from the data that cybercriminals abused the pandemic as a malicious social engineering technique.

According to [22] annual threat report, attackers are using a browser injection method to target home workers who are browsing the web. The malware is said to infect the user when unknowingly complacent workers are browsing the Internet from home, subsequently covertly bypassing security to access corporate networks. The attacks increased at the height of the pandemic from March 2020 onwards. APT groups: APT36 from Pakistan and an APT linked with China used spear phishing to install malware to capture machine information using screen capture tools. It was also reported that these COVID-19 scams were able to install remote-access Trojans to access these machines and most likely use at a later date [29]. This data shows the level of influence and leverage COVID-19 had to be maliciously used against remote workers. The Foreign Secretary Dominic Raab [30] warned of cyber actors such as criminals and APT groups targeting individuals using phishing techniques to scam remote workers. The NCSC and CISA further warned against this threat which could be a technique to leverage access into the National Health Service (NHS) database.

Their incentive is most likely traversing the network and searching for integral information on the UK's approach to COVID-19. Moreover, the NCSC suggests that the criminals behind the COVID-19 phishing scams will be targeting remote workers for financial gain [31]. The NCSC further report of advanced phishing methods using World Health Organisation (WHO) Director-General Dr Tedros Adhanom Ghebreyesus as the originator. The emails were observed by the NCSC with the finding to be of a malware called Agent Tesla. Using this spear-phishing technique, attackers can then access PII and steal sensitive information to further sell on the Dark Web [32].

In addition to phishing techniques, the NCSC report of an increase is user credential exploit. Attackers used password spraying techniques to attempt to compromise NHS workers credentials and use these to access the corporate database [31]. It is likely that if successful, the attackers would be looking to steal personal identifiable information (PII) to sell on the Dark Web for financial gain. It is not known what type of attackers these are; however, criminals or APT are most likely since their motive is financial. Moreover, since remote workers have been utilising communication platforms such as Zoom and Microsoft Teams, attackers have been planting phishing emails with embedded malware. The NCSC and CISA report that attackers can then hijack the software, eavesdrop on the conversation or takeover the conference calls [31]. Reports suggest that attackers were using this vulnerability and falsely impersonating other persons to access these online conferences [33].

Nation-State. These are highly skilled professionals who act upon direction from a country whose incentive is to disrupt or harm a foreign entity [34]. In the second quarter of 2020, the Australian Government reported an increase of 67 million cyberattacks compared with the prior quarterly report [35]. The Australian government further report in their cyber strategy report that these attacks, along with others, are originating from nation-states. It is not reported which nation-states are behind these attacks; however, reports suggest that the attacks were attempts on customer data [35]. The USA financial department also saw a steep rise, cybercrimes rose by 238% between February and April 2020 [36]. Again, the report suggests the attacks are from nation-state actors and criminals who exploited databases using COVID-19 phishing scams. The report concludes that the USA was inadequately prepared due to the Trump administration disbanding various top cyber positions, and thus not having the correct professionals in-place.

Figure 4 shows data from [37] on nation-state actors, criminal, and unknown attackers. The data suggest, that since the start of the pandemic, criminals with financial motive have dominated the cyberattacks; however, nation-state attacks still stand with 13% of the attacks meaning organisations must still stay vigilant to the nation-state threat.

Insider. An Insider threat is a person who is a current or former employee, or someone that has access to an organisation and poses a significant malicious threat [38]. Moreover, an insider can be someone who accidentally abuses their power within the network, more commonly known as a negligent insider [38]. An example of a negligent insider can be an employee who has been granted too high privileges and accidentally deletes critical data. A report by Fortinet [38] state that 54% of

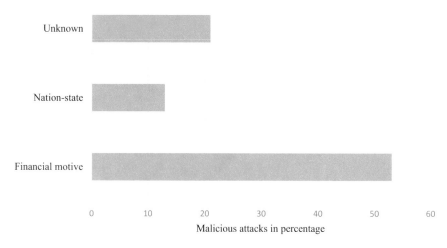

Fig. 4 Global data on malicious attacks by threat actors [37]

negligent insider attacks are believed to be from lack of training. Increasing information security through governance and practical training should be an organisation objective.

Generally, there are two types of incentives that malicious insiders have, these can be categorised into:

1. **Digital**: A digital threat is an insider whose incentive would be to maliciously access the network and extract sensitive information, mass data, and traversing the network to access information not associated to that person [39].
2. **Behavioural**: A behavioural threat is an insider who attempts to circumvent network security to violate the system to disrupt or harm business operations. Warning signs would be a known disgruntled employee whose behaviour is unusual, especially who is active during non-working hours [39].

The report by [38] shows the most likely motivations from a malicious insider perspective. The data was collected during 2019 from 400,000 professionals specifically on the insider threat. Figure 5 shows that fraud was the main factor for an insider, the criminal incentive is to leverage fraud and use this for financial gain. Additionally, monetary gain holds the same meaning, with the underlying incentive being financial gain. On the contrary, sabotage was at 43% which, most likely originates from malicious employees whose aim is to cause harm to a specific organisation. Figure 5 shows a full list of an attacker's motives based on 400,000 individual's opinions.

The increase of the aforementioned cyber threats indicates a need for a strong cyber hygiene culture individually and more importantly in corporate environment. The next section discusses cyber hygiene and the best practices to reduce the chances of cyber-attacks.

Fig. 5 Malicious insider
motives against
organisations [38]

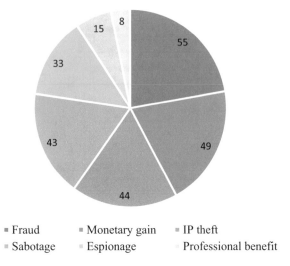

▪ Fraud   ▪ Monetary gain   ▪ IP theft
  ▪ Sabotage   ▪ Espionage   Professional benefit

4 Critical Analysis of Individual Cyber Hygiene

Cyber Hygiene is not much dissimilar to personal Hygiene. Often these two concepts are compared closely with each other. According to [40] personal hygiene refers to maintaining health to prevent diseases. There are several factors involved in preventing the risk of spread, for example, environment, hand hygiene, sterilisation of equipment, disposal of waste etc. Similarly, Cyber Hygiene is related to the safety of data. The factors such as maintaining functional devices and updating necessary protective software are considered to prevent the risk of data loss or corruption. Like personal hygiene, cyber hygiene needs to be adopted as a regular practice.

Cyberspace is often interpreted as the internet (a network of networks). According to the cybersecurity intelligence (2020) Cyberspace is regarded as managing physical space, however, with the advent of the internet, the term is more related to symbolic and figurative space that exists within the scope of the internet. Words like Cyber-security, cyber-crime, cyber-war, cyber hygiene etc. are derived from it. These all are preventive measures to keep the data safe either stored or in transmission, from external or internal cyber-attacks.

The frequency of Cyber-attacks is rapidly increasing in all sectors and an increase in demand for the improvement of cyber hygiene is required [41]. According to the European Union Agency for Network and Information Security [42], cyber-attacks are increasing in the corporate world. The UK government report indicated that around 80% of cyber-attacks are because of the poor cyber hygiene within the victim organisation, in other words, the individual system user. A survey conducted by the UK government agency, [43] also indicated that almost half of the business (46%) reported having cybersecurity breaches or attacks in the last 12 months. In a report from [44], it was estimated that the total number of connected devices by the end of 2020 will be around 200 billion. This estimate was collected with the help of

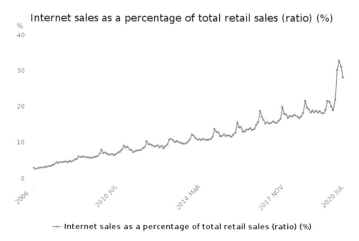

Fig. 6 Rate of internet sales by office for national statistic

Symantec security centre based on year on year trends, however, this did not include the COVID-19 pandemic factor.

Based on various research, it was identified that the online activity across the UK has increase by around 129%. This increase has resulted in the growth of the internet retail sales market. According to the Office for National Statistic the percentage of internet retail sales from total sales pre Covid-19 was 19% (February 2020), but this spiked to 32.8% (May 2020) post COVID-19 (illustrated in Fig. 6). According to the same report, the annual increase in online sales was approximately 2%. While post COVID-19 there was an increase of approximately 19%. Similarly, in April 2020, 46.6% of staff work from home or remote locations. Since Covid-19 many organisations were forced to start remote working at a cost of their survival. Hence, the introduction of emergency systems to operate online have introduced poor cyber hygiene practices. Newly joined users in this domain, with no or very little knowledge, have increased the chance for cyber-attack.

In recent history, banks were facing serious cybersecurity issues due to their desire for digital transformation. Banks in the UK launched many campaigns to update security procedures and educate individual users for better cyber hygiene on government instruction to safeguard the public interest. Attackers can easily sweet talk individuals into either transferring money or give away their access details. Except for private organisations, some governments initiated different activities like a review of cyber hygiene practices and preventive measures by the European Union Agency for Network and Information Security [45]. The report concluded that there is no one solution that fits all scenarios and this is because of other variables in mind such as local infrastructure, culture etc. Hence different hygiene solutions were proposed for organisations in different countries to protect themselves.

In February 2013, the National Institute of Standards and Technology (NIST) was directed to work with stakeholders to develop a framework for reducing cyber risk on

US Presidential order. The first version of Cybersecurity frameworks was released a year later in February 2014, later modified in 2018. The new framework consists of three main components i.e. Core, tiers, and profiles. These frameworks and security measures prevent form problems like Data loss, security breach, misplaced data etc.

For building the best practices culture of strong cyber hygiene following are some common measures discussed in different literature. These hygiene practices are guidelines, they define what an individual should do and do not define what they must do.

I.   Update system software and hardware

Post 2010, due to the increase in cyber-attacks and systems getting more integrated the software & hardware manufacture diverted to focus on security along with the maximising the performance. Continuous software updates are released, and better secure hardware equipment are made available. These updates do not necessarily mean that you are entirely safe from attacks. The hackers are crafty and find loopholes or backdoors to infiltrate. These updates also come with a cost. This might be affordable by organisations but for an individual, this might be challenging. In some cases, software updates are not free, also continuous update needs more memory space. Similarly, replacing or updating hardware is also costly and hence mostly this is not done by individuals. However, this does not mean that one should stop applying available or possible updates.

II.   Sending and receiving emails

Emails are one of the biggest threats in cybersecurity. These are the source of importing malicious applications programs (malware) into the system and individuals must understand and review what drops in their inbox. However, attackers are nifty and structure emails or phishing websites that look legitimate. An individual can further secure email using two-factor authentications or by email encryptions to secure leak of information.

III.   Strong access policies

One of the main requirements of remote working is access to organisational systems and data provided through authenticated access. These credentials can be used on a personal device over a home network or sometimes on a public network. In these cases where either an individual is sharing a network or device, the risk of cyber-attack is greatly increased. Use of preventive measures like VPN on a public network would make it difficult for attackers and facilitate tracking. One very important security measure recommended is to maintain Cyber Hygiene is "Strong Password". Although the strong password policy had made passwords difficult to break but easy to discover. To reduce the risk of forgetting password the user normally writes the password down in a location where they can find it easily. Mostly it is either on a desktop file or one a mobile phone. A quick sniff through recent document can easily reveal the login details. Also, some users will save the credentials in browsers. This could be easily exported and accessed or even used in case of shared devices.

IV. **Cyber hygiene plans**

Cyber hygiene practices should be part of a routine, a habit to ensure the safety of data. Hence it is important to draw up plans to prevent, identify, respond, and recover. These plans should be revisited on regular intervals or in case of any system changes. For example, an individual might consider changing passwords to prevent malicious activities and unauthorised access especially default password on applications and hardware devices such as routers provided by ISP's. Performing software updates should also be in the review list. For more advance and secure environment, an individual or organisation may want to implement security framework like NIST framework.

5 Critical Review of Cloud Security and Cyberattacks Against Remote Workforce

During the COVID-19 pandemic, homeworking was used as a mean to contain and slow down the propagation of the virus. In April 2020, 46.6% of people in employment worked from home [45] and 56% of employees used their personal computers to work remotely [46]. This abrupt change resulted in companies deploying digital tools and processes to make remote work fast and reliable without reducing employee's self-management abilities.

Among the tools deployed by companies to facilitate remote working, cloud-based collaboration tools (Google Suite, Lotus, Slack, Microsoft Office 365 etc.…) were mostly used. Cloud-based collaboration tools provide advantages such as simple deployment and maintenance, easy accessibility, storage and sharing capabilities, increased mobility, real-time communication, secure data transfer and management etc.…

After the fast deployment of cloud-based collaboration tools, an increase of cyber-attacks on cloud-based systems was recorded. In 2020, up to 24% of data breaches were related to cloud assets and 73% involved email or web application servers [47].

This section reviews the main layers of cloud computing, explains the various surfaces used by attackers to perform attacks and describes the Man-In-the-Middle (MiTM) and the Distributed-Denial-of-Service (DDoS) attack.

5.1 Cloud Computing Layers

Cloud computing provides services by relying on three main models: Infrastructure as a Service (IaaS), Platform as a Service (PaaS) and Software as a Service (SaaS). Each layer offers specific services and can be accessed using various methods. Figure 7 gives an overview of the different layers and how users interact with them.

Fig. 7 Cloud layers

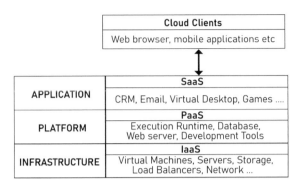

Infrastructure as a Service (IaaS) provides users with computational resources such as virtual machines, file-based storage, firewalls, IP addresses, Virtual Local Area Networks (VLANs), software bundles and other resources [48]. On a practical level, with IaaS users have control over operating systems and have a wide range of possibilities available. Well-known IaaS services are Amazon Web Service EC2 and S3 [50].

Platform as a Service (PaaS) provides users with "a computing platform typically including operating system, programming language execution environment, database and web server" [48]. With PaaS, users can develop and deploy applications in a controlled environment and reduce the constraints linked to buying and managing software and hardware devices. However, users have no control over other resources such as the network, servers, or storage. Well-known services based on PaaS model are Google App Engine and Microsoft Azure [50].

Software as a Service (SaaS) provides users with the ability to use applications running on the cloud. The applications are installed and maintained by the cloud providers and clients can access them mainly through web browsers however, some of those applications require the user to download and install the software. In this scenario, users do not have control over the cloud infrastructure and are limited to the application.

Most of the layers and their applications can be accessed by clients using web browsers and do not require any preinstalled software to be utilized.

5.2 Attack Surfaces

Attack surfaces refer to a set of sections during a user's interaction with the cloud. During a typical use of the cloud, three main parties interact: the service user, the service, and the cloud provider. These three element's interaction results in the following surfaces: *Service-to-User* surface, *User-to-Service* surface, *Cloud-to-Service* surface, *Service-to-Cloud* surface, *Cloud-to-User* surface, and *User-to-Cloud* surface. Figure 8 gives a detailed description of the interaction process between the three parties.

The most vulnerable surfaces are the *Service-to-User* and *User-to-Service* surfaces. Since the interaction between these parties takes place through a simple server-to-client interface, it allows attackers to execute a wide range of attacks that are not cloud-specific. On the *Service-to-User* and *User-to-Service* surface, users access services through web-browsers hence attacks possible in this case can range from browser-based attacks to more traditional attacks. Some of the popular browser-based attacks used in this case are SSL certificate spoofing, Phishing attacks, Cache Poisoning, Cross-user Defacement and more traditional attacks performed are buffer overflow, SQL injection, privilege escalation, spoofing, MiTM etc.

Cloud-to-Service surface and *Service-to-Cloud* surface refer to the interface used by the service and the cloud system to interact. Although it is hard to differentiate both instances (cloud service and cloud system), their interaction generates vulnerabilities that can be used to perform attacks. On the *Cloud-to-Service* surface, attacks revolve on ways the cloud system can impact its services and with administrator privileges, the damages from attacks on this surface can be irreversible. Attacks on this surface can impact service availability, data privacy, data integrity and much more. On the *Service-to-Cloud* surface, the impact is the same, the attacker tries to use the service instance to impact the server, attacks usually involve DDoS and attacks on cloud system hypervisor. To perform a DDoS attack on this surface, attackers use the service to request more processing resources to create resource exhaustion.

During a standard interaction between a user and the cloud system, there usually is a service in the middle to give access to the cloud. In this case, the *Cloud-to-User* and *User-to-Cloud* surface refer to the interface provided by the cloud system to users to control services, manage settings, add/remove services and perform other changes. Attacks on these surfaces involve phishing attacks and other user-directed attacks, the aim is often to manipulate users into giving access to attackers by mimicking cloud providers or other standard phishing methods.

Fig. 8 Users interaction with the cloud

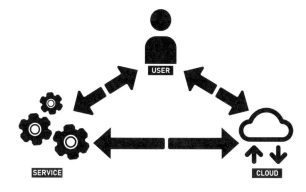

5.3   Distributed-Denial-of-Service (DDoS)

"A Distributed-Denial-of-Service is an Internet-based assault that is delivered from multiple sources to one destination" [50]. A successful DDoS attack render networks inaccessible, stops network devices functions and sometimes impact Internet Service Providers (ISP) resulting in a loss of service on the client's end [50]. In a typical DDoS attack, they are four standard components: an attacker, master hosts, zombie hosts and a victim host [51]. In this scenario, the attacker is the individual performing the attack, the master hosts are computers and IT devices running the master program, the zombie hosts are computers running the agent program (bots) and the victim host is the target of the DDoS attack. When the attack is initiated, master programs send commands to the agent programs through a client/server medium. Using this method, more than thousands of agent programs can be initiated and perform synchronised attacks on the target. Figure 9 illustrates a typical DDoS attack on the cloud.

In this scenario, an attacker uses VM machines to create a Denial-of-Service. Using various bots and compromised machines over the internet, the attack is initiated, and bots send a lot of fake traffic to the target. However, cloud services function such as when VM machines are overloaded (requires more resources), the cloud system goes into an auto-scaling mode and assigns more CPU (resources) to the VM. Except for the auto-scaling method, the cloud system has several methods to deal with overloaded VM. VM can either have more resources assigned or be migrated to higher resource capacity servers and much more. In case there is no control system in place, resources keep on being allocated to the VM. This process empowers the DDoS. Ultimately, the cloud system runs out of resources (resource exhaustion) and result in a Service Denial [52].

A more advanced and reliable method to perform DDoS is using BotClouds. With Botnets being hard to gather and maintain, the BotCloud method resolves this issue by providing bots that are always available and completely under the attackers' control. Botnets are easy to set and attackers can legally purchase a large group of

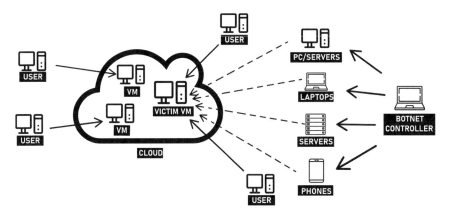

Fig. 9   DDoS attack in cloud

Table 1 Direct and indirect effects on DDoS attack on the cloud [52]

Direct effects	Indirect effects
Service downtime	Attack mitigation costs
Economic losses due to unavailable services	Energy consumption costs
Business and revenue losses	Reputation and brand image loss
Dependent services downtime	Collateral damages in cloud
Scaling driven economic losses	Smoke-screening attacks

machines from Cloud Service Providers (CSP) and install bots on each machine to form a cloud-based botnet (BotCloud) [53].

A successful DDoS can have serious impacts on CSP (Cloud Service Provider), users, businesses, and other instances. The effects can be classified into two categories: direct effects and indirect effects. Table 1 shows the various impact DDoS can have at various levels.

5.4 Man-in-the-Middle Attack (MiTM)

"A man-in-the-middle attack refers to an attack where the attacker goes between communicating devices and snoops the traffic between them" [54]. In this attack, the attacker connects to both devices and relay traffic between them giving an impression of communication going normally while it goes through a third-party device (the attacker's system) eavesdropping on the communication. Once the attack is successful, the attacker "can either listen to the conversation and or pose as one of the recipients of the communication" [55]. Furthermore, the attacker can intercept and modify legitimate communication between communicating devices [56].

For a Man-in-The-Middle attack to be successful the first requirement is to be able to intercept and relay traffic between devices. The challenge is to be able to insert an illegitimate user into the data stream without being detected. For that, the attacker needs to convince both devices that his machine is their intended recipient. Attacks like session hijacking or ARP poisoning can sometimes be used to create trust on both sides of the communication.

During a typical interaction between a user and the cloud, the user is connected to a wireless router or an access point either through the Wi-Fi using devices such as tablets, laptop or smartphone or through a wired connection on a laptop or desktop. The router receives data from the cloud and transmits it back to the user. Figure 10 gives an overview of the communication process between remote workers and the cloud.

As previously mentioned, users access cloud services using web-browsers. Using various tools, it is possible to perform a Man-in-The-Middle attack and monitor

Fig. 10 Communication
between cloud and remote
workers

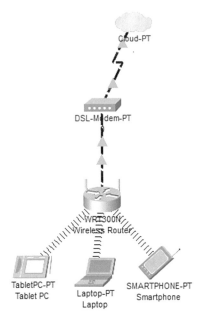

incoming traffic from a user. In this scenario, the user uses a Windows 10 machine
and the attacker uses Kali Linux to perform the attack. The following tools are used:
Nmap, Ettercap-graphical and TCPDUMP. Ettercap is a preinstalled tool in Kali
Linux used to perform various versions of MiTM attack and Nmap is a network
discovery and security auditing tool used for scanning.

- The first step is the reconnaissance phase. It is critical to gather information on
  the victim such as IP address, open ports, services running on the machine etc.
  Using Nmap with the *-sP* attributes, the attacker scans a range of IP address to
  identify potential victims and type of services running on the system using the
  following command "*nmap -sP 192.168.1.0/24*".
- The next step is to use Ettercap to first perform an ARP poisoning attack, that will
  give Man-in-The-Middle access. This step is important since it allows the attacker
  to corrupt ARP tables of each instance (router and target). To do that, the first step
  is to start *Sniffing* process in Ettercap, next set *Targets*: *Target 1* as the network's
  default gateway (*192.168.1.0*) and *Target 2* as the IP address of the previously
  identified target (IP address of Target) and then launch *ARP Poisoning*.
- Finally, while the ARP poisoning is running, use a traffic sniffing tool like *Wire-
  shark*, *TCPDUMP* and others. The traffic sniffing tool can capture traffic coming
  from the target.

Using this process, attackers can gather targets' critical information, gain unauthorised access and/or cause irreversible damages. It is then critical to implement
processes and methods to improve users and cloud security. Various means are
employed by Internet Service Providers, Cloud Service Providers, and IT experts

to counter those attacks. The next section not only discusses efficient countermeasures to deploy to counter DDoS and MiTM attack but also provide guidance on general security measures users should practice and be aware of.

6 Recommendations to Counter Malicious Cyberattacks

The recent rise in the adaptation to a remote working environment poses a potential threat to cloud services and data security. It is needful that employees and business owners adhere to precautions and safety mechanisms that are either ethical or technical. These procedures are termed as countermeasures or controls towards potentials cyber-attacks. This section will address potential countermeasures towards DDOS attack and MITM attack as well as other relevant security controls that need to be considered within a remote working environment.

6.1 Mitigation Against DDoS Attack

In a cloud environment where resources are shared by many users, there is a great possibility for a DDOS attack to occur where legitimate users are denied access. The attack scenario demonstrated in the previous section causes service unavailability by exploring possible vulnerabilities within the given network as well as its security framework. The purpose of this mitigation is to ensure that network communication is not blocked, or legitimate users are not denied access to services. One of the key challenges in dealing with a DDOS attack is to differentiate between a request from an attacker and that of a legitimate user, though the former is derived from a large number of distributed machines [57]. A targeted user can minimize the incoming threat by using specially built network equipment or cloud-based security service. The mitigation procedure will be considered as routing (traffic routing across multiple data centres), detection (detection of attack signature as it occurs), response (filter malicious traffic at the network edge), and adaptation (adaptation of machine learning to attack patterns) [58, 59]. However, the mitigation procedure in this section will emphasis on filtering malicious network traffic from overwhelming the main server.

6.1.1 Deploying Web Application Firewall

By intelligently dropping malicious bot traffic and absorbing the rest of the traffic, the DDoS defence network responds to an incoming established threat. The deployment of a Web Application Firewall (WAF) on layer 7 (application layer) attacks, protects web applications by filtering and monitoring HTTP requests between web applications and the internet. A DDOS attack on layer 7 could be targeted toward the CPU intensive or the business logic code of an application which involves making

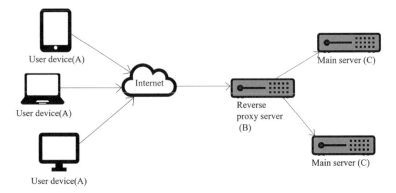

Fig. 11 Reverse proxy server workflow

requests of large files or adjusting file sizes. The enforcement of the latency trigger on each object to intercept traffic to the compromised object while transmitting traffic to objects that do not experience response-time degradation [60].

WAF serves as a shield between the web application and the internet. WAF operates as a reverse proxy. A reverse proxy intercepts a request made by the client to a server. The interception takes place at the network edge where a request is sent to the main server by the reverse proxy for a response. In this scenario, a request from the client is been monitored and reviewed in line with the policies that make up the architecture of the reverse proxy [61].

A typical reverse proxy server workflow is illustrated in Fig. 11. In the scenario illustrated, all request from the user devices(A) to the internet goes straight to the reverse proxy server(B). The reverse proxy then sends the request to the main server(C). The required response is then passed on from the reverse proxy server(B) to the user devices(A).

WAF serving as a reverse proxy, a set of rules and policies are established. This provides a defence mechanism against vulnerabilities by filtering malicious traffic. A WAF's benefit derives in several ways depending on the speed and ease with which policy amendments can be enforced, making room for an efficient response to varying attack vectors. In the occurrence of a DDOS attack, the rate-limiting can immediately be implemented by amending WAF policies [62]. A web service with tones of requests can adapt to a WAF as a means of balancing the traffic load and evenly distribute the request to servers available. This also ensures that the web platform does not reveal the IP address of its origin server(s), making it difficult for attackers to administer an attack against it. Instead, the attack will be directed to the reverse proxy server which is well equipped with the needed security infrastructure to prevent overwhelming of the main server. Figure 12 shows a WAT application scenario.

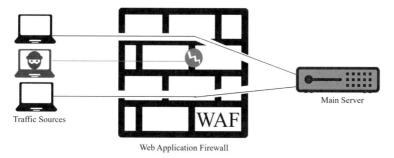

Traffic Sources

Main Server

Web Application Firewall

Fig. 12 WAT application scenario

6.2 Mitigation Against Man-in-the-Middle Attack

The attack demonstrated regarding man-in-the-middle is geared towards intercepting data flow and gaining unauthorized access to data in transit or data at rest. To outline relevant mitigation procedures towards such an attack within a remote working environment, it relevant to consider the major factors that lead to the attack. In most cases, an insecure communication link is exploited by attackers which leads to network hijacking and network stripping. It is needful to tailor the mitigation procedure in line with achieving a secured and well-encrypted connection to prevent network intrusion or minimize the level of attack impact. The use of multi-factor authentication focus on ensuring that only legitimate users have access to network resources. However, authentication alone does not eliminate the potential occurrence of a MITM attack [63]. It is needful that network operators or organizations deploy a more reliable network encryption protocol. This ensures end-to-end encryption between the user and the network, hence making data transmission secured.

6.2.1 End-to-End Encryption

End-to-end encryption is simply encrypting data during creation and only decryption at the point of use [64]. This means data transmission remains encrypted until been accessed by the legitimate recipient. End-to-end encryption can be between a user's application and a data processing server. It can also be between two users on the same application either being a mail service or data transfer service. It also satisfies interaction between a user and multiple devices, especially in situations where a user needs to make use of several devices in carrying out work duties. This also ensures that in each step of the data transmission, an encrypted data object is been handled instead of the data itself.

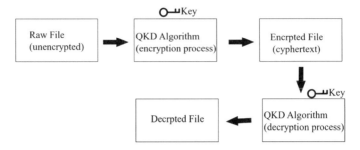

Fig. 13 Basic illustration of QKD concept

6.2.2 Quantum Key Distribution for End-to-End Encryption

The *Quantum Key Distribution (QKD)* algorithm can be adopted for end-to-end encryption as indicated in [65]. QKD makes use of random bits as tools in the encryption and decryption process, while every message is been converted into a photon. Photon only handles the encryption of data by making use of tokens which expires after a period [66]. Deploying QKD involves the value of the generated bit which is been determined by the photon either being 1 or 0 which is been generated at the sender's point and calculated for decryption at the recipient's end [67]. An interception from a third party will result in the cryptographic file being destroyed and returned to the sender. The QKD mechanism operates on the unpredictable principle of quantum physics [68] making it impossible for a third party to gain access or eavesdrop on transmitted data as well as to determine data characteristics. A private key which is included in the process of data transition is required by the recipient to gain access to the data. The private key generated is used in accessing and converting data which in return ensures data privacy is maintained. The working mechanism of QKD is illustrated in Fig. 13.

If by any means an attacker gain access to transmitted data, they only get the encrypted data and not the actual data. This means there is a cryptographic control of data at every stage of the transmission. The end-to-end encryption ensures that data is not exposed hence preventing an attacker from enjoying any unauthorized privileges either through eavesdropping or session hijacking.

6.3 Security Controls and Related Threat Elimination

Although there are various sophisticated countermeasures that can implemented to mitigated or limit impact of cyber-attacks, there are also simple security controls that implemented correctly have a huge impact on potential threats. Table 2 shows some important controls and the threats it eliminates.

Table 2 Security controls and threats affected

Security control	Eliminated threat
Deploying the use of virtual private network	Preventing the exploitation of user information such as IP address and location by ensuring anonymity
Monitoring of administrative changes by reviewing logs regarding password change, account creation, and deletion	The monitoring process provides an insight into network activities to determine anomalies that might lead to unauthorized intrusion
Adapt the use of Alphanumeric passwords with symbol combination which might be difficult to predict	The implementation prevents impersonation and identity theft
Limiting administrative access to make sure not all privileges are given to users	Prevention of unauthorized access and data falsifying
Installing a network intrusion detection system such as SNORT	The setup provides first-hand knowledge on any potential unauthorized intrusion within a given network by providing detailed logs and analysis on network performance
Ensuring effective BYOD management through the implementation of policies in that regard	Establishes security awareness among workers regarding the use of personal devices in accessing company resources. A well-Structured policy could prevent the use of malicious devices as well as session management

7 Conclusion

The research done in this paper highlight the way cloud systems provide essential infrastructure and services in the context of remote work. The benefit derived from cloud systems cut across users, developers, and service providers. As initially discussed in this paper, cloud services and its related technology faces numerous threats due to the high volume of demand and deployment of tools on cloud resources. It is of no doubt that remote work has come to stay amid COVID-19, as well as the potential increase of related cyber threats. Organisations and business owners are constantly advised to maintain cyber hygiene and to implement policies that safeguard their employees, resources, and operations while using a cloud system. Potential cyber-attacks administered against cloud systems and its related attack vectors were addressed in the research. The attack procedures expose how sophisticated cyber-attacks have become and the vulnerabilities that are been exploited on a cloud system. These attacks result in services unavailability, identity theft, unauthorised access and deprived of data privacy. As a means of addressing these issues, the research paper outlines relevant mitigation and security controls that are both technical and ethical. The attacks demonstrated hijacks data communication and floods network connection with malicious traffic. This results in Man-in-the-Middle and Distributed Denial of Service. By deploying a Web Application Firewall, network traffic is filtered to prevent cloud servers from being overwhelmed. A Quantum Key

Distribution is recommended as an encryption protocol that provides end-to-end encryption for data communication to prevent eavesdropping.

Irrespective of all the technical mitigation outlined, the paper also provided ethical controls that need to be adhered to by employees while working from home. It is recommended that organisations provide security awareness and to also implement rigorous BYOD policies that govern employees on using devices in carrying out work duties. These policies are to serve as a guide for employees to stay safe in their operation while working from home.

Amid the COVID-19 pandemic, remote work is inevitable and the deployment of cloud resources while at home has become part of the daily work duties. Ensuring a safe cloud environment is paramount for both employers and employees as well as information resources.

References

1. Frenzel LE (2018) Networking: wired and wireless. Electron Exp (Second Edition) 2:217–242
2. Rouse M, Irei A, Scarpati J (2019) Router in home networking. [Online]. Available: https://sea rchnetworking.techtarget.com/definition/router. Accessed 14 Sept 2020
3. Pratt L (2019) Only 30% of the UK workforce experienced working from home in 2019. 27 March 2020. [Online]. Available: https://employeebenefits.co.uk/30-uk-work-home-2019/. Accessed 14 Sept 2020
4. Office for National Staistics (2020) Coronavirus and the social impacts on Great Britain. Accessed 4 Sept 2020
5. Office for National Statistics (2020) "Coronavirus and the social impacts on Great Britain. 9 April 2020
6. Office for National Statistics (2020) "Coronavirus and the social impacts on Great Britain. 7 May 2020
7. Office for National Statistics (2020) Coronavirus and the social impacts on Great Britain. 5 June 2020
8. Office for National Statistics (2020) Coronavirus and the social impacts on Great Britain. 10 July 2020
9. Office for National Statistics (2020) "Coronavirus and the social impacts on Great Britain. 7 Aug 2020
10. AVIVA (2020) Tech Nation: number of internet-connected devices grows to 10 per home [Online]. Available: https://www.aviva.com/newsroom/news-releases/2020/01/tech-nat ion-number-of-internet-connected-devices-grows-to-10-per-home/
11. Schmidt F (2013) Tapping the world's fiber optic cables. 30 June 2013. [Online]. Available: https://www.dw.com/en/tapping-the-worlds-fiber-optic-cables/a-16916476. Accessed 14 Sept 2020
12. Baddar SA-H (2017) Chapter 8: how on earth could that happen? An analytical study on selected mobile data breaches. Adaptive Mobile Comput 153–183. ISBN 978-0-12-804603-6
13. Allison PR (2015) The security dangers of home networks [Online]. Available: https://www. computerweekly.com/feature/The-security-dangers-of-home-networks. Accessed 04 Sept 2020
14. NCSC (2019) Home and mobile working [Online]. Available: https://www.ncsc.gov.uk/col lection/10-steps-to-cyber-security/the-10-steps/home-and-mobile-working. Accessed 07 Sept 2020

15. Jacoby D (2014) IoT: how I hacked my home. The story of a researcher who wanted to see how vulnerable he actually was [Online]. Available: https://securelist.com/iot-how-i-hacked-my-home/66207/. Accessed 05 Sept 2020

16. Sivaraman V, Chan D, Earl D, Boreli R (2016) Smart-phones attacking smart-homes. WiSec 18–22. ISBN 978-1-4503-4270-4/16/07

17. Avast (2019) Avast smart home: security report [Online]. Available: https://cdn2.hubspot.net/hubfs/486579/avast_smart_home_report_feb_2019.pdf. Accessed 06 Sept 2020

18. Lee C, Zappaterra L, Choi K, Choi H (2014) Securing smart home: Technologies, security challenges, and security requirements. In: IEEE conference on communications and network security, pp 62–72

19. PandaSecurity (2020) WPA vs WPA2: which WiFi security should you use? 08 April 2020 [Online]. Available: https://www.pandasecurity.com/mediacenter/security/wpa-vs-wpa2/. Accessed 14 Sept 2020

20. Waliullah M, Gan D (2014) Wireless LAN security threats & vulnerabilities: a literature review. Int J Adv Comput Sci Appl 5(1):176–183

21. Rafferty L, Iqbal F, Hung PC (2017) A security threat analysis of smart home network with vulnerable dynamic agents. Comput Smart Toys 127–147. ISBN: 978-3-319-62071-8

22. Fortinet (2020) Global threat landscape report: a semiannual report by FortiGuard labs [Online]. Available: https://www.fortinet.com/content/dam/fortinet/assets/threat-reports/threat-report-h1-2020.pdf. Accessed 13 Sept 2020

23. Avast (2019) Avast threat landscape report [Online]. Available: https://press.avast.com/hubfs/media-materials/kits/2019-Predictions-Report/Avast%202019%20Threat%20Landscape%20Report.pdf?hsLang=en. Accessed 10 Sept 2020

24. Coker J (2020) Cyber-criminals exploiting remote working by attacking RDP ports 07 May 2020 [Online]. Available: https://www.infosecurity-magazine.com/news/criminals-exploiting-remote/. Accessed 10 Sept 2020

25. ESET (2019) It's time to disconnect RDP from the internet [Online]. Available: https://www.welivesecurity.com/2019/12/17/bluekeep-time-disconnect-rdp-internet/. Accessed 14 Sept 2020

26. Stockley M (2019) RDP BlueKeep exploit shows why you really, really need to patch 01 July 2019 [Online]. Available: https://nakedsecurity.sophos.com/2019/07/01/rdp-bluekeep-exploit-shows-why-you-really-really-need-to-patch/. Accessed 13 Sept 2020

27. NCSC (2020) Weekly threat report 5th June 2020 [Online]. Available: https://www.ncsc.gov.uk/report/weekly-threat-report-5th-june-2020. Accessed 10 Sept 2020

28. Fintechnews (2020) The 2020 Cybersecurity stats you need to know 20 Aug 2020 [Online]. Available: https://www.fintechnews.org/the-2020-cybersecurity-stats-you-need-to-know/. Accessed 10 Sept 2020

29. Check Point (2020) Cyber attack trends: 2020 mid-year report 06 2020 [Online]. Available: https://www.checkpoint.com/downloads/resources/cyber-attack-trends-report-mid-year-2020.pdf. Accessed 13 Sept 2020

30. GOV.UK (2020) UK condemns cyber actors seeking to benefit from global coronavirus pandemic 05 May 2020 [Online]. Available: https://www.gov.uk/government/news/uk-condemns-cyber-actors-seeking-to-benefit-from-global-coronavirus-pandemic. Accessed 13 Sept 2020

31. NCSC, CISA (2020) Advisory: COVID-19 exploited by malicious cyber actors 08 April 2020 [Online]. Available: https://www.ncsc.gov.uk/files/Final%20Joint%20Advisory%20COVID-19%20exploited%20by%20malicious%20cyber%20actors%20v3.pdf. Accessed 13 Sept 2020

32. Francoeur B (2020) Breach Report—March 2020, 31 March 2020 [Online]. Available: https://www.gflesch.com/elevity-it-blog/breach-report-march-2020. Accessed 14 Sept 2020

33. Pyman T (2020) Cyber-criminals are exploiting coronavirus crisis by targeting victims via Zoom or Microsoft Teams and using email scams including false offers of face masks, 08 April 2020 [Online]. Available: https://www.dailymail.co.uk/news/article-8200883/Cyber-criminals-targeting-victims-Zoom-Microsoft-Teams-offering-face-masks-email-scams.html. Accessed 14 Sept 2020

34. Shakarian P, Shakarian J, Ruef A (2013) Introduction to cyber-warfare. Syngress, Waltham

35. Cockburn G (2020) NAB flags cyber attacks during the pandemic have intensified, 14 Sept 2020 [Online]. Available: https://www.theaustralian.com.au/news/latest-news/nab-flags-cyber-attacks-during-the-pandemic-have-intensified/news-story/8cedc744da49f4bf4c766cfd1410dfa7. Accessed 20 Sept 2020

36. USA Primer (2020) 2020 thematic brief: US cybersecurity efforts, 09 Sept 2020 [Online]. Available: http://thirdway.imgix.net/pdfs/2020-thematic-brief-us-cybersecurity-efforts.pdf. Accessed 14 Sept 2020

37. Clement J (2020) Share of malicious data breaches worldwide per threat actor in 2020, 27 Aug 2020 [Online]. Available: https://www.statista.com/statistics/256656/threat-actors-malicious-data-breaches/. Accessed 14 Sept 2020

38. Fortinet (2019) Insider threat report [Online]. Available: https://www.fortinet.com/content/dam/fortinet/assets/threat-reports/insider-threat-report.pdf. Accessed 14 Sept 2020

39. Petters J (2020) What is an insider threat? Definition and examples, 17 June 2020 [Online]. Available: https://www.varonis.com/blog/insider-threats/. Accessed 14 Sept 2020

40. W. H. O. (WHO) Hygiene, World Health Organisation (WHO), [Online]. Available: https://www.afro.who.int/health-topics/hygiene. Accessed 08 Sept 2020

41. Vishwanath A, Neo LS, Goh P, Lee S, Khader M, Ong G, Chin J (2019) Cyber hygiene: the concept, its measure, and its initial tests. In: Decision support systems. Elvesier

42. E. U. A. F. N. a. I. Security (2016) Enisa, Dec 2016 [Online]. Available: https://www.enisa.europa.eu/publications/cyber-hygiene/at_download/fullReport. Accessed Sept 2020

43. C. M. &. S. Department for Digital (2020) Cyber security breaches survey 2020, 26 March 2020 [Online]. Available: https://www.gov.uk/government/publications/cyber-security-breaches-survey-2020/cyber-security-breaches-survey-2020#contents. Accessed 8 Sept 2020

44. Security R (2019) Why is cyber hygiene important?, RSI, 4 Sept 2019 [Online]. Available: https://blog.rsisecurity.com/why-is-cyber-hygiene-important/. Accessed 10 Sept 2020

45. O. F. N. Statistics (2020) Coronavirus and homeworking in the UK. Office For National Statistics

46. Morphisec (2020) Morphisec's 2020 WFH employee cybersecurity threat index. Morphisec

47. Verizon (2020) 2020 data breach investigations report. Verizon

48. Rodmunkong T, Wannapiroon P, Nilsook P (2014) The challenges of cloud computing management. Int J Signal Process Syst 2:160–165

49. Puthal D, Sahoo BPS, Mishra S, Swain S (2015) Cloud computing features, issues and challenges. In: International conference on computational intelligence & networks, Bhubaneshwar

50. Anderson B, Anderson B (2010) USB-based virus/malicious code launch. In: Seven deadliest USB attacks. Elsevier, pp 65–95

51. Foo B, Glause MW, Howard GM, Wu Y-S, Bagchi S, Spafford EH (2008) Intrusion response systems: a survey. In: Information assurance. Elsevier, pp 377–415

52. Somani G, Gaur MS, Sanghi D, Conti M (2017) DDoS attacks in cloud computing: issues, taxonomy, and future. In: Computer communications. Elvesier

53. Badis H, Doyen G, Khatoun R (2014) Understanding botclouds from a system. IEEE, Troyes

54. Knapp ED, Langill JT (2015) Chapter 7 hacking industrial control systems. In: Industrial network security. Elsevier, pp 171–207

55. Oriyano S-P, Shimonski R (2012) Client-side attacks and defense. In: Mobile attacks. Elsevier

56. Prowell S, Kraus R, Borkin M (2010) CHAPTER 6 Man-in-the-Middle. In: Seven deadliest network attacks. Elsevier

57. Darwish M, Ouda A, Capretz LF (2016) Cloud-based DDoS attacks and defenses. Department of Electrical and Computer Engineering University of Western Ontario, Ontario

58. Wang N, Ho KH, Pavlou G, Howarth M (2008) An overview of routing optimization for internet traffic engineering. IEEE Communication Surveys

59. Zekri M, Kafhali SE, Aboutabit N, Saadi Y (2017) DDoS attack detection using machine learning techniques in cloud computing environments. In: 2017 3rd international conference of cloud computing technologies and applications (CloudTech)

60. Koyfman M (2015) Intelligent layer 7 DoS and brute force protection for web applications. F5 Networks, Inc., p. 4

61. Randhe VS, Chougule AB, Mukhopadhyay D (2014) Reverse proxy framework using sanitization technique for intrusion prevention in database. Department of Information Technology, Maharashtra Institute of Technology, Pune
62. Dermann M, Dziadzka M, Hemkemeier B, Hoffmann A, Meisel A, Rohr M, Schreiber T (2008) Best practice: use of web application firewalls. OWASP German Chapter
63. Wang L, Wyglinski AM (2014) Detection of man-in-the-middle attacks using physical layer wireless security techniques. Wiley Online Library
64. Bai W, Pearson M, Kelley PG, Mazurek ML (2020) Improving non-experts' understanding of end-to-end encryption: an exploratory study. In: European workshop on usable security
65. Abirami N, Nivetha MS, Veena S (2018) End to end encryption using QKD algorithm. Int Open Access J 951
66. Kuhn DR (2015) A quantum cryptographic protocol with detection of compromised server. National Institute of Standards and Technology
67. Richard J, Alde DM, Dyer P, Luther GG, Morgan GL, Schauer M (2006) Quantum cryptography
68. Bruß D, Lütkenhaus N (1999) Applicable algebra in engineering communication and computing. arXiv

The Magic Quadrant: Assessing Ethical Maturity for Artificial Intelligence

Andi Zhobe, Hamid Jahankhani, Rose Fong, Paul Elevique, and Hassan Baajour

Abstract This paper discusses the need for measuring ethics for organisations that use and develop artificially intelligent software. The primary objective for this paper is to bridge the gap between artificial intelligence and ethics through the development of an ethical maturity framework that can be globally adopted and implemented through the design stages of AI that considers ethics through the lifecycle of the technology to support ethical evolution. This is a discussion that's missing in the realm of AI and is much needed in the rapidly evolving world of AI to regain control and confidence in the technologies we use.

Keywords The magic quadrant · AI · Ethics · Digital societies

1 Introduction

The use of artificially intelligence within systems continues to be increasingly ubiquitous however ethical considerations within the topic appears to be less familiar among researchers within in Artificial Intelligence [34].

We first aim to understand general perception of ethics AI. To do this, a literature review will be conducted to identify current ethical considerations made and consequences AI when those considerations are not made. In addition to the review, we will carry out a questionnaire which will provide first-hand data asking specific questions which we can analyse to support the need for a greater focus on ethics within artificial intelligence.

It is accepted that measuring ethics poses a number of challenges. It is considered difficult to quantify and quite relative to the subject. Throughout research, we will establish methods used to measure ethics and hope to assess current technologies said to incorporate AI, how data is handled by an intelligent system and link it to normative, meta and applied ethics as well as their subcategories and how these collide with legislations and regulations.

A. Zhobe · H. Jahankhani (✉) · R. Fong · P. Elevique · H. Baajour
Northumbria University, London, UK
e-mail: Hamid.jahankhani@northumbria.ac.uk

© The Author(s), under exclusive license to Springer Nature Switzerland AG 2021   313
H. Jahankhani et al. (eds.), *Cybersecurity, Privacy and Freedom Protection in the Connected World*, Advanced Sciences and Technologies for Security Applications, https://doi.org/10.1007/978-3-030-68534-8_19

It is considered achievable to develop a framework that can be used to assess the ethical maturity of an organisation in respect to artificially intelligent systems. The limitation to this may be that it cannot be globally adopted due to the natural ethical diversity across the world. I would consider a framework that can be utilised as an assessment in the development of AI a realistic outcome of this project.

2 Digital Societies

The World undergoes major digital revolution with the emerging technology – AI in twenty-first century. Such disruptive changes not only improve the quality of our living standards, but also create challenges to policy-makers in regards to privacy, digital security, safety, transparency and keen competitions [27]. United Nations calls for global "digital cooperation" on the provision of a better digital future for all [33].

In order to build an inclusive digital economy and society, there is a need of improving the accessibility of the digital infrastructure, ehealth-care and e-governmenet services. The digital divide is the gap that exists between individuals who have access to modern IT and those who is unable to access. The International Digital Economy and Society Index (I-DESI) 2018, using a similar methodology to the EU DESI index, compares the average performance towards a digital society and economy of EU countries with 17 non-EU countries. I-DESI measures performance in five dimensions including broadband connectivity, human capital (digital skills), use of Internet by citizens, integration of technology and digital public services [10]. The digital divides in terms of the DESI within the European Unions or I-DESI at the international level are still large. In terms of gender, there are still digital inequality in particular to those Eastern Europe countries in according to the "Women in Digital" scoreboard [9]. The scoreboard assesses EU countries' women's inclusion in digital economy in the areas of Internet use, Internet user skills as well as specialist skills and employment based on 13 indicators.

There is also a need for the World to apply the existing frameworks of human rights in the digital age. More and more decisions were made by AI nowadays and these certainly improve our lives a lot. However, it is questioned that whether such technologies also create new ways of not respecting or even violating human basic rights. For instance, the deepfake technologies increase the chance of online harassment to women [31], and the development of personal care robot deepens social isolation together with deception and loss of dignity [16]. The rights of children also require high attentions in the digital age as most of young children go online without the presence and guidance of their parents. There is insufficient protection to the information that the children could access through the internet and social media. They are the most vulnerable to online bullying and sexual exploitation. The widely spread of violence, hate speech and fake news in social media destroy the human virtues such as love, respect and trust. There is also a trend for the people feels "happier and more comfortable" to interact with machine [32]. Is there an evolution of the meaning of a society from a group of people live together and help each other to a more digial meaning—a group of people interact with emerging technologies?

3 What Is Identity?

In trying to define Identity, Zhu and Badr [35] quote Aristotle's law of identity in logic which rather cryptically states "each thing is identical with itself". They then present an interpretation of Leibniz's Law which provides a more practical definition representing identity in terms of the properties of an object. Leibniz states that for two entities to be identical, they must have all the same properties and conversely, if two entities have all the same properties then they are identical. Zhu and Badr then go on to examine that statement and conclude that reliance on identification attributes alone to identify an object leads to an incomplete solution. Unfortunately, this incomplete solution is mirrored in most traditional digital Identity Management Systems in existence.

Muhle et al. [26] describes a digital identity as follows. "A digital identity can be simply described as a means for people to prove electronically that they are who they say they are and distinguish different entities from one another". Whilst this definition is non-controversial and is in line with accepted thinking, it lacks accuracy and depth. To illustrate, if we extend this definition to the physical world, then a physical identity can be described as a means of proving physically that a person is who they say they are. This is traditionally achieved using a passport or some other identity papers issued by a trusted authority. However, a passport isn't a "physical identity" it is a credential that proves a person's physical identity and authenticates their claim that they are who they say they are. It then follows that what is referred to as Digital identity is sometimes a digital credential that authenticates some claim that a person or entity makes and is sometimes a digital identifier that distinguishes that person or entity from others in the same domain or context. In his seminal blog of 2016, now considered to be the foundation of the Self-Sovereign Identity concept, Allen [2] supports this argument by lamenting the confusion caused by the conflation of state-issued credentials such as an identity card with the fundamental concept of identity. He highlights that having one's credentials revoked doesn't mean that one loses one's identity and cites this as a crucial factor in the need for Self-Sovereign Identity. Hence, digital identity must be considered an umbrella term, that encompasses identifiers, attributes associated with those identifiers, and any claims associated with those identifiers.

Kuperberg [19] expresses digital credential in terms of identity representations, describing these in two different types consisting of owning a physical item such as an identity card or knowing a secret such as a password. These factors are commonly referred to as "something you have" and "something you know" respectively. In doing so, he omits what is typically referred to as a third authentication factor, "something you are", which relates to biometric-based authentication. Whilst this may seem like an oversight, his reference to numerous implementations based on biometric authentication, shows that he clearly sees biometrics primarily as a method of protecting access to the private authentication information held within the first two factors e.g. fingerprint protected smart card or a faceID protected IOS credential store.

Ferdous et al. [12] refers to the laws of identity as defined by Cameron and uses this as a basis to define Self-Sovereign Identity laws. Cameron's laws are presented in the form of seven principles covering:

1. User control and consent
2. Minimal disclosure for a constrained Use
3. Justifiable parties (disclosure only to parties with legitimate uses)
4. Directed identity (the ability to restrict exposure of identity information to the intended targets)
5. Pluralism of operators and technologies (the ability of multiple identity technologies run by multiple providers to interact)
6. Human integration (Ensuring we define reliable, unambiguous and secure communication mechanisms between the user and user interface when exchanging identity information)
7. Consistent experience across contexts (expressing clearly to the user what identity information is acceptable and what information has been applied in a given application context whilst ensuring the experience is consistent across contexts).

These laws are still respected in 2019, with state-of-the-art identity technologies including Self-Sovereign Identity seeking to justify their worthiness through adherence to these 7 principles, and Cameron himself stating in 2018 that "Identity on the Blockchain must be subject to the Laws of Identity.". Whilst it is evident that Cameron's laws contributed to the thinking that led to the GDPR, Ferdous et al. [12], consider these laws relevant but incomplete in the context of Self-Sovereign Identity. Some of the wording surrounding the definition of Cameron's laws is overly complex e.g. "Pluralism of operators and technologies" could be summarised by the principle of "Interoperability" and inevitably some of the examples used to illustrate these concepts could do with an update reflecting today's technology landscape, but the principles defined back in 2005 have clearly stood the test of time.

Ferdous, et al. defined digital identity using a mathematical model called the Digital Identity Model or DIM [12]. Whilst this model supports Ferdous' intention of producing a model which is both methodological and comprehensive, it is not easily accessible to those without a mathematical background and such does not have the widespread appeal required to base a reference model on.

4   Ethics and Artificial Intelligence

Driven by technological advances and public interest, Artificial intelligence (AI) is considered by some as an unprecedented revolutionary technology with the potential to transform humanity [4]. With such a bold statement and potential impact, it is deemed imperative that we consider how organisations are measuring their capability or if the consideration is being made at all. Especially with the use of technology in the public eye, potentially having access to vast amounts personally identifiable information.

AI, whether we like it or not plays a major factor in our daily lives, with varying degrees of opinions amongst researchers. AI has always been a strange field [1] and its rapid growth in adoption has diluted its definition among adopters which each seemingly having their own interpretation thus unique implementation. In an attempt to define AI, we must dissect the term. Artificial in the oxford dictionary is defined as "made or produced by human beings rather than occurring naturally, especially as a copy of something natural" and intelligence from the same source is defined as,"the ability to acquire and apply knowledge and skills" [3]. Often referred to as the father of AI, Dr. John McCarthy also thought to have also coined the term AI, defines the collective,Artificial Intelligence, as the science and engineering of making intelligent machines, especially intelligent computer programs. It is related to the similar task of using computers to understand human intelligence, but AI does not have to confine itself to methods that are biologically observable [24].

Ethics even outside of the scope of information technology can be quite complex. There are three main categories of ethical theories as McCartney and Parent go on to explain [25].

Normative theories tell us why we do things that in some instances may appear counterintuitive to what we think an ethical decision would be. These theories are often called ethical systems due to that they provide a system allowing people to determine ethical decisions that individuals should make.

Typology of theories within normative ethics:

	Situation ethics	Rule ethics
Teleological ethics	Consequences of individual action	Adherence to rule with good consequences
Deontological ethics	Situation dependent principles	Adherence to rule that is right in itself

The table above presents the basis for evaluating an action according to the four theory categories. It also presents important representatives for the category [6]. One immediate challenge this poses is it discusses consequences however consequences are widely subjective and dependent on individual morals or law governing the location.

Meta ethics doesn't address desired behaviour however is related to the study of ethical theory itself. Applied ethics considers how we apply normative theories to often addressed by organisational specific issues.

With a large number of organisations claiming to be implementing AI in their technologies and processes, we need to be clear on any ethical considerations made in design stages for ethical decision making in AI. A part of identifying this process is also identifying what qualifies a system as artificially intelligence. McCarthy, considered the Father of AI, states his lack in surprise at the slow development of human-level AI as well as "the demand to exploit what has been discovered has led many to mistakenly redefine AI" [22]. With this in mind, the pursuit for clear definition is also considered throughout this dissertation and the importance of having a clear definition of AI.

McCarthy has been very active in the realm of AI and has already discussed the logical AI system, uncovering that Logical AI is more intellectually ambitious than approaches to AI based on evolution. This is due to the requirement of understanding enough common sense and about the requirements for decisive system to put the basic facts in a logical computer program. Evolutionary AI may be capable of reaching human level intelligence without any human understanding on how this was achieved [23]. With this in mind, it would appear that an AI system making its own ethical considerations without strict restrictions to preserve ethics can quite easily become unethical. Does this mean that artificial intelligence needs to measure against human intelligence or does AI need to measure against an ideal? Furthermore, who decides the measure of ideal?

One problem with artificial intelligence is we have a different measure and definition of intelligence. Some definitions measure success in terms of fidelity to human performance, whereas some others measure against an ideal concept of intelligence, which we will call rationality. A system is rational if it does the "right thing," given what it knows [30].

The creation of thinking machines raises a host of ethical concerns relating to ensuring that such machines do not harm humans and other morally relevant beings, and to the moral status of the machines themselves [13]. With this in mind it is believed that we can begin to hold machines accountable for actions, which means they must be capable understanding the concept of ethics.

The use of the term autonomous is usually coined with AI which leads to invalid conclusions about the way machines can be kept ethical [8], a hair-raising statement that would suggest an AI self-governing system could surpass the line of ethical decision making if the focus is autonomy.

According to Forbes, 75 out of 176 countries globally are actively using AI technologies for surveillance purposes, including smart city/safe city platform [28] and the UK are no exception. Where ethics become important is in areas such as we saw in relation to Facebook and Cambridge Analytica which resulted in over 86 million records being share by Facebook to Cambridge Analytica. The lack of ethics shown is worrying by technology leaders.

The National Health Service (NHS) have plans to venture into AI with aims of becoming world leaders in digitising national healthcare [7]. Initial concerns with the reports are the lack of ethics within their plans and the current ethical issues which have only been exacerbated by COVID-19. One example of this is the handling of Personally Identifiable Information (PII) is often transferred over the counter. PII is considered to be information which can be used to distinguish or trace an individual's identity and includes things like date of birth and addresses [18]. Historically, this information would always be shared verbally across a counter, with social distancing, PII is being disclosed at a 2-m distance which substantially increase the chances of a person or persons becoming victim of cybercrime.

Bradley Malin had discovered in his research through trail matching that 87% of Americans can be uniquely identified by combining only their birth date, gender and zip/postal code [21]. This is the exact information that's required for the health representative to access your records and confirm your appointment which can easily be overheard and repurposed for nefarious purposes.

The NHS is due to their recent publications announcing their digital transformation goals of becoming innovators in the healthcare industry through the use of artificial intelligence to make better decision [15]. While the National Health Service have identified several challenges in achieving this, ethics doesn't appear in their agenda. Their lack of tackling fundamental ethics when handling client data within hospitals doesn't bode well for an intelligent system designed from unethical practices.

Social networks are amongst the technology industries pioneering and, in many cases, boasting their AI ability to deliver. However, the uptake of social networking sites has raised some alarming questions regarding their potential to facilitate deception, social grooming and the creation of defamatory content, amongst others sharing the platform. Social networking sites require extensive and active user involvement, which develops an illusion of control when, in fact, the operation of the social networking sites and the outcomes of their use and the data collected are not at all transparent [20].

A compelling, history making example of this was the Cambridge Analytica & Facebook collaborative scandal which took place in 2016 that used a social network to steer a United States presidential election. Cambridge Analytica, through the use of machine learning, which is a subset of AI, began with finding a way first to obtain raw data about millions of users. With a platform that as reported by Statista has over 2.6 billion active monthly users, where better to go digging for data than Facebook [11]. Once the data is collected intelligent systems were used by Cambridge Analytica to develop profiles about the users through all of the collected data. They could curate an advertisement that would be most convincing to the profiles with the understand that different profiles would be better persuaded with varying messages in support of the same candidate or issue.

Before any of this can happen, the raw collection of the data requires approval. Aleksandr Kogan, a psychologist researcher at Cambridge University created a survey which asked respondents about their personality of participants by paying each one a few dollars, to get those participants. The survey asked for respondents' consent to access of their Facebook data, including the pages that they liked. This not only gave them access to insights on the participants profiles but due to the features of Facebook, also gave Cambridge Analytica information of the friends who hadn't opted in of those participants who had opted in [29].

In a research conducted by Capgemini, a leading consultancy company that operates internationally, in 2019 in an attempt to identify whether interactions of AI enabled systems are ethical, Capgemini had concluded that ethics in AI will always be a work in progress with continued development throughout the evolution of AI. The report below distinctly represents that among the 4447 consumers of a survey, 76% of respondents were not happy with the current regulations in place surrounding

AI which would insinuate consumers sense a lack of lack of readiness in artificially intelligent systems for handling consumer data and decision.

With these in mind, it is clear that ethics play a huge part in the successful adoption of artificial intelligence however it's also apparent that it is one of the biggest challenges.

AI algorithms could present issues in the financial sector should a lack of satisfactory ethical considerations be made. In a publication by Ramsey and William M, they look at the consequence of an unethical AI system processing mortgage applications and the challenges in those systems. S lawsuit is brought against the lender due to an application being allegedly rejected due to their race. Despite the bank rejecting such claims and rendering them impossible stating the algorithms are blinded to race despite approval rate for black applicants steadily dropping. The report goes on to explain the complexity of identifying the cause, labelling it "almost impossible" to explain why or how the system could evolve into utilising race as a factor in the decision-making process [13].

The tech industry has seen a massive shift in the last two decades as have the consumer expectations. We have become consumers of outcomes rather than physical products. I will use a fitbit to explain, practical and simple design not hugely stylish but it's not the way it looks that makes the fitbit so popular, it's the notion of a healthier lifestyle.

Social media, however, is free. Contrary to popular belief, these organisations won't sell your data, your data isn't the product.

Companies are making conscious decision to make AI more diverse, examples of this include technologies like Siri, in intelligent voice controlled personal assistant being able to communicate in a number of different voices that we might associate with different genders as well as different accents.

There is further room for improvement though, especially in the realm of AI and how it can help to improve diversity. Earlier this year, a globally trending movement, Black Lives Matter was seen across the world with people of all races attending the streets with aims at tackling diversity and injustice. BLM could vastly benefit from AI implemented ethically to tackle racial bias in everyday lives. A study conducted by Daugherty, Paul R; Wilson, H James and Chowdhury, Rumman looks at how AI can be used to overcome harmful biases rather than perpetuate. For this to happen, the authors reached the conclusion that in order to achieve diversity, this must be tackled at the design stage whereby the humans teaching artificial intelligence include and adapt their data to be more diverse which suggests these biases may simply be an unconscious reflection of the humans that are developing the code also adding Software development remains the province of males—only about one-quarter of computer scientists in the United States are women 3—and minority racial groups, including blacks and Hispanics, are underrepresented in tech work, too [5].

Evidence indicates that machine learning algorithms could contribute to oppression and discrimination due to historical legacies of injustice reflected in training data directly to economic inequality through job displacement, or through a failure to reflectively account for who benefits and who does not from the decision to use a particular system. Furthermore, AI technologies pose an existential risk to humanity by altering the scope of human agency and self-determination, or by the creation of autonomous weapons [17].

5   The Magic Quadrant

For years, we have assessed leading cloud providers' maturity, this is often referred to as the magic quadrant in a report that is routinely conducted by Gartner. A study conducted and published by Gartner based on a company's current performance, growth and capabilities placing candidates into one of four quadrants: Niche Player, Challenger, Visionary or Leader.

This gives consumers or organisations the power of visibility into the strength of cloud providers against their competitors in a booming market, with 2020 seeing the highest growth in IaaS due to consolidated Data centres with public cloud service adoption growing 17% on its previous year. "At this point, cloud adoption is mainstream." (Gartner Forecasts Worldwide Public Cloud Revenue to Grow 17% in 2020, [14]).

For the purpose of this research same principals and research of an organisation to identify what quadrant they find themselves in, the organisations learnings from being in a quadrant and how it'll change their approach to ethics in relation to the use or development of Artificial Intelligence.

The reason this approach is used is because AI can't be measured as a scale between for example; good and bad. As there are many facets for improvement to consider, we need further knowledge from these results in order to improve our capabilities in developing ethical artificial intelligence.

6   What Is an Ethical Maturity Model?

There appears to be a gap, or at the very least an uncommon model to AI using or developing organisations that can be used to asses an organisations readiness for AI or, ethical maturity in order to develop ethical and intelligent systems.

The purpose is to see if we can create visibility and a communicable outcome to organisations with regards to ethical AI.

An ethical maturity model emerged from a questionnaire with the purpose being to identify the organisations ethical readiness for AI and their ability to successfully utilise or develop AI technologies while maintaining ethics. Each question in the questionnaire fall within two categories to establish a scale in the form of a quadrant, Vision and Capability all of the questions carry the same weight though the responses will have a value between 1 and 5.

6.1   The Categories and the Quadrants

An equal set of question that fall into the two facets will fall into a quadrant once the results are calculated which is used to assess an organisations vision and capability, placing the organisation on a quadrant of the two dimensions gauging their ability into a category of a starter, an equipped, an aspiring or leader in ethical maturity thus resulting in a position either higher up the vision on the x axis or further along the y axis representing their capabilities.

For that reason, we will also need to group the questions equally between visionary question or capability questions for the initial model.

The questions was evaluated for their position on vision and capability and an icon will be used to indicate the organisations position in the quadrant resembling the Magic Quadrant produced by Gartner.

A starter represents an organisation who doesn't have a great deal of confidence in their ability and isn't resourced to adequately execute ethical AI therefore would be discouraged at present from doing so.

Aspiring would represent an organisation with a strong vision and confidence however not adequately resourced.

An organisation who is equipped have the right resources but lack transparency within the organisation which impacting confidence, innovation and vision.

A leader represents an organisation who have set high standards of transparency with an ability to execute successful ethical AI projects on those high standards.

The results from a set of questions was used to calculate the capabilities with the realm of the row labels listed in the table above.

The assessor will have the opportunity to respond to each question on a scale between 1 (Strongly disagree) and 5 (Strongly agree). Once every question is answered. Then we were able to accumulate the results for each category giving us a percentage for each category to assess the capabilities against vision of the organisation.

One thing to note is that the higher the weight, the greater the impact on the end result, for the purposes of discovery I will ensure the weight is consistent, meaning there are the same number of questions for vision as there are for capability. Greater research will be needed to identify which category has a stronger bearing or which questions carry a heavier weight thus increasing the impact on the outcome.

6.2   The Question

A set of questions have been devised which was used to collect outcome values for calculations which will then be placed on quadrant to aid an organisation in visualising their current stance and capability in successfully implementing ethical AI.

Vision questions:

1. Our organisation has a mature vision for ethics in implementing AI projects and products
2. Our organisation has formed its AI strategy using input from stakeholders inside the organisation that represents our customers and wider society
3. Our organisation has taken appropriate steps to communicate this strategy both internally and externally
4. Our organisation recognises that ethical risk accompanies the use of AI and suitably manages this risk
5. Our development teams are enabled and empowered to raise concerns of AI and can expect these concerns to be taken seriously.
6. The users of AI technology within the organisation are aware of the effect their actions have on a technology living up to its ethical obligations
7. The organisation has the right cultural change strategy and plan to drive the expected outcomes
8. Our organisation's IT investment is aligned with business value with regard to opportunities and threats of ethical considerations
9. Our organisation has an active strategy to reduce risk from ethical consideration in our IT systems on an ongoing basis
10. Our organisation has invested sufficiently in technologies to understand data science model outcomes
11. Our organisation has published Vision or similar statements concerning areas of Corporate Social Responsibility.

Capability questions:

1. Our organisation has a suitable policy document that outlines how to implement an AI project ethically according to the company's guidelines
2. Our organisation is aware of and makes use of leading standards and best practice to guide the development and use of AI technology in an ethical manner
3. The organisation has clearly defined AI & Ethics guiderails supporting its policy or strategic intent for the ethical use of AI technology
4. During an AI development or deployment project, there are resources within the organisation who can lend their assistance with ethics related problems and concerns
5. Our organisation works to "Ethical by Design" concepts with processes containing key steps that reflect upon and assess adherence to our ethical obligations with respect to Artificial Intelligence technology
6. Our organisation would be able to quickly identify when an AI project falls short of its ethical obligations
7. Our organisation has the right capabilities to understand how AI technologies that have been "released" are well monitored to ensure they continue to meet their ethical obligations

8.   The organisation fully understands and engages with stakeholders affected by the AI technology they are developing/adopting throughout the development and/or deployment of the technology
9.   We have a mature data science development platform that enables our data science resources to understand and explain ethical implications of the models they develop
10. We have robust data access measures for your data scientists
11. We understand the types of data used for data science and make clear the considerations in using various data sets with respect to ethical implications.

With these twenty-two questions, split evenly amongst the two categories, we used a test case to aim in identifying how this can improve an organisations visibility in the AI ethics and make this communicable across their organisation allowing them to set adequate targets for improvement.

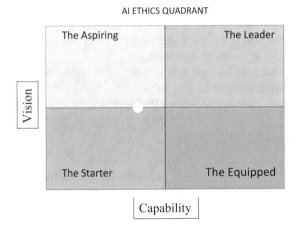

Visual aid provided by the chart can give the director of Company a clear visual representation that can be shared and communicated with other teams and employees within the organisation, particularly the technology departments and those in direct contact with AI.

7 Conclusion

While we have been able to create a model to assess an organisations ethical maturity towards AI, we haven't done enough to define AI, despite discovering numerous times through various references that the term Artificial Intelligence is not clearly understood. Since intelligence is subjective, seeking a specific definition the umbrellas the technologies that incorporate intelligence may not be the right approach.

However, it can be concluded that the definition of AI isn't the issue, transparency is. And with this ethical maturity model, we have been able to support an organisation in not only understanding its capability to deliver ethical AI but also make these finding communicable within their organisation in a way that is easy to ingest.

References

1. Allen JF (1998) AI growing up: the changes and opportunities. AI Mag 19(4):13–23
2. Allen C (2016) The path to self-sovereign identity. https://www.lifewithalacrity.com/2016/04/the-path-to-self-soverereign-identity.html. Accessed 15 Sept 2019
3. Artificial, Intelligence (2020) Oxford online dictionary. https://en.oxforddictionaries.com/definition/money
4. Brock J, von Wangenheim F (2019) Demystifying AI: what digital transformation leaders can teach you about realistic artificial intelligence. Calif Manag Rev 61(4):110–134
5. Daugherty P, Wilson H, Chowdhury R (2019) Using artificial intelligence to promote diversity. MIT Sloan Manag Rev 60(2)
6. Ekvall T, Tillman A, Molander S (2005) Normative ethics and methodology for life cycle assessment. J Clean Prod 13(13–14):1225–1234
7. England N (2019) NHS England. NHS aims to be a world leader in artificial intelligence and machine learning within 5 years [online]. https://www.england.nhs.uk/2019/06/nhs-aims-to-be-a-world-leader-in-ai-and-machine-learning-within-5-years/. Accessed 23 June 2020
8. Etzioni A, Etzioni O (2017) Incorporating ethics into artificial intelligence. J Ethics 21(4):403–418
9. European Commission (2020b) Women in digital scoreboard 2019—country reports, European Commission, Europe [online]. https://ec.europa.eu/digital-single-market/en/news/women-digital-scoreboard-2019-country-reports. Assessed 1 Nov 2020
10. European Commission (2020a) Creating a digital society. European Commission, Europe [online]. https://ec.europa.eu/digital-single-market/en/creating-digital-society. Assessed 1 Nov 2020
11. Facebook: Active Users Worldwide (2020) Statista [online]. https://www.statista.com/statistics/264810/number-of-monthly-active-facebook-users-worldwide/. Accessed 12 July 2020
12. Ferdous M, Chowdhury F, Alassafi MO (2019) In search of self-sovereign identity leveraging blockchain technology. IEEE Access 7:103059–103079
13. Frankish K, Ramsey W (2018) The Cambridge handbook of artificial intelligence. Cambridge University Press, Cambridge, pp 315–316
14. Gartner (2020) Magic quadrant for cloud infrastructure and platform services [online]. https://www.gartner.com/doc/reprints?id=1-1ZDZDMTF&ct=200703&st=sb. Accessed 15 Oct 2020
15. Gould M (2020) Regulating AI in health and care—NHS Digital. NHS Digital [online]. https://digital.nhs.uk/blog/transformation-blog/2020/regulating-ai-in-health-and-care. Accessed 12 July 2020
16. Hosseini S, Goher K (2017) Personal care robots for older adults: an overview. Asian Soc Sci 13:11. https://doi.org/10.5539/ass.v13n1p11
17. Krafft P, Young M, Katell M, Huang K, Bugingo G (2020) Defining AI in policy versus practice. In: Proceedings of the AAAI/ACM conference on AI, ethics, and society
18. Krishnamurthy B, Wills C (2009) On the leakage of personally identifiable information via online social networks. In: Proceedings of the 2nd ACM workshop on Online social networks—WOSN '09
19. Kuperberg M (2019) Blockchain-based identity management: a survey from the enterprise and ecosystem perspective. IEEE Trans Eng Manag (99):1–20. https://doi.org/10.1109/TEM.2019.2926471

20. Light B, McGrath K (2010) Ethics and social networking sites: a disclosive analysis of Facebook. Inf Technol People 23(4):290–311
21. Malin B (2005) Betrayed by my shadow: learning data identity via trail matching. J Priv Tech
22. McCarthy J (2007) From here to human-level AI. Artif Intell 171(18):1174–1182
23. McCarthy J (2000) Concepts of logical AI. Logic-based artificial intelligence, pp 37–56
24. McCarthy J (2004) What is artificial intelligence? www.formal.stanford.edu/jmc/whatisai/wha tisai.html
25. McCartney S, Parent R (2020) 2.1 major ethical systems [online]. https://opentextbc.ca/ethics inlawenforcement/chapter/2-1-major-ethical-systems/. Accessed 12 July 2020
26. Mühle A, Grüner A, Gayvoronskaya T, Meinel C (2018) A survey on the essential components of self-sovereign identity. Comput Sci Rev 30:80–86
27. OECD (2018) OECD science, technology and innovation outlook 2018: adapting to technological and societal disruption. OECD Publishing, Paris. https://doi.org/10.1787/sti_in_outlook-2018-en
28. Press G (2019) Artificial intelligence (AI) stats news: AI is actively watching you in 75 countries. Forbes [online]. https://www.forbes.com/sites/gilpress/2019/09/18/artificial-intell igence-ai-stats-news-ai-is-actively-watching-you-in-75-countries/. Accessed 23 June 2020
29. Rathi R (2019) Effect of Cambridge analytica's Facebook ads on the 2016 US Presidential election. Medium [online]. https://towardsdatascience.com/effect-of-cambridge-analyticas-fac ebook-ads-on-the-2016-us-presidential-election-dacb5462155d. Accessed 12 July 2020
30. Russell S, Norvig P (1995) Artificial intelligence. Prentice Hall, Englewood Cliffs, N.J.
31. Sample I (2020) What are deepfake—and how can you spot them? Guardian UK [online]. https://www.theguardian.com/technology/2020/jan/13/what-are-deepfakes-and-how-can-you-spot-them. Assessed 1 Nov 2020
32. Sydell L (2018) Sometimes we feel more comfortable talking to a robot [online]. https://www.npr.org/sections/alltechconsidered/2018/02/24/583682556/sometimes-we-feel-more-comfor table-talking-to-a-robot?t=1604349931906. Assessed 1 Nov 2020
33. United Nations (2020) The age of digital interdependence. Report of the UN Secretary-General's High-level Panel on Digital Cooperation [online]. https://www.un.org/en/pdfs/Dig italCooperation-report-for%20web.pdf. Assessed 1 Nov 2020
34. Yu H, Shen Z, Miao C, Leung C, Lesser VR, Yang Q (2018) Building ethics into artificial intelligence. In: Proceedings of the 27th international joint conference on artificial intelligence (IJCAI'18), pp 5527–5533
35. Zhu X, Badr Y (2018) Identity management systems for the internet of things: a survey towards blockchain solutions. Sensors 18(12)

A Systematic Literature Review of the Role of Ethics in Big Data

Jade Roche and Arshad Jamal

Abstract The aim of this research is to identify the ethical standards and practice that exists for big data and what the gaps are, through reviewing academic literature about various industries such as health, research, and social media. The aim is to provide a roadmap for future research on this topic within academia, policy makers and law makers. Big data is a relatively new concept and has been used by many different types of organisations on a large scale over the last decade, which has impacted individuals as consumers, citizens, and employees. Big data has provided insights into consumers and the public, at an unprecedented scale but standards of managing this data have not been implemented at the same speed. Through a Systematic Literature Review (SLR), an analysis of existing academic research into big data and ethics, and how it has been applied within industries will be critically analysed. By utilising academic databases and applying an exclusion and inclusion criteria, this will locate relevant good quality papers for the SLR. The SLR was narrowed down to 14 papers, which focused on different industries and elements of big data ethics. By using this broad approach, reoccurring themes that exist universally when managing big data and ethical issues appear, such as privacy concerns, accountability, and definitions regardless of data type and purpose. Big data has proved controversial with how it has been used by some organisations, while simultaneously positive in other areas. The topic of using big data ethically has arisen socially, politically, and academically, and the purpose of this SLR is to determine how far this conversation has progressed and what existing practice is. Areas for further research include the impact of new technologies and concepts such as IoT, Smart Cities and their relationship to big data and ethics. Law makers should lead the way of progressing this topic and introducing frameworks and best practice and soon policy makers will follow.

Keywords Big data · Ethics · Privacy · Accountability

J. Roche · A. Jamal (✉)
Northumbria University, London, UK
e-mail: arshad.jamal@northumbria.ac.uk

J. Roche
e-mail: jade.roche@northumbria.ac.uk

H. Jahankhani et al. (eds.), *Cybersecurity, Privacy and Freedom Protection in the Connected World*, Advanced Sciences and Technologies for Security Applications, https://doi.org/10.1007/978-3-030-68534-8_20

1   Introduction

Big data has become a fundamental element for some businesses, regardless of industry, and has provided insights into people and behaviour at an unprecedented rate. With increasing publicity over data management scandals such as Cambridge Analytica, election interference and Snowden disclosures, the public are aware of how data is used. The use of big data is everywhere, and effects individuals in their capacity as consumers, citizens, and employees but the understanding of what big data is, is not extensive [1]. The gathering of metadata about people's online interactions has made commercial organisations and public bodies aware of different traits of individuals and what appeals to them, meaning more is known about people's behaviour.

Throughout this paper, the term 'big data ethics' is used to describe the practice of applying ethical considerations or decision-making about how large datasets are used and the impact the use of this dataset has on individuals and society. The question of using data ethically is being retrospectively applied to big data already in use, and is often considered alongside other data issues such as data governance, cyber security and data privacy [2]. This can lead to data ethics not being prioritised or the risks not adequately documented and addressed. Big data is used across different industries such as commercial, political and health and increasingly the use of this data often reflects the privileged in societies view and can disadvantage or leave out poorer parts of society and their experiences [3]. This can have substantial and unintended consequences such as the A-Level results algorithm in the UK in 2020 [4].

There is a regulatory gap in managing big data, across all industries. Big data that does not contain directly identifiable personal data, or data that can be combined with other data to identify an individual, is not under the scope of the General Data Protection Regulation (GDPR) [5]. Therefore, the use of big data is unregulated, and organisations can use it to make decisions about their operations and about the people who use their services. There are some industries that have ethical requirements such as research ethics or computer ethics, but these do not match the wide reach in which big data is used. In addition, anonymised data can be shared widely and as a result data broker companies which buy and sell data and in the US the industry are worth around $200 billion [6]. Data may be gathered for a specific and legitimate reason, is then shared or processed for another reason, and this can happen multiple times and the origins of the data may be lost but constantly reused for commercial purposes, with competing organisations now collaborating with data ecosystems [7]. Organisations could argue there is nothing wrong with their use of big data, and it a lot of cases, this may be correct. However, the lack of scrutiny leaves the practices open to negative consequences for individuals and a lack of transparency about how decisions are made about them.

In this paper by using a SLR method, it will be examined what ethics is applied to big data, what standards and practice exist, what the definitions are, and who is accountable for overseeing big data ethics. Section 2 of this paper will outline the research methodology used for finding academic papers, such as search criteria,

Research Question	Context
What is "big data ethics"?	Review the existing definitions and terminology when discussing big data ethics and are they consistent in different industries and jurisdictions.
Who is responsible for big data ethics?	Who is responsible in an organisation for ensuring big data ethics and is there external regulations or codes of conduct the organisation is accountable to?
How can big data ethics be achieved?	What steps does an organisation need to take to demonstrate they considered big data ethics? Are there industry frameworks that can be applied? How widespread is there adoption?

Fig. 1 Research questions

selection process for academic papers and how papers were selected for the SLR. Section 3 sets out the core reoccurring themes and discussions of the papers selected, which include defining big data and ethics, privacy concerns, where accountability lies, examples of where ethics has been applied, and the type of ethical theory that can be applied. Section 4 further discusses the thematic findings of the SLR and explores the gaps in existing literature and explores grey literature on this topic. Section 5 provides recommendations for future research on this topic, and actions policy makers could take to implement big data ethics now.

Aims and Research Questions

Research questions along with the context are conceptualised in Fig. 1 below.

2 Research Methodology

The methods of conducting this SLR is influenced by Kitchenham, however there are some instances where Kitchenham methodology does not work, particularly as this topic is qualitative, whereas Kitchenham's is based on quantitative data [8]. To conduct a SLR which produces unbiased results, a defined criteria of search terms and criteria were formed. The search criteria consist of inclusion and exclusion metrics and a quality assessment of the papers shortlisted. Due to the large number of results generated by the search terms and metrics, the journal articles were then assessed based on title and abstract to determine if they were relevant to the topic. If the title and abstract was relevant the whole paper was read to determine if it should be included in the SLR.

2.1 Search Terms

The search terms used on their own and combined were "big data ethics", "big data", "digital ethics" or "ethical" and "data". The terms on their own could provide a substantial return of journals found, however by combining the terms, more relevant results were returned. By using the search criteria and the inclusion and exclusion criteria laid out in the following section, the results were refined to only include relevant papers located through the search terms.

2.2 Search Criteria

The Figure 2 outlines the exclusion and inclusion criteria of the systematic literature review.

Big data and research surrounding this topic has substantially evolved in the last decade so the decision to have an 8-year search was necessary as papers from before this time may be have been outdated. As it is, the oldest paper selected for the SLR is 2014. To ensure relevant good quality papers were sourced, reputable databases were used to locate the papers. Grey literature such as blog posts from non-reputable sources were excluded to ensure the quality of the research and that only grey literature of impact in the topic area was included.

Inclusion criteria	Exclusion criteria
The academic papers must be sourced using databases IEEE, Scopus or ScienceDirect.	Any paper published before 2012 are to be excluded.
Grey literature must be from a reputable source i.e. government, regulatory, professional body, or news source.	Papers that are not in English.
The paper must contain information about ethical standards, big data processing or implication of big data use on society.	Papers focusing solely on personal data processing or very specific or niche elements of big data are to be excluded.

Fig. 2 Inclusion and exclusion criteria

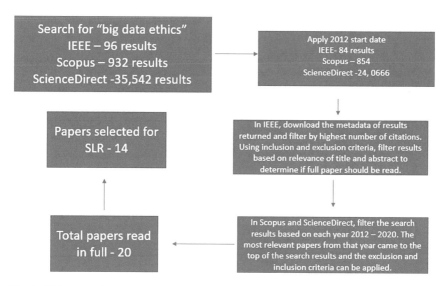

Fig. 3 SLR paper selection process

2.3 Search Results

Through IEEE, a search was conducted using the words 'big data ethics' with a start date of 2012. This yielded 83 results. A further search was conducted using the terms 'data' and 'ethics' with a start date of 2012 which yielded 472 results. Using the extraction function in IEEE the results were exported to excel and sorted by highest citations to lowest. Based on title and abstract, papers were selected for review. In Science Direct the terms 'big data ethics' unfiltered returned 35,542 results and when the criteria of papers since 2012 was applied, this returned 24,066 results. In Scopus the terms 'big data ethics' returned 932 results and since 2012 there were 854 results. The results were sorted within the databases based on relevance and searches were done by filtering down to each year since 2012. By applying the exclusion and inclusion criteria and reviewing papers based on title and then reviewing the abstracts, papers for a full review were identified. In total 20 papers were read in full and these were narrowed down to 14 based on relevance and quality. Figure 3 provides a map of how the papers for the SLR were selected.

2.4 Quality Assessment

To qualify for inclusion in the SLR and ensure good quality papers were selected, the answers to the questions laid out in Fig. 4 must be yes. The topic of big data ethics is qualitative and to use criteria that worked within other SLRs such as Kitchenham quantitative approach, did not fit this topic, therefore a contextual assessment needed

SNO	Article Author (s) and title	Topic/Focus/Question	Study Significance	Methodology (Data Collection and Analysis Methods)	Main Themes and Findings	Key Contributions	Limitations	Further research
1	Towards Ethical Data Ecosystems: A Literature Study Minna M. Rantanen, Sami Hyrynsalmi, Sonja M. Hyrynsalmi	Examination of the ethical aspects considered of data ecosystems in academic literature through systematic literature study.	Determining the existing research into data ecosystems and examining the different elements of the academic literature.	This study's focus is on existing work of primary studies on ethical governance of data ecosystems.	Privacy, Accountability, Ownership, Accessibility, Motivation with managing of datasets.	Sets out existing literature and for data governance and ethics and what additional research is needed.	It focused on primary studies only ecosystems, however the main themes are relevant for the topic.	Ethical themes identified but acknowledges that the discussion is fragmented in content and domains.
2	Big Data ethics Andrej Zwitter	Reviews theories of ethics and how they apply or can be adopted in big data.	The study examines the meaning of big data and the ethical themes such as privacy that need to be examined alongside managing big data.	By examining ethical theory, the paper could apply the moral responsibility of big data owners towards those who the data is about and impacts.	Examining ethics, moral responsibility, big data governance and societal impact.	Provides themes of what needs to be improved to compliment big data ethics.	Due to the age of the paper, examples are out of date, but it is still relevant as the issues raised in 2014 still exist, even for privacy issues since the GDPR.	The paper written in 2014 foresaw problems that existed with big data in the early days and addressed the potential problems that can arise, and the steps organisations can take to mitigate the issues with big data.

Fig. 4 Quality assessment criteria

to be formed. The questions laid out in Fig. 4, address this balance. Quality assessments in the sense of contextual papers, as opposed to statistical driven assessments, are difficult to define and their quality can be open to interpretation based on the reader [9]. By using a data extraction form in the following section and in Appendix II, this provides evidence of robust assessment of the papers, their relevance and contribution to this SLR.

2.5 Selection and Data Extraction

An initial assessment was done based on the article title to determine its relevance. Once the title was deemed relevant, the abstract was read to determine if it could add value to the research topic. If the abstract proved relevant, the whole paper was read. Once the paper was checked that it met the quality assessment, the next step was to document the selection process for the papers included in the SLR. An example is shown below in Fig. 5 of two of the papers in the summary data extraction form and the full form for all 14 is in Appendix II. Additional contextual literature such as academic books and grey literature were also reviewed to determine if big data ethics had progressed across industries and if it mirrored academic research.

Fig. 5 Example of data extraction form

Number	Question
1.	How relevant is the paper to the title and abstract? Does it stay on topic?
2.	Has the paper put forward new ideas or expanded the discussion of the topic?
3.	Was the paper clear and comprehensive on the topic?
4.	Did the paper use strong evidence such as references based on their arguments?
5.	Does the paper demonstrate knowledge of the topic through complexity and depth of discussions?
6.	Are the papers unbiased with balanced arguments?

2.6 Descriptive Analysis of Papers Selected for SLR

The papers reviewed were all based on secondary data research. One of the limitations of the SLR papers was that there was no primary data collection and little evidence to demonstrate public and industry opinion or understanding of the topic. On balance, the papers generally had a negative view of big data use and its impacts on society and examples such as Cambridge Analytica appeared repeatedly. Due to the commodification of big data and the benefits it has brought to companies, politicians, healthcare, and other industries, this means it is here to stay. Therefore, literature which include examples of big data being used ethically and the impact would be helpful to demonstrate if ethics works and how industry can implement it. A benefit of not researching a specific sector, was that it shined a light on the broad interpretation and low maturity level of big data ethics across different industries providing the reality of big data ethics in a general sense. In Fig. 6, this shows the years the papers are were published showing a spread-out timeline and simultaneously that recent papers are more common and relevant to this topic. In Fig. 7, this demonstrates what country the papers originate in, further demonstrating the global nature of this question, but interestingly it is explored more in the US where their privacy legislation is not as robust the as EU. In Sect. 3, the reoccurring themes, and findings from the SLR will be explored.

Fig. 6 SLR papers by year
of publication

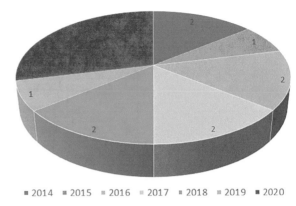

■ 2014 ■ 2015 ■ 2016 ■ 2017 ■ 2018 ■ 2019 ■ 2020

Fig. 7 Publication location
of papers

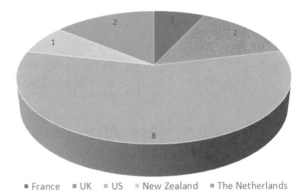

■ France ■ UK ■ US ■ New Zealand ■ The Netherlands

3 Findings and Descriptive Themes from SLR

Given the scale of results found after the search criteria was implemented, this demonstrates that the questions raised around big data use are not new. A core observation is that big data ethics can have different priorities or perspectives depending on industry. A common conclusion of the papers within the SLR is that more research is needed on the topic. Figures 8 and 9 outline respectively how frequent the reoccurring topics appeared in the SLR and which papers touched on the main themes identified.

3.1 Definitions and Terminology

The term big data is often used in an oversimplified manner and it can be difficult to explain the different elements of big data and the impacts it has on industry and society, for example, the filter bubbles people are exposed to online which is undermining common sense in some situations [10]. Richards and King also argue that the

Topic	Topic overview	Volume of papers	Authors
Definitions of terminology within big data	Papers that attempt to define or provide reference to definitions of "big data", "data ethics" and "big data ethics"	8	Mittelstadt, Ibiricu, Rantanen, O'Leary, Herschel, Markham, Vayena, Richards.
Types of ethics	The different models/frameworks that could be considered when using big data.	7	Mahieu, Herschel, Wylie, Ibiricu, Yallop, Markham, O'Leary.
Examples of big data ethics.	Example of ethics being applied to big data and what big data ethics frameworks could look like.	7	Herschel, Ibiricu, Zwitter, Markham, Vayena, Richards, O'Leary.
Accountability for big data ethics	This includes discussion on ownership and accountability within an organisation, industry, and at regulatory level for big data ethics.	6	Mittelstadt, Ibiricu, Wylie, Markham, Richards, Rantanen
Privacy	Discussions on data protection, privacy concerns because of big data and the role of legislation and its relevance to big data ethics.	11	Mahieu, Herschel, O'Leary, Richards, Yallop, Zwitter, Ibiricu, Rantanen, Wenhong, Vayena, Deng,

Fig. 8 Breakdown of themes identified in SLR

term big data focuses on the size of the data but it does not take into consideration the decisions made about individuals even if the decisions are made from smaller datasets [11]. In the US, questions asked for Mark Zuckerberg in 2018 by the US senate, shined a light on the gap between elected officials and their understanding of big data and data ethics, when compared with practitioners within industry, which shows that law-makers are reactive and do not fully understand the industry they

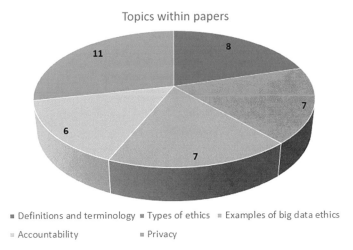

Fig. 9 Topics explored in SLR papers

are regulating for [12]. Currently, any definition and implementation of ethics to big data would be industry or company led, and not through widespread regulation like data protection. From the papers reviewed there was not a universal accepted definition of big data and in some cases, the papers suggested definitions based on their research. In terms of defining "big data ethics" this is more difficult, and authors did not generally attempt to offer a definition. The focus was on defining 'big data' and 'ethics' separately and examining a possible alignment between the two.

3.2 Privacy

Data Protection legislation within the EU focuses on personal data, however big data normally does not include directly identifiable individuals, so how big data is processed does not come under the scope of the legislation [13]. Throughout the SLR, data protection legislation and expectations of privacy are discussed repeatedly by various papers. Spina draws attention to the big data driven economy that is continuing to grow as technology evolves and innovates but this brings up issues such as social media content and marketing, where decisions need to be made that impact society, but they are not covered by data protection as groups are targets and not an identifiable individual [10]. Due to the threat of large fines from the GDPR many senior managers ask the question 'are we compliant' and focus on meeting data protection requirements, however to move towards big data ethics, the question should change to 'are we doing the right thing' [14]. The consensus amongst the papers were that data protection legislation fell short when it came to big data and it did not provide the regulatory oversight required for how it is used. In some papers which predate 2018, but discuss privacy issues, the arguments are still relevant as the law did not overly change when managing big data.

3.3 Accountability

A reoccurring term that is used in various articles is the concept that the owner of the big datasets has 'power'. Owning and controlling big data and deciding who else gets access to the data automatically means power for those in control of the data. If Facebook sells the data on what people 'like' this information is powerful for Facebook and who they sell it to, but it also means power on how it is subsequently used. When a commercial organisation is unrestricted on how it can use a money-making tool, there would be natural resistance to implementing additional rules that could reduce income. A reoccurring theme within the papers was who is accountable for implementing big data ethics with various conclusions. Wylie argues that the responsibility lies with all of us, however the author digs deeper to conclude that 'serving society through good data work lies with practitioners- all of them, including people who produce, curate, analyse and/or interpret data' [12]. In contrast, Markham argues that responsibility towards managing big data ethically is difficult to pin down due to the complexity and integration of big data. If data are extracted by multiple departments for different reasons and as technology evolves, who is responsible for ethics both within business and society becomes complex [15].

3.4 Types of Ethics

Big data ethical questions are not new, however big data underestimated the impact on individuals' abilities to make informed decisions about their wants and needs, particularly when it comes to products and campaigns marked to them' [2]. Ethical questions with big data are complex and often overlooked in the production of big data but elements of ethics used from other areas could be drawn on to implement ethics on big data for example research ethics or building in ethical considerations into decision-making [15]. Ibriricu put forward that ethics involves applying moral behaviour to business standards and that ethics should be integrated at design of technology and processes [16]. Another ethical category that big data could come under is computer ethics which is focused on professional conduct through policies and codes of conduct, thus putting responsibility on individuals and employees [17]. Generally, the papers explored ethics based on consequential ethical theory, however Herschel and Miori examine existing ethical theories and evaluated on how this might fit with big data practices [13]. A related element of big data that also needs to be considered is the ethics of algorithms which generally uses big data for making assumptions about individuals and decisions previously made by humans are being replaced by algorithms. Algorithms reflect the values of the designers and can be inherently biased as a result [18]. The GDPR partially deals with this where individuals have a right to object to automated decisions and can request a human intervention to review the decision [19–21].

3.5 Examples of Big Data Ethics

Individuals and technology have become more interconnected, but ethics has not developed at the same rate and varies depending on industry. Medical research has a mature approach to data, as ethical practices have been used in that industry for a long time, however in other industries, ethics is virtually non-existent [22]. Big data ethics that has worked well in industries such as the medicine and research. In research, any project must put the needs to the individual first, rather than putting the researcher's motivations or arguments for the greater good first [1]. Before embarking on a research project, researchers will generally need ethical approval which involves demonstrating transparent participant consent forms, as well as methodology for how the research will take place [23]. When considering health and big data a balance needs to be found between utilising data for the common good, and respecting individuals right to privacy [24]. This is relevant with the current global pandemic, Covid-19. Different governments globally have different approaches to using big data to minimise spread of the virus as well as respecting the levels of privacy their citizens expect and treating the data ethically. In somewhere like the UK, thus far use of tracking apps has been consent based, whereas elsewhere such as China, citizens do not have the same level of privacy expectations and have little choice when it comes to national big data initiatives [3]. In addition, as demonstrated through systems like track and trace, when big data that is centralised, managed and good quality, this can help improve public health. However, if there are issues with a dataset it can have significant consequences on individuals for example being told to self-isolate unnecessarily could result in financial loss, or loss of access to education for children. The following section will discuss the findings from the themes of the SLR and exam of grey literature and what further considerations for future research are needed.

4   Discussion of Findings

The aim of this research was to understand the current landscape of big data ethics, what this means in practice and how it is applied within industries. Many individuals both in their professional and personal capacity associate data protection legislation with regulating how organisations manage data. However, the GDPR, only governs personal data, which leaves a gap in the regulation of big data. In addition, if big data does contain personal data under the GDPR definition, an organisation may still be able to identify an appropriate lawful basis for their processing [19–21]. Following the publicised scandals such as the use of big data in elections, these highlighted the impact of misusing big data and a gap in regulations and governance. Therefore, this means it is up to organisations to decide how their data is used and shared, without much restriction.

An aim of this research was to understand what big data ethics meant. However, the SLR demonstrated that there is a lack of agreed and consistent terminology for common terms such as big data and ethics as numerous papers who had slightly differing views on what these terms meant. Unlike personal data where there is a UK and EEA wide definition in law, with extensive guidance from the ICO, the same does not exist for big data from an official or regulatory body [19–21]. Without clear definitions this has led to some organisations and industries implementing their own version of ethics to big data, but this would be inconsistent with limited oversight.

Through the SLR, another aim of the research was to understand who is responsible for the implementation and management of big data ethics within an organisation, and what external forces is this accountable to. This was discussed inconsistently across the papers and there was no single approach or framework identified as being commonly used. Most papers seem to have a consensus that there should be some form of regulatory oversight before meaningful action and change in practice will happen. This is demonstrated through Data Protection evolution, where in the UK before the introduction of the GDPR which had substantial fines, many organisation started to seriously consider their data protection obligations for the first time with senior management backing, regardless that the new legislation mirrored existing legislation which dated back to 1998 and a lot of the requirements should have been done anyway [25].

Interestingly, the UK government have a data ethics framework which is principle based, and therefore can be applied to any dataset. The framework was last updated in 2018 and introduced under Matt Hancock, who is currently the UK Health Secretary, leading the UK's response to managing Covid-19 and using data to help inform decision making [26]. The data ethics framework was not discussed in the papers selected for review and there was little grey literature about the framework available, aside from the government websites and no specific academic literature on the framework was found. This suggests that the framework is not extensively used by industry, nor is it hugely publicised. In addition, in 2019 ISO published the standard ISO 20545:2019 which addresses information technology and big data [27]. This was not mentioned in the articles under the SLR and there was limited grey literature on the standard. This demonstrates that there are the beginnings of a move towards big data governance which could translate to ethical practices however it is not mature, and adaption seems to be limited. In the following section concluding remarks and recommendations for further research, policy makers and law makers are laid out.

5 Conclusion and Recommendations

There are gaps in the governance and implementation of big data ethics, as well as societal understanding of what big data is, and how it affects consumers, citizens, and employees. The existing practices are not fit for purpose and this is demonstrated through a lack of basics being pinned down. Universal definitions were not evident in the literature reviewed, however there was reputable grey literature that attempted to address this gap. The academic literature did not appear to accept or adopt the government and ISO frameworks, although the frameworks are relatively new so more time for research into these frameworks may be required. Given the large amount of results that the search criteria yielded, this demonstrates this topic is important, fast-moving, and evolving. This is also a global issue as demonstrated by the fact the papers of the SLR stemmed from countries across the world, and not specific to one industry or part of the world.

To progress big data ethics, the following are recommendations for further research, and steps for policy makers and law makers to consider.

Due to the fast evolution of technology and how it is used, means big data ethics is not in isolation. Ethical considerations also need to be made on the use of algorithms, IoT and Smart Cities. Further research is required on these technologies and their relationship with big data and explore if an ethical framework could be applied across all these types of technologies.

There is an opportunity for big data ethics to tap into pre-existing frameworks such as information governance, or data management best practice. By tapping into other frameworks, which are trendy within some industries, it does not mean costly new projects and lengthy implementation periods [28]. Potentially the data governance framework could complement big data ethics and where senior accountable officers are given responsibility over specific datasets to ensure it is managed appropriately. These responsibilities could expand by adding ethical considerations towards the collection and processing of a dataset. This is recommended for policy makers and senior managers to adopt this approach as an interim measure before a wider framework is commonly available and utilised.

Big data ethics is relevant in current affairs especially in considerations and discussions on how big data can help with the Covid-19 pandemic. As the public become aware of big data is used and its impact, now would be appropriate to education the public on this topic, like with data protection in 2018. The GDPR and the hype surrounding it, demonstrated that the public could learn their rights, and businesses and organisations would take rules around data seriously, because of the reputational impacts, operational disturbances, and regulatory fines that could be imposed if they failed in their responsibilities. If the government took the lead on this matter in a meaningful way, the public and industry would follow.

Overwhelmingly the use of big data has benefited the lives of the public, however that does not mean there should not be ethical considerations made on the impact of specific decisions about big data use, as there can be monumental consequences such as swaying elections, or excessive monitoring of citizens. Ultimately it is evident more work is needed on this topic to determine what big data ethics should look like or how big data should be regulated to serve the interests of the public. Further research is needed into this area, but adoption by industry of established frameworks like ISO or commitment to more transparency and accountability will help make ethics more common and business as usual.

References

1. Richterich A (2018) The big data agenda: data ethics and critical data studies. University of Westminster Press, London
2. Zwitter A (2014) Big data ethics. Big Data Soc 1(2)
3. Chen W, Quan-Haase A (2018) Big data ethics and politics: toward new understandings. Soc Sci Comput Rev 38(1), 3–9
4. BBC (2020) BBC NEWS, technology [Online]. https://www.bbc.co.uk/news/technology-538 36453. Accessed 13 Sept 2020
5. EU (2020) General data protection regulation [Online]. https://gdpr-info.eu/. Accessed 28 Sept 2020
6. LA Times (2019) Shadowy data brokers make the most of their invisibility cloak [Online]. https://www.latimes.com/business/story/2019-11-05/column-data-brokers#:~:text=The%20d ata%20broker%20industry%20is,files%20on%20millions%20of%20Americans. Accessed 28 Sept 2020
7. Rantanen MM, Hyrynsalmi S, Hyrynsalmi SM (2019) Towards ethical data ecosystems: a literature study. In: 2019 IEEE international conference on engineering, technology and innovation, Valbonne
8. Kitchenham B (2007) Guidelines for performing systematic literature reviews in software engineering. Software Engineering Group, Durham
9. Denyer D, Tranfield D (2011) The Sage handbook of organizational research methods. Sage Publications Ltd., London
10. Spina A (2017) A regulatory Mariage de Figaro: risk regulation, data protection, and data ethics. Eur J Risk Regulat 8(1):88–94
11. Richards NM, King JH (2014) Big data ethics. Wake Forest Law Rev
12. Wylie C (2020) Who should do data ethics? Patterns 1(1)
13. Herschel R, Miori VM (2017) Ethics & big data. Technol Soc 49:31–36
14. Yallop AC, Aliasghar O (2020) No business as usual: a case for data ethics and data governance in the age of coronavirus. Online Inf Rev
15. Markham AN, Tiidenberg K, Herman A (2018) Ethics as methods: doing ethics in the era of big data research. Social Media+Society 4(3)
16. Ibiricu B, van der Made ML (2020) Ethics by design: a code of ethics for the digital age. Records Manage J
17. O'Leary DE (2016) Ethics for big data and analytics. IEEE Intell Syst 31(4):81–84
18. Mittelstadt BD, Allo P, Taddeo M, Wachter S, Floridi L (2016) The ethics of algorithms: mapping the debate. Big Data Soc 3(2)
19. Information Commissioner's Office (2020) Lawful basis for processing [Online]. https://ico. org.uk/for-organisations/guide-to-data-protection/guide-to-the-general-data-protection-regula tion-gdpr/lawful-basis-for-processing/. Accessed 9 Sept 2020
20. Information Commissioner's Office (2020) Rights related to automated decision making including profiling [Online]. https://ico.org.uk/for-organisations/guide-to-data-protection/ guide-to-the-general-data-protection-regulation-gdpr/individual-rights/rights-related-to-aut omated-decision-making-including-profiling/. Accessed 7 Sept 2020

21. Information Commissioner's Office (2020) What is personal data? [Online]. https://ico.org.uk/for-organisations/guide-to-data-protection/guide-to-the-general-data-protection-regulation-gdpr/key-definitions/what-is-personal-data/. Accessed 29 Aug 2020

22. Mahieu R, van Eck NJ, van Putten D, van den Hoven J (2018) From dignity to security protocols: a scientometric analysis of digital ethics. Ethics Inf Technol 20(3), 175–187

23. Health Research Authority (2020) Applying to a research ethics committee [Online]. https://www.hra.nhs.uk/approvals-amendments/what-approvals-do-i-need/research-ethics-committee-review/applying-research-ethics-committee/. Accessed 2020 Sept 13

24. Vayena E, Salathé M, Madoff LC, Brownstein JS (2015) Ethical challenges of big data in public health. PLoS Comput Biol 11(2)

25. Davies CLJ (2018) GDPR: implementing the regulations. Bus Inf Rev 35(2)

26. Gov.UK (2018). Data ethics framework [Online]. https://www.gov.uk/government/publications/data-ethics-framework/data-ethics-framework. Accessed 29 Aug 2020

27. International Organization for Standardization (2020) BSOL online [Online]. https://bsol.bsigroup.com/Bibliographic/BibliographicInfoData/000000000030341487. Accessed 10 Sept 2020

28. Hagmann J (2013) Information governance—beyond the buzz. Records Manage J 23(3):228–240

Transforming Higher Education Systems Architectures Through Adoption of Secure Overlay Blockchain Technologies

Foysal Miah, Samuel Onalo, and Eckhard Pfluegel

Abstract The adoption of Distributed Ledger Technology (DLT) has been growing tremendously in recent years following the introduction of Bitcoin in 2009. However, the usefulness of DLT is not limited to the financial sector, and this paper investigates the viability of DLT architectures for use in Higher Education (HE). This sector faces challenging financial constraints, and one way to address this problem is to adopt emerging DLT technologies as architectures for HE systems. This article presents the ASTER Open Source system, a hybrid DLT integration within the context of a student submission system for assignment grading purposes. ASTER addresses many concerns of traditional system architectures such as centralisation, system downtime, and decoupling; all of which are mitigated through the use of blockchain technology. The advantages and drawbacks of such a new approach are discussed, including the aspect of security concerns relating to student work being submitted to a public ledger.

Keywords Blockchain security · Security overlays · Decentralised ledger · Technologies · Higher education systems

1 Introduction

From both an institutional and student perspective, traditional learning management systems rely heavily on centralised infrastructure, and even though these solutions function well, there are limitations that require re-evaluation. Whilst the focus here

F. Miah · S. Onalo · E. Pfluegel (✉)
Kingston University London, Kingston upon Thames KT1 2EE, UK
e-mail: e.pfluegel@kingston.ac.uk

F. Miah
e-mail: foysal@kingston.ac.uk

S. Onalo
e-mail: k1450301@kingston.ac.uk

© The Author(s), under exclusive license to Springer Nature Switzerland AG 2021
H. Jahankhani et al. (eds.), *Cybersecurity, Privacy and Freedom Protection in the Connected World*, Advanced Sciences and Technologies for Security Applications, https://doi.org/10.1007/978-3-030-68534-8_21

343

is on HE systems, the above postulation can be applied to most systems currently in operation.

The first obstacle for an institution is cost. Universities in the UK and elsewhere are under extreme pressure to reduce costs with drastic measures being taken to ensure continued operation across the board. As mentioned previously, the current solutions function well, however, the cost of maintaining a centralised system is high, typically requiring dedicated staff to manage system issues and relentlessly update the software to keep in line with changing external factors. While the recent uptake of cloud infrastructure technology, presenting features such as Software as a Service (SaaS) and Platform as a Service (PaaS), have dramatically reduced overheads, this approach introduces a different set of problems.

Second, there are the concerns of the student to address. Submitting assignments to a centralised system means the student is reliant on the software vendor to keep their submission safe, not to mention their personal data, which is particularly important with the recent data protection changes decreed by the European Commission. Institutions and vendors must adhere to GDPR guidelines [5] to ensure personal data is kept secure or they could face severe fines. Submissions to a centralised system are prone to intermittent outages, especially during critical submission times, mainly due to insufficient infrastructure resourcing. The obvious solution would be to increase resource availability during peak times. However, the increased resources would remain idle during off-peak periods, the cost of which would need to be considered by the institution and by extension, passed to the student. Finally, insufficient security measures are also a problem with personal data being the primary target for cyber criminals. Even though the strict regulations imposed by GDPR legislation have obligated vendors to improve their security protocols, such measures incur costs that are passed down to the student to bear.

With the introduction of a decentralised submission system, the above issues would no longer be relevant. Such a proposed system would be released as Open Source software maintained by a community of developers; the infrastructure would be formed as a peer-to-peer network with the students, institutions, and the public running client software as processing nodes. This architecture would require little to no staff to maintain. A further potential benefit could be an additional revenue stream for universities, by selling off the currency that is generated for submission processing. In times of a pandemic, austerity, and an uncertain financial climate, this type of technology could potentially help bring running costs down dramatically while maintaining infrastructure integrity. Research regarding currently active blockchain networks has found no attempt so far in developing a blockchain that is explicitly targeting the HE sector. There is an existing blockchain that deals with the lengthy time it takes to publish a research paper to related journals [7], but this does not deal with assignment submissions by students.

The main contribution of this paper lies in secure blockchain technology. We present the design and development of a secure student submission system with a novel blockchain architecture based on security overlays. This has led to the creation of an Open Source prototype solution named ASTER [4], providing an end-to-end decentralised and secure system, mitigating the potential security risk of using a public, insecure blockchain as far as the confidentiality of blockchain data is concerned.

The paper is structured as follows: in Sect. 2, we review traditional systems and architectures including pertinent aspects of Blockchain technology. In the next section, our system design and implementation are presented. This is followed by a description of the security in Sect. 4. Section 5 concludes the paper.

2 HE Systems Architectures

In this section, we review salient aspects of Higher Education system architectures and explain the advantages of DLT versus a centralised approach.

2.1 Removing Central Points of Failure

Time after time, there are reports identifying corporations that have had their systems breached in one way or another. The global governments then put legislation in place to ensure these breaches do not become commonplace. However, when a breach is found, and a company is fined, it is not the company that suffers in the long run. Fines, especially against the largest companies, can be recouped simply by raising the prices of their products affecting the consumer. Not to mention the costs involved in ensuring security is kept up to date, which are invariably passed on to the consumer. These attack vectors exist because of one fundamental architectural design flaw—the centralised nature of traditional systems.

Within the HE context, our ASTER system aims to remove all central points of failure, this is achieved by the very nature of DLT design, by replicating the network across a global cluster of nodes. By doing so the network is protected from security issues such as DDoS attacks thus eliminating availability issues plaguing traditional architectures.

Figure 1 shows a comparison between a traditional cloud system and a potential architecture utilising DLT.

DLT addresses these security concerns by design as there is no single point of failure to exploit. An attacker would need to target every single node on the network simultaneously to have any effect; any such attempt would be extremely cost-prohibitive.

2.2 Infrastructure and Software Architecture

Since the beginning of computing, traditional systems have been designed around the idea of centralisation, housing data and applications on a network server onsite.

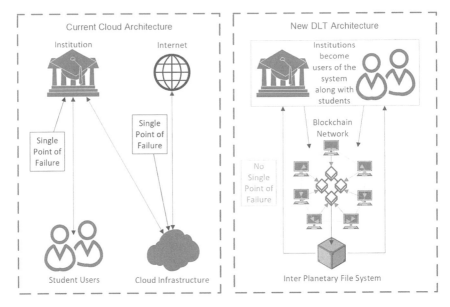

Fig. 1 High level architecture comparison

As time has gone on, methods such as multi-point failover processes have been introduced to mitigate data loss and cloud solutions have been adopted primarily to allow for service continuation and cost reduction.

The HLD below depicts a possible scenario that could be adopted in the form of a cloud-hybrid solution. The diagram depicts the application layer between the users and the LMS, the cloud platform, which would host the blockchain and the communication between the IPFS storage layer (Fig. 2).

2.3 Addressing System Downtime

Any system analyst maintaining an enterprise system will confirm that any form of system downtime creates added pressure to their workload. Even scheduled downtime is always a cause for concern. There is some semblance of control if the entire architecture is on-prem, but this is becoming more of a distant memory, especially when considering cloud solutions, where the baton of ownership is being passed on to a 3rd party with their supposedly iron-clad promise of adherence to accompanying SLAs.

No matter the operational process in place to address downtime, there is one inevitability. Traditional and current systems are always going to be prone to some level of system downtime which is why no service provider guarantees 100% uptime. A simple online search will point out services offering 99.x% uptime, and it has

Fig. 2 High level diagram

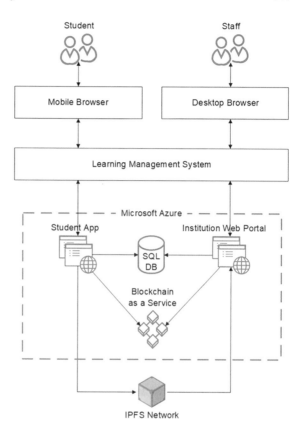

become commonplace for a service to be measured for reliability with the number of successive 9 s. But this is where DLT disrupts the status quo—by offering 100% uptime. One might argue that performance may be an issue, and it may very well be, but the system would never suffer downtime—ever. The only time the system could be down is if every node stopped using the service.

3 System Design and Implementation

In this section, the design and implementation of our system will be presented.

3.1 General Architectural Considerations

Current architecture methods used are very archaic, even when considering using infrastructure in the cloud. These methods are still modelled around the idea of centralisation. A typical architecture might involve an application layer, middleware, services, data layer and a platform layer. Each of which will typically sit on a server making up the full stack. In a more modern design, these layers may be separated to ensure reliability and maintainability, but still, sit on servers. When it comes to cloud service design, all that is happening is that these servers are no longer owned and maintained by individual companies, but leased out by cloud service providers.

Currently, the vast majority of universities may maintain their infrastructure and equipment, or lease services from cloud service providers. Figure 3 shows how a typical architecture might look against the proposed architecture.

Our proposed architecture would replace the Infrastructure as a Service, with a DLT infrastructure, containing Off-Chain Data (in the case of the ASTER system, this would represent IPFS), On-Chain Data (the IPFS reference data on the EVM), the Blockchain Network (client nodes confirming the transactions taking place) and the Blockchain Transaction Ledger (the confirmed transactions). Stripping away the old infrastructure layer would mean universities would be saving on costly equipment and leasing costs as well as expensive on-going maintenance costs.

3.2 Potential Architecture Types

Several types of architecture can be designed to implement ASTER. There are many frameworks available to assist with this, and the difficulty here is choosing the correct tools for a useful implementation. This section will illustrate three methods of potential architecture designs: *Ethereum-Based*, *Independent Blockchain*, and finally, *Hybrid Solution*.

Ethereum Based Architecture An Ethereum based approach would entail storing files on the Ethereum blockchain, this can be achieved by creating an Ethereum contract between the university and the student, which would allow students to submit their work directly to the Ethereum blockchain. Once the work is submitted, the student will have proof the submission has taken place, as this will be reflected in their Ethereum wallet and the ethscan [3] explorer, where all transactions are logged. This approach has numerous advantages. It consists of a straightforward method to store student submission and would bring an effortless setup only requiring the creation of an Ethereum contract. This solution seems an obvious choice until the costs are explored. Due to the monetisation of Ethereum, mitigate this cost for a university may prove challenging and may not be ethically practical. Also, the resulting system will still be reliant on legacy architecture where there is a potential point of failure. Furthermore, a significant disadvantage is the lack of data confidentiality. Entire student submissions would be stored openly on the public transaction database, with obvious potential detrimental consequences to students and lecturers alike.

Independent Blockchain The development of an independent, purpose-built blockchain would allow for a currency that is not monetised outside of its intended environment but used for assignment submission and content creation. The

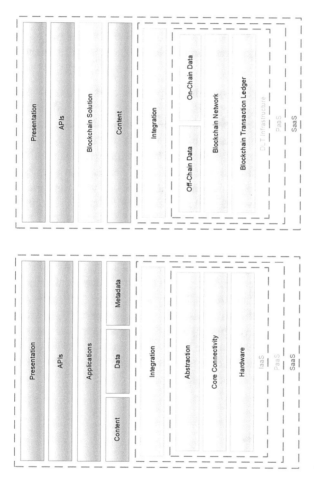

Fig. 3 Architectural comparison

blockchain would require students and possibly the university to mine currency by processing the submissions and content to the blockchain. This mined currency would then be used for submission fees. Any additional currency accumulated could then be used to purchase other services from an institution, such as graduation tickets, merchandise, and other university products or services.

Reviewing the advantages of this architecture, it can be stated that while there is no real monetary value associated with the cryptocurrency, students may be able to use the currency to exchange for other university items. Furthermore, there would be little to no maintenance costs arising. On the other hand, this would require the development of a complete blockchain network, which is a serious challenge. The system will still be reliant on legacy architecture where there is a potential point of failure.

Hybrid Solution With a hybrid system, it is possible to use two separate systems to achieve the end goal, such as using IPFS and Ethereum. IPFS is a distributed storage system which allows any user to store any type and size of data [6]. Currently, there is no native feature within IPFS that establishes who or when a file has been submitted to its network, which means using the platform on its own is not viable, however, by decoupling the data from the user submission details, it is possible to design a system which stores the submissions on the IPFS network and the submission details on the Ethereum network. Both of which would create an immutable record. As discussed above, storage on Ethereum is highly cost-prohibitive; however, this hybrid system would only be storing up to 1 KB of data on the Ethereum network, which equates to 54 pence per KB per submission. With an average of 6 submissions per year, at the Ethereum price stated earlier, submissions would cost £9.72 over the duration of a student's 3-year undergraduate course. However, this indicative cost is subject to market price fluctuations. The hybrid solution exhibits several positive aspects. It is highly cost-effective, has little to no maintenance costs and can be implemented using rapid development, as the storage system is already established. The inconveniences of this solution are that transaction costs are subject to Ethereum market price fluctuations and that the system will still be reliant on legacy architecture where there is a potential point of failure.

3.3   The ASTER System

The ASTER proof of concept system is based on the hybrid architecture described previously. ASTER utilises an Ethereum smart contract to store the IPFS address created on file submission. ASTER can be thought of as a hybrid dApp which will use Metamask to transact between EVM and IPFS. The architecture consists of a mobile client front end and a web portal for administrative tasks. A data controller handles the data processing between the client front end and data storage. Finally, the data storage layer utilises IPFS, which will store the student submissions ready for lecturers to mark.

The mobile front end has been created using Xamarin forms allowing for a single codebase to be shared across various mobile platforms. The data controller is implemented as a web service, with C# being the coding language. The web portal for lecturers is coded using ReactJS. The storage layer uses the IPFS network storing all submissions to a public network of active storage nodes, with transaction data being stored on the Ethereum network. In order to simulate a transaction being processed, the Ethereum contract will be created using Solidity and submitted to the Rinkeby test network.

Finally, there is a need for a database to allow for credentials to be managed. The database is created using the data first approach using Microsoft Entity Framework and deployed to the Microsoft Azure Cloud Platform. A data controller API angles the business logic between the database and client application. The Azure platform provides commercial cloud computing services across many data centres across the globe. To simulate a typical HE back-end, Microsoft Azure is configured to host the web application and the data API, as well as the database back-end.

4 Implementing Security

Security is a fundamental requirement for the proposed system and its application area. Student submissions need to be protected concerning integrity and confidentiality against both internal and external attackers. In this section, we commence by illustrating the existing security mechanisms of blockchains. The need for additional security in the form of data confidentiality will be highlighted, and a novel mechanism for providing this security requirement and its role within ASTER will be presented. This continues our research on security protocols [9] and security overlays [10].

4.1 Standard Blockchain Security Features

It is vital to understand that while cryptography is used in particular, specific areas of blockchain technology, it does not provide complete and comprehensive security. However, the security qualities provided rival many centralised systems by a substantial margin. The use of these features is quintessential to the successful implementation of ASTER.

Secure Hash Functions A fundamental operation of the blockchain system is the block hashing process; this process is responsible for verifying every new block added to the public ledger and uses cryptography to achieve verification. Various cryptographic methods are in use within different blockchain implementations, the most popular being Secure Hashing Functions. Rapid integrity verification is achieved through the use of sophisticated data structures.

Public-key Cryptography Asymmetric (Public)-key Cryptography was first suggested in 1976 by Whitfield Diffie and Martin Hellman [2], their idea was to introduce a public and private key pair and is the underlying principle of industry-standard encryption and digital signature algorithms such as RSA [11] or DSA [11] used today. Public-key Cryptography is used to tackle the following main security challenges within the blockchain process: to validate the authenticity of a transaction and to provide ownership anonymity.

On the initiation of a transaction, to ensure ownership of the data contained, a cryptographic signature must be passed along as part of the transaction. As the cryptographic signature is created using a private/public key combination unique to the owner, the blockchain network and the client nodes within the network can then confirm the origins of the transaction thus validating the transaction as authentic.

Since the majority of blockchain implementations are public, ownership anonymity becomes a high priority, as anyone can interrogate the blockchain ledger, and if the transaction data are not anonymous, the ownership is easily identifiable. The blockchain process handles this aspect by allowing the transaction originator and recipient to create a wallet address using asymmetric encryption.

4.2 The Need for Data Confidentiality

The use of encryption is not necessarily available when creating data stored in a blockchain. However, in many systems nowadays, this necessity arises, partly due to the exposure of online systems to attacks, partly due to more sensitive data and transactions present. Major blockchain providers such as Hyper Fabric Ledger [1] and Multichain [8] have responded to this need by releasing permissioned blockchains, where access control can be managed using a central entity. Encryption of data is provided as an additional feature, sometimes as a paid premium feature. This approach contradicts the original philosophy behind Blockchain systems such as Bitcoin, as it is deviating from the idea of decentralisation. It also requires time and overhead for managing these permissions and might require the setting up of a Public Key Infrastructure.

4.3 Virtual Private Security Overlays

In our previous research [10], we have suggested an alternative security approach, based on security overlay architectures. The basic idea is to apply a suitable secure information dispersal scheme such as *secret sharing* [12] in order to diffuse sensitive data on several blockchains. This achieves transaction confidentiality as long as a threshold number of individual blockchains is not inspected simultaneously. In particular, if this idea is applied to public blockchains, the resulting architecture may be seen as a blockchain with additional security properties and is referred to as *Virtual Private Blockchain* (VPBC) in analogy to a Virtual Private Network in traditional network security. In this approach, confidential transaction content is replaced with "fake" pseudo-content the precise choice of which will strongly depend on the specific application scenario. The transaction recipient will be able to retrieve the original data by combining a set of fake transactions, using a suitable method. Depending on how the transaction data is structured and what the specific blockchain application prescribes in terms of security requirements, an additional out-of-band channel might be required. The main advantage of this approach is that it does not rely on encryption, as it implements confidentiality through covertness. In addition, it is very flexible and can be based on any number of individual blockchains and transactions.

4.4 ASTER Security Approach

The main difference of ASTER to the VPBC approach is the restriction to a single Private Blockchain, in this case, the Ethereum system. Data diffusion will be achieved through multiple transactions, and the arising need for a secure out-of-band channel is implemented based on email. The motivation behind this design decision is the fact that one of the earlier versions of ASTER was already implemented based on Ethereum; and that the existence of an email channel between students and lecturers is a realistic assumption.

A secret sharing scheme with parameters m and n is also called a (m, n) threshold scheme and it has the property that given data (the *secret s*) can be divided into n parts (the *shares*) in such a way that m shares are sufficient to reconstruct s.

Consider an intended transaction with sensitive transaction data d, requiring protection. This will be shared as n shares $d_1,...,d_n$ using fake transactions and an email message if required. This will be explained in the following example: assume the submission of a student assignment. The transmitted information is the student name, ID number and the actual assignment document and a reasonable decision would be to consider the ID number ID (for data privacy reasons) and assignment document A (in order to prevent cheating) confidential. Hence, the data d are the concatenated latter two pieces of information.

In order to create suitable fake student assignments, one can proceed as follows, where without loss of generality we will discuss the individual pieces separately: denote ID_i the ith share of the ID number. Rather than including this share information in the fake assignment, we can send the values $ID_i + R_i$ and $R_1 \oplus R_2 \oplus ... \oplus R_n$ using the email channel where the R_i are random numbers. Including the fake assignment documents requires additional care, as typically shares in a secret sharing scheme appear as random values. Unless it would be argued that documents would be encoded a (potentially proprietary) binary encoding scheme, the following approach could create plaintext documents: slightly abusing notation, we will use the same R_i to denote a new set of numbers to be determined. The aim is to create a fake set of assignment documents B_i. If we consider the set of equations $A_i \oplus R_i = B_i$ ($i =$

$1, \ldots, n$) we can solve for the R_i and proceed as in the case of the student ID numbers, using an email.

A mechanism to explain the resulting proliferation of assignments submissions needs to be in place. This could be achieved by simply having an artificially large number of students enrolled for the assignment. In case of suspicion raised, this could be explained as having distance learning students, students from the previous cohort retaking the assignment, and so on.

5   Conclusion

This paper has investigated Distributed Ledger Technology and how the concept could be applied to improve the current student submission system implementation that is in use. We have investigated the viability of DLT integration within the context of a student submission system for assignment grading purposes and defined a detailed design of a prototype and chosen specific technologies to integrate with the ASTER prototype system. This prototype system has been developed which showcases the use of two unrelated blockchain technologies to submit and store student assignment submissions, with a legacy backend configured on the Azure platform simulating student data that would be in use by a university. An innovative mechanism to establish data confidentiality, an aspect often neglected in current blockchain technology, has been designed.

The ASTER system currently has limitations and will require additional work to become a fully functioning and production-ready application. The prototype was aiming to demonstrate the ability to produce an application that would connect to an already existing system and whilst the API is able to generate lists of assignments, students, courses, modules and lecturers, it is not possible to create lists assigned to particular users, but this can be achieved by revisiting the LINQ code.

Due to timing constraints, the Xamarin forms application could not be built and would have been an added benefit for those wishing to use ASTER on a mobile platform, however, the ASTER front-end client can be used on a mobile device as it is responsive. However, the main research question has been proven, which was to create a hybrid application that will allow student assignments to be stored on a decentralised system and also making use of distributed ledger technology. The benefit of using such emerging technologies is also highlighted successfully in the prototype, particularly where the cost of submitting a document or collection of files of any size is a fraction of a penny.

ASTER is potentially looking at re-defining how the IT infrastructure within HE currently operates, a move like this would usually require a cultural change across the institution, however it may be possible to cushion the change impact by introducing the architecture gradually. Starting with ASTER which deals with the core of HE business—the dissemination, collection and grading of student papers. Targeting this particular system initially will ensure buy-in from academics as well as the student body. With a successful implementation through the institutions existing

VLE, additional services can be provisioned incrementally. Introducing DLT within the HE sector would also provide the much needed, positive exposure which has been marred by groups and individuals misusing the technology and tainting it with the perception of untrustworthiness.

References

1. Cachin C et al (2016) Architecture of the hyperledger blockchain fabric. In: Workshopon distributed cryptocurrencies and consensus ledgers, vol 310
2. Diffie W, Hellman M (1976) New directions in cryptography. IEEE Trans Inform Theory 22(6):644–654
3. Etherscan.io (2019) Ethereum (ETH) blockchain explorer. https://etherscan.io/
4. FoysalM (2019) ASTER. https://github.com/FoysalM/ASTER
5. Ico.co.uk (2018) Penalties, https://ico.org.uk/for-organisations/guide-to-data-protection/guide-to-law-enforcement-processing/penalties/
6. Ipfs.io (2019) Datasets. https://awesome.ipfs.io/datasets/
7. Mackey TK, Shah N, Miyachi K, Short J, Clauson K (2019) A framework proposalfor blockchain-based scientific publishing using shared governance. Front Blockchain 2:19. https://doi.org/10.3389/fbloc.2019.00019. https://www.frontiersin.org/article/
8. Multichain (2019) Stream confidentiality. https://www.multichain.com/developers/stream-confidentiality/
9. Obinna O, Pfluegel E, Clarke CA, Tunnicliffe MJ (2017) A multi-channel steganographic protocol for secure SMS mobile banking. In: The 12th international conference for internet technology and secured transactions (ICITST-2017). IEEE, Cambridge
10. Onalo S, Deepak GC, Pfluegel E (2020) Virtual private blockchains: security overlays for permissioned blockchains. In: The fifth international conference on cyber-technologies and cyber-systems CYBER 2020. IARIA XPS Press, Nice, France
11. Rivest RL, Shamir A, Adleman L (1978) A method for obtaining digital signatures and public-key cryptosystems. Commun ACM 21(2):120–126
12. Shamir A (1979) How to share a secret. Commun ACM 22(11):612–613

Centralised IT Structure and Cyber Risk Management

Kamran Abbasi, Nick Petford, and Amin Hosseinian-Far ⓘ

Abstract Against the backdrop of organisational needs to derive value from IT Organisations through agility, efficiencies and cost effectiveness, many organisations have adopted a decentralised IT organisational structure, enabling individual business units the autonomy to implement, operate and govern technology. The increase risk that poses organisations through cyber-attacks, raises the question of how IT security could effectively provide the level of organisations governance to counter cyber threats in a decentralised organisational model. In exploring the challenges in the decentralization of IT security, we highlighted that the accountability of such activities would become diluted, with each business unit managing security in their own methods and practices or lack of, while unable to take full accountability due to the complex independencies of modern system architectures, often resulting in a lack of ownership, accountability and reporting of security at an organisational group level. This ultimately increases the overall security risk to the organization. We further highlighted that while centralization of IT security at a group level would be more effective, a hybrid model of IT security at two-levels with strategy and policy at the central governance level and a degree of autonomy and decision at the IT Operational level could also be considered.

Keywords IT · Information security · Cybersecurity · Centralisation · Decentralisation · IT organization · IT value

K. Abbasi · N. Petford · A. Hosseinian-Far (✉)
University of Northampton, Northampton NN1 5PH, UK
e-mail: Amin.Hosseinian-Far@Northampton.ac.uk

K. Abbasi
e-mail: Kamran.Abbasi@Northampton.ac.uk

N. Petford
e-mail: Nick.Petford@Northampton.ac.uk

1   Introduction

As businesses compete with one another for the competitive edge and dominant market share, it has become evident that IT can play a crucial role in enabling firms to meet their strategic objectives [1]. Firms may have to increase their investments in Information Technology (IT) to remain efficient, innovative, agile, and compete against their market competitors.

As Pajic et al. [2] highlight the increasing use of information technology has resulted in firms needing to evaluate the productivity impact of IT investments through IT value measures [2]. However, IT value has been a continuous discussion for organisations. Lei and Huifan suggest that organisations are challenged to determine the overall organisation performance generated by IT capabilities [3].

In today's organisations, data is considered as a treasured asset, which with appropriate data analytics techniques, can enhance business decision makings [4]. Lowry and Wilson [5] argues that modern business organisations increasingly depend on their IT departments, he further goes on to suggest that IT Organisations are not merely expected to provide supporting services but more so becoming strategic partners and providing value-added services, moreover aligning its objectives and priorities with those of the departments and organisations overall strategy [5].

Lowry and Wilson's view relies on the assumption that IT performance will be optimised to meet the businesses demands and needs. Often IT performance is criticised for the lack of service quality and agility to meet the requirements of the broader organisation [5], as Whyte et al. suggest that IT organisation often failed to support businesses efficiently and in particular to change business attitudes and satisfy user needs [6].

This can result in organisations moving towards outsourcing their IT Services to third party organisations or decentralisation of the internal IT organisation and its capabilities. Moreover, it is against the backdrop of organisations move towards decentralisation of their IT capabilities and the increased risk of security breaches and the implications of them to an organisation's reputations and revenue that this paper aims to review some of the critical considerations of a Centralised IT security capability.

2   IT Security in IT Organisations

As the growth of online channels such as e-commerce and mobile commerce continues to increase and become a key revenue generator and strategic objective for most organisations, the need for robust IT Security governance has also become apparent. While previously IT security was often seen as a reactive measure, afterthought or over-head cost, the growing pressures to keep data and systems safe from customers, stakeholders and government regulators has forced organisations to elevate proactive and robust security measures [7], which includes people, technology and process considerations [8].

According to Hooper and McKissack [9], in the past ten years, cybersecurity breaches have cost organisations worldwide billions of dollars. Most notably, technology firms such as eBay, Adobe Systems, AOL and Sony Interactive Entertainment's PlayStation Network have suffered heavy losses, resulting in widespread media reporting and served to attract organisation and public awareness of the potential damages of security breaches [9]. For this paper, security is defined as the protection against undesirable disclosure, destruction, or modification of data in a system and also the protection of systems themselves [10]. There are three key elements which underpin this definition, and these are vulnerabilities, exploits and threats.

- Vulnerabilities—these are bugs, weaknesses or flaws found in the design of the system architecture or processes which allow attackers to comprise these vulnerabilities to execute nefarious activities such as un-authorised access to data, Phishing or denial of service attacks (DDoS) [11],
- Exploits—these are actions which are executed by attackers on the identified vulnerabilities using various tools and techniques, often for purposes of self-satisfaction or financial gain [10, 11],
- Threats—these refer to the impending risk of an exploit that may be executed on identified vulnerabilities. Threats enable organisations to put in place countermeasures to mitigate and nullify the vulnerabilities and potential attack.

Whilst the importance of IT Security for an organisation is apparent [8–10], Organisations have in recent years been faced with the dilemma of centralising or decentralising their IT capabilities.

Brynjolfson, in his paper titled 'information assets, technology and organisation' [12] explained how information technology had the potential to significantly affect the structure of organisations. Almost 26 years on there remains a debate on how best to formulate the IT Organization within the context of the wider organisation. A continued 'merry-go-round' has witnessed the early popularity of centralisation to decentralisation in the 1980s and then re-centralisation of the 1990s [13]. In recent years with the growing disruptive digital phenomena and organisation drivers to promote innovation and agility [14], businesses are again seeking to ask the question whether to centralise or decentralise their IT organisations.

King [15] makes the basic assumption that centralised IT benefits the organisation by economies of scale while decentralised IT benefits by economies of scope [15]. Centralisation of IT versus decentralisation of IT refers predominantly to three key aspects. Firstly, the control of autonomy of decisions making in the organisation. Centralised organisations largely concentrate the decision into a single business unit, person or a group of individuals, while decentralisation primarily means devolving the decision-making authority and autonomy to individual departments and business units. This is supported by Richardson et al. [16] report from a 1987 study by Przestrzelski suggesting decentralisation can be broadly defined as "a dynamic, participative philosophy of organisational management that involves selective delegation of authority to the operational level" [16].

In the context of IT Security, individual business units would now have the freedom to make their own security-related decisions, such as the procurement and delivery of software, hardware, security governance and processes, controlled use of administrative privileges, and vulnerability assessment and remediation activities. Secondly, the physical location of resources. Centralisation often has resources in one place, while decentralisation spreads resources across multiple locations within the organisation. Thirdly, capabilities and functional activities. In centralisation, control and governance of functional capabilities would be driven from a central competency centre, while in decentralisation the functional capabilities would be disseminated across single or multiple business units.

In traditional organisations IT security has often fallen under the CIO organisation providing centralised security governance, and compliance, Hooper and McKissack [9] argue that while this arrangement made sense, the downside resulted in IT security often being diluted in the plethora of other capabilities that IT was responsible for, not only in relation to priority but also budget allocation, with IT security often fading into the background unless there are had been a major security breach [9]. Whilst the authors do not advocate the decentralisation of IT security to individual business units within an organisation, they do however pose the question of where best fits a central IT security capability within an organisation. The authors highlight that while placing a central IT security function under the CIO could have benefits of synergies between both functions and efficiencies resulting in greater value for the organisation, this could also result in inhibitors for the security capability to highlight security threats, vulnerabilities and exploits of the CIO function. Whilst, separating the two functions out also comes with the challenge of diluted accountability as much of the security governance and principle are reliant on the underpinning IT systems and processes.

3   Centralisation and Decentrlisation of IT Security

For most organisation, the risk of IT security breaches remains high, ensuring business continuity, threat avoidance, quick incident resolution and disaster recovery. In a decentralised model, the accountability of such activities becomes diluted with each business unit managing security in their own methods and practises or lack of, while unable to take full accountability due to the complex independencies of modern system architectures, often resulting in a lack of ownership, accountability and reporting of security at an organisational group level. King [15] explores aspects of both centralising and decentralising. He suggests that centralisation of control preserves top management prerogatives, capitalising on economies of scale and to preserve organisational integrity in operations. The economies of scale arise from exploiting the full potential of technologies that cause the output to increase more rapidly than costs. The costs of duplicating overhead and facilities can be avoided, and organisational protocols are easier to enforce, while decentralisation allows lower-level managers discretion and authority in decision making, while also fostering a culture of innovation of new opportunities and responsibility for their decision making, possibly improving their performance. However, decentralisation of control may lead to problems of accountability and decision making if lower-level managers lack key competencies and are not held accountable for decisions [15]. King's point on key competencies and accountability is particularly pertinent to IT Security. Khallaf and Majdalawieh examined whether the CIO's competency is a determinant of IT security performance measurement. The study highlights that CIOs' knowledge in IT acquired through their education or work experience improves the performance of IT security [17]. The study reaffirms King's [15] viewpoint that by decentralising capabilities there may be a loss of key skills and competencies, with IT security itself being a complex domain which requires experience, knowledge with security architectures, processes and governance professionally designed [18]. To explore this view further, we explore some of the most common attacks cybersecurity vulnerabilities that organisations face today and how they would be complicated in a decentralised security landscape. Several studies have attempted to classify, characterise and provide recommendations to tackle cyber and cyber-enabled threats and security implications e.g. [19, 20]. A study by Humayun [10] Identified and analysed common cybersecurity vulnerabilities. The findings highlighted that Denial of service (DoS) was the most commonly addressed vulnerability (37%). The second most common vulnerability was Malware (21%), and finally, the third most common was Phishing (9%) [10]. We can see from the authors' research that all three vulnerabilities constituted to 67% of cybersecurity threats that organisations encounter today. In order to understand this better, we describe some of their key characteristics.

3.1 Malware

Malware is a shorthand term used for malicious software. In this attack, software programs are deployed on to user computers or servers to gain unauthorised access. The intent behind these types of attacks is to compromise organisational network devices in order to gain control of the host systems and networks for malicious aims [10]. A variety of malware types exists such as Viruses, Trojan Horses, Worms, Ransomware and Spyware. One of the most recent trends, Ransomware, a type of Malware has over the last five years gained prominence [21]. Ransomware is where a victim of an attack is blackmailed. According to Cartwright and Cartwright [21] there approximately hundreds, if not thousands, of ransomware strands in the wild [21].

Two such examples of Ransomware have been the cyber-attack that affected more than sixty NHS trusts in the United Kingdom, with 200,000 computers affected globally. The impact of this resulted in many facilities unable to access patient records which led to delays in surgeries and cancelled patient appointments [22]. The second of the attacks was that of South Korean web-hosting firm Nayana paying a $1 million ransom in 2017 clearly demonstrates how lucrative Ransomware can be for attackers [21].

3.2 *Denial of Service (DOS)*

Denial of service attacks have been around for many years, and they are triggered by a flood of network requests to an organisation's servers and networks, ultimately bringing the infrastructure down and enabling attackers to access vulnerabilities during the infrastructures recovery phase for bringing services back up. Large organisations have not been immune from DOS attacks, Yahoo, Amazon.com, eBay, CNN.com, Buy.com, ZDNet, were all subjected to total or regional outages of several hours caused by distributed denial-of-service (DDoS) attacks [23].

3.3 *Phishing*

Phishing is one of the most common forms of cyber-attack. Phishing works by attackers deceiving people with socially engineered approaches of downloading Malware or surrendering sensitive data such as passwords, personal information or bank details.

Curtis et al. [24] highlight that whilst technologies have evolved with organisations deploying tools such spam filters to effectively detect and deter known phishing campaigns, attackers continuously find new ways to evade these technologies such as through sophisticated and personalised e-mails ("spear-phishing") that take advantage of human limitations and biases and persuade people to respond [24].

Considering the impact that the previously described vulnerabilities can cause to organisations, rather than decentralise IT security processes, governance and capabilities, it is apparent that organisations should strategically align their IT Security and Business in way that it meets business needs, goals and strategies [25]. In the case of the NHS malware attack, it was identified that due to a lack of centralised security investment that many of the Windows operating systems were more than 15 years old and were no longer updated or supported by Microsoft [22].

Kearns and Lederer [26] highlight the while IT Investment plans are often planned in isolation it is the utmost importance that IT and business investment plans are aligned on the strategic objectives of the organisation in order to obtain effectiveness [26]. Furthermore, El Mekawy et al. [25] suggest that Information security processes (ISP) are an integrated part of IT strategy and business operations [25].

4 Information Security Processes

Centralised Information security can enable organisations to implement security risk-assessment processes. According to Laliberte [27], conducting risk-assessments are not only a good idea but can help organisations determine where organisations should invest their efforts both financially and effort to reduce its security exposure [27]. Laliberte [27] further argues that more importantly, risk assessments help to identify the key assets they need to protect and the threats and vulnerabilities those assets face. By assessing the likelihood of an incident and the effect of the incident actually occurring, the organisation can make a more informed decision about how and to what extent it should proceed to protect that asset. In essence, the risk assessment covers six key phases:

1. Asset identification;
2. Threat assessment;
3. Vulnerability assessment;
4. Risk determination;
5. Identification of countermeasures;
6. and finally, Remediation planning.

Oppliger [18] goes further to posit that the output of the security risk assessment goes further than just remediation planning, It forms the basis of the security policy, strategy and architecture at a technical, organisational and legal level [18].

Whilst Security risk assessments make sense, a study by Hooper and McKissack [9] found that the use of formal assurance techniques based on risk and security metrics at a central level did not always provide effective insights and communications tools to senior executives [9]. The survey resulted in these key findings:

- 75% of respondents indicated that metrics were important or very important to a risk-based security program.
- 53% didn't believe or were unsure whether the security metrics used in their organisations were properly aligned with business objectives.
- 51% didn't believe or were unsure whether organisations metrics adequately conveyed the effectiveness of security risk management efforts to senior executives.

With these challenges already existing at a centralised IT security model, it would only be compounded by decentralising IT security in how to formulate, capture, measure, consolidate and action the overall security posture of an organisation, resulting in an increased risk of security breaches.

Lowry and Wilson [5] posits that centralised organisations that meet or exceed the service qualities of their business partners, the organisation, in turn, is far greater to derive IT related benefits. Conversely, if IT quality is low the organisation's ability to innovate and respond to market conditions will be hindered, leading the business to alternative IT models such as decentralisation. Magnusson [13] further support this notion from research on a case study of a large Swedish organisation, where he notes that a level of IT Support quality had resulted in some departments having to abdicate from IT altogether, decreasing their usage and even matters of organisation compliance [13]. While the literature supports that there is a relationship between IT perception of Service quality, there is a contradictory element, whereby although acknowledging the lack of IT service quality, some organisations may refrain from outwardly recommending decentralising. This may be down to a market context, whereby the concern of the available skills in less developed economies may act as an inhibitor to decentralise the IT organisation or that there is a lack of agreement on the key organisational objectives that drive the centralisation/decentralisation of IT.

King [15] sets out organisational measurements/objectives of IT that drive the discussion on centralisation and decentralisation. This study adapts Kings models to incorporate Security aspects for an organisation:

- The need to provide IT security capability to all organisational units that legitimately require it.
- The need to contain the capital and operations costs in the provision of computing services within the organisation.
- The need to satisfy special computing needs of user departments.
- The need to maintain organisational integrity in operations that are dependent on computing, i.e., avoid mismatches in operations among departments.
- The need to meet information requirements of management and security of the data.
- The need to provide computing services in a reliable, secure, professional, and technically competent manner.
- The need to allow organisational units sufficient autonomy in the conduct of their tasks to optimise creativity and performance at the unit level, while not putting the organisation at a risk of security breaches.
- The need to preserve autonomy among organisational units, and if possible, to increase their importance and influence within the larger organisation, however, key capabilities with the required high level of governance such as IT Security remain centrally governed.
- The need, wherever possible, to make the work of employees enjoyable as well as productive.
- The need to counter security threats, vulnerabilities and exploits.

5 Conclusion

In summary, whilst a complete decentralising of IT Security capabilities across the organisation would create lack of governance, diluting accountability, increasing cost and skills while increasing the risk of IT security breaches there are rational arguments for both centralisation and decentralisation of the IT security function. Magnusson [13] highlights that centralisation and decentralisation may not be 'opposites or alternatives' but as mutually dependent. The model that Magnusson refers to the hybridisation of IT at two-levels with strategy and policy at the central governance level and a degree of autonomy and decision at the departmental/business unit level. This model also supports findings of Richardson et al. [16] that high performing organisations included those with simultaneous decentralisation and centralisation at two levels of the organisation [16]. Furthermore, in relation to IT Security, Hooper and McKissack [9] support the notion of a hybrid configuration, with an introduction of a CISO (Chief Security officer) reporting to the CEO.

The configuration would be a split between operations and the more strategic level. For example, the IT department would be in charge of the day-to-day technical security operations while the CISO would operate independently and be responsible for the strategic aspects of the organisation's security posture. In conclusion, the impacts of IT security breaches for organisations are both vast in terms of financial and reputational damage. Organisations should keep consistency through centralisation of IT Security with two options (1) Complete centralisation of IT security at an IT level; or (2) a hybrid configuration with a CISO reporting to the CEO as a strategic security capacity and IT performing the day-to-day security operations underpinned by the Strategy of a CISO. Rather, Organisations should refrain from devolving IT Security responsibilities in a decentralised manner to individual business units which will only lead to dilution of responsibility, accountability of security capabilities and governance across the organisation increasing the risk of security breaches and attacks.

References

1. Cane A (1992) Information technology and competitive advantage: lessons from the developed countries. World Dev 20(12):1721–1736
2. Pajić A, Pantelić O, Stanojević B (2014) Representing IT performance management as metamodel. Int J Comput Comm Control 9(6):758–767
3. Lei C, Huifan W (2017) Design and construct IT Performance architecture: settle the IT productivity paradox from critical realism. Procedia Eng 174:537–542
4. Campbell J, Chang V, Hosseinian-Far A (2015) Philosophising data: a critical reflection on the 'hidden' issues. Int J Organizat Collect Intell (IJOCI) 5(1):1–15
5. Lowry PB, Wilson D (2016) Creating agile organisations through IT: the influence of internal IT service perceptions on IT service quality and IT agility. J Strateg Inf Syst 25(3):211–226
6. Whyte G, Bytheway A, Edwards C (1997) Understanding user perceptions of information systems success. J Strat Inf Syst 6(1), 35–68
7. Farsi M, Daneshkhah A, Hosseinian-Far A, Chatrabgoun A (2018) Crime data mining, threat analysis and prediction. In: Cyber criminology. Springer, Berlin, pp 183–202
8. Herath T, Herath H, Bremser WG (2010) Balanced scorecard implementation of security strategies: a framework for IT security performance management balanced scorecard implementation of security strategies. Inf Syst Manage 27(1):72–81
9. Hooper V, McKissack J (2016) The emerging role of the CISO. Bus Horiz 59(6):585–591
10. Humayun M, Niazi M, Jhanjhi NZ, Alshayeb M, Mahmood S (2020) Cyber security threats and vulnerabilities: a systematic mapping study. Arabian J Sci Eng 1–19
11. Abdullah SM, Ahmed B, Ameen M (2018) A new taxonomy of mobile banking threats, attacks and user vulnerabilities. Eurasian J Sci Eng 3(3):12–20

12. Brynjolfsson E (1994) information assets, technology and organisation. Manage Sci 40(12):1645–1662
13. Magnusson J (2013) Intentional decentralisation and instinctive centralisation: a revelatory case study of the ideographic organization of IT. Inf Res Manage J (IRMJ) 26(4):1–17
14. Cozmiuc DC, Petrisor II (2020) Innovation in the age of digital disruption: the case of Siemens. In: Disruptive technology: concepts, methodologies, tools, and applications. IGI Global, pp 1124–1144
15. King JL (1984) Centralized versus decentralized computing: organizational considerations and management options. Comput Sur 15(4):319–349
16. Richardson HA, Vanderberg RJ, Blum TC, Roman PM (2002) Does decentralization make a difference for the organization? An examination of the boundary conditions circumbscribing decentralized decision-making and organizational financial performance. J Manage 28(2):217–244
17. Khallaf A, Majdalawieh M (2012) Investigating the impact of CIO competencies on IT security performance of the U.S. Federal Government Agencies. Inf Syst Manage 29(1):55–78
18. Oppliger R (2007) IT security: in search of the Holy Grail Oppliger. ACM [Online]. https://cacm.acm.org/magazines/2007/2/5725-it-security/fulltext. Accessed 2 Sept 2020
19. Jahankhani H, Al-Nemrat A, Hosseinian-Far A (2014) Cyber crime classification and characteristics. In: Cyber crime and cyber terrorism investigator's handbook. Syngress, pp 149–164
20. Hosseinpournajarkolaei A, Jahankhani H, Hosseinian-Far A (2014) Vulnerability considerations for power line communication's supervisory control and data acquisition. Int J Electron Secur Digit Forensics 6(2):104–114
21. Cartwright A, Cartwright E (2019) Ransomware and reputation. Games (Mdpi) 10(2):26
22. Collier R (2017) NHS ransomware attack spreads worldwide. CMAJ 189(22):786–787
23. Dayanandam G, Rao TV, Babu DB, Durga SN (2019) DDoS attacks—analysis and prevention. In: Innovations in computer science and engineering. Singapore, Springer, pp 1–10
24. Curtis SR, Rajivan P, Jones DN, Gonzalez C (2018) Phishing attempts among the dark triad: patterns of attack and vulnerability. Comput Hum Behav 87:174–182
25. El Mekawy M, AlSabbagh B, Kowalski S (2014) The impact of business-IT alignment on information security process. In: International conference on HCI in business, pp 25–36
26. Kearns GS, Lederer AL (2000) The effect of strategic alignment on the use of IS-based resources for competitive advantage. J Strateg Inf Syst 9(4):265–293
27. Laliberte S (2004) Risk assessment for IT security. Bank Account Fin 17(5):38–43

Blockchain and Artificial Intelligence Managing a Secure and Sustainable Supply Chain

Elias Pimenidis, John Patsavellas, and Michael Tonkin

Abstract Supply chain management is often the most challenging part of any business that manufactures, sells goods, or provides services nowadays. Regardless of whether the operations are mostly physical or online, managing supply chains relies entirely on being able to manage shared information securely, efficiently and effectively. Managing the information within the context of a closely-knit supply chain offers the benefits of extra resilience and ability to recover quickly from major disturbances. The authors propose here the development of a blockchain enabled and Intelligent Agent supported supply chain community that will provide a secure, intelligent, responsive and sustainable operational partnership.

Keywords Supply chain management · Blockchain · AI · Chain of custody

1 Introduction

The global economy is highly dependent on China and more particularly supply chains across the world are dependent on Chinese input and drive. China's share of global trade in some industries exceeds 50%—in the global trade of telecommunications equipment, for example, China's share (by volume) was 59% in 2018. Because of Covid-19, it is likely that this period of globalisation will not only come to a halt, but it will reverse. Some multinationals were forced under the conditions to relocate their supply chains away from China to other parts of Asia and even closer to their core operations, in Europe and the Americas. Such moves will lead to building

E. Pimenidis (✉) · J. Patsavellas · M. Tonkin
University of the West of England, Bristol, UK
e-mail: elias.pimenidis@uwe.ac.uk

Cranfield University, Bedford, UK

J. Patsavellas
e-mail: john.patsavellas@cranfield.ac.uk

M. Tonkin
e-mail: Michael2.Tonkin@live.uwe.ac.uk

quasi-independent regional supply chains, allowing global companies to provide a hedge against future shocks to their network [10].

Supply chains are difficult to set up and even more difficult to move, especially in the automotive sector. As more firms make a shift to such operation paradigm, the shift to regionalised supply chains will be the predominant outcome of this crisis. Optimising transportation and storage for risk mitigation though is not an easy, safe and inexpensive venture and companies will need advanced technologies to support such significant changes and mitigate the relevant risks. Smart ways of managing supply chains will need to be engaged to provide security of transactions and operations, and intelligent management that will support the creation of sustainable local and regional supply chains.

Blockchain driven and Artificial Intelligence managed supply chain could be the answer to securing the operations of a local/regional supply chain and providing the intelligence required to instil enhanced efficiency in operation. The above will yield sustainability into the supply chain, allowing it to counter the increased costs of shifting away from Asian markets and absorbing the higher wage costs of western economies [2, 4].

Blockchain offers a secure ledger for sharing documents. This is the founding stone of the development and operation of a digital community that supports a supply chain in any industry. Such digital communities are not a new concept as they date back from the dot.com era of the early years of the 21st century, called Valued Added Communities (VAC) [6]. Partners in the digital community share information of transactions allowing other partners to complete complementary transactions as they collaborate in fulfilling a partner's requirements and at the same time meeting their own objectives. The level of detail shared between members of the VAC depends on the type of partner, but all participants in the supply chain receive notification of every transaction completed between members of the VAC offering transparency that supports sustainability.

2   Supply Chain in an Uncertain World

Disruption is an everyday part of life arising from a myriad of circumstances. Normal service can be restored relatively quickly in some cases depending on the impact, its duration and the effect it might have on society and the economy. Supply chain flow usually remains constant and continuous, with just the odd blip, now and again. This was the view up until the early part of 2020 and before the Covid-19 pandemic. This unique situation has placed a different kind on supply chains and has forced companies to rethink their strategies and their approach to sustainability of supply chains.

2.1 What Is Supply Chain?

In the realm of manufacturing, a supply chain is the process of the flow of goods from the upper echelons of value creation to the end customer consumption. It is a form of symbiotic connection in which customers and suppliers work together to achieve the best interests of each other, buying, converting, distributing and selling goods and services to create specific final products and to add value to their organisations.

Through the control of information flow, logistics and capital, intermediate products and final products are prepared from the procurement of raw materials and supplied to customers by distribution networks. All such systems contribute and are part of the supply chain. Failure of one system can affect the normal operation of the supply chain. Having alternatives available to pick up the disturbed work or services can lead to seamless operations and efficiency in performance [5].

All of the above depends on secure, transparent, and intelligent management of information. This is where Blockchain can contribute to support sustainable supply chains [1].

2.2 A Secure and Sustainable Digital Supply Chain

Political, economic, social, technological, environmental and legal factors have constant, profound, often unexpected, and dramatic impacts, both positive and negative, on supply chains and the wider domain interests they serve. Certainty is an elusive commodity. An agile, data-driven supply chain ecosystem will be better prepared to react and mitigate such impacts.

Sharing assets and capabilities across supply chains will increasingly provide the foundation to achieve the greater flexibility necessary to cope with an uncertain future. Collaboration within the supply chain ecosystem is vital.

Sustainability is the ability for a business to operate successfully without compromising the ability of future generations to meet their own needs. To be sustainable, supply chains must become more flexible and responsive whilst incorporating increased resilience and traceability. Once again, technology and innovation will be the enablers with engineers at the heart of the ecosystem, delivering supply chain success [1, 5]. Blockchain can contribute to enhanced traceability and Artificial Intelligence algorithms can provide the technological edge to make supply chains, flexible, responsive and efficient, while maintaining the transparency of transactions that support all members.

2.3 *Value Added Communities*

The quest for flexible and sustainable supply chains based on transparency and sharing of information is not new to the business world.

In the past successful businesses evolved into colossal organizations that incorporated a large number of business functions. These shared little information or direct interaction and were managed centrally through a complex web of activities that contributed little towards customer satisfaction or to the organization's core objectives. The stand-alone mega organization is too dysfunctional for the modern business world. E-businesses were the first to realize this at the dawn of the 21st century. They had made information the key driver of all activities; retaining only the core functions of their business and focusing only on those activities that contribute directly to the attainment of competitive advantage. In this way, e-businesses have become more flexible and responsive to customer requirements, creating additional value for their customers and achieving a larger customer base.

Notwithstanding some spectacular failures at the turn of the 12st century, the dot-com era has enhanced the experience in this area and has shown that the ability to create added-value is the key factor of success in the modern competitive environment. To reap fully the benefits of Internet technologies and extract value for both the business and the customer, e-businesses realised that they needed to further exploit the wealth of information they gather by sharing it across businesses that can be seen as complementary within the supply chain. This led to the formation of mutually collaborative online communities known as "value-added communities".

Value-added communities (VACs) are groups of businesses that function at the various points of the supply chain and are connected electronically to enable optimal response to customer demand. At the same time this electronic network should offer maximum return for the "community" as a whole. This is done through the establishment of a series of communicating computer systems that support the key activities of each of the participating businesses. Customer demand is used as the empowering input for all the above systems. Through the electronic business facility of the trading organization (brand-owning company), information is processed, filtered and forwarded through the relevant networks to other computer systems such as MRP, MRPII, ERP, that each may support the function of one of the members of the "community". Thus planning and coordination of activities within the "community" can be performed according to evolving market trends and continuously revised on a real time basis [6].

E-business development though and especially the above concept of VACs involves a considerable level of uncertainty and risk. Developing a VAC involves integrating a number of different business functions, belonging to different organizations, which may be linked to diverse and conflicting objectives, or differing levels of commitment to the evolution and functioning of the VAC [8].

In the early days of value added communities, communication was based on slow internet connections, email systems and simple text messaging, with information shared on a peer to peer network like architecture as shown in Fig. 1 below.

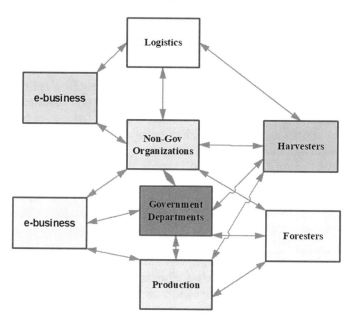

Fig. 1 Value added community structure [7]

Covid-19 is beginning to reshape trade fundamentally by accelerating the trend towards shortening supply chain reaction times. Intelligent agents operating on behalf of various members of a blockhain supported VAC can utilise the shared information in negotiating the completion of transactions. The security and privacy offered by blockchain technologies supports the transparency of such transactions and the sharing with the partner members [3].

3 A Distributed Ledger and Blockchain Technology

In the business context, a ledger, or general ledger, is defined as a central repository of the accounting information of an organization in which the summaries of all financial transactions during an accounting period are recorded.

In the common business environment, a digital ledger is stored in a central server, and distributed access is provided with read and/or read/write privileges. To assure security, there is some sort of access control mechanism that authenticates users, enables secure access, and enforces access restrictions (for example, read-only).

In a system with ongoing transactions and a heavy volume of read and write access, the central server model can be inefficient.

An alternative is a secure distributed ledger, which consists of an expandable list of cryptographically signed, irrevocable records of transactions that is shared by a distributed network of computers.

Every participant has the same copy of the ledger. Each participant may propose a new transaction to be added to the ledger and when consensus that the transaction is valid is reached, it is added to the register.

Trust is central in a distributed ledger and it involves two concepts:

Security protocols and mechanisms, generally based on Public-Key Cryptography, ensure that the creator of each transaction is authenticated and validated.

Transaction creators prove they are entitled to make a transaction by satisfying the particular conditions associated with this application. Meeting these conditions involves the use of a secure digital signature [9].

Distributed Ledger Technology (DLT) is based on peer-to-peer (P2P) network technologies enabled by the Internet, but internet-based transfers require ensuring that an asset is only transferred by its true owner and ensuring that the asset cannot be transferred more than once, i.e. no double-spend. The asset in question could be anything of value by Nakamoto in 2008 proposed a novel approach of transferring "funds" in the form of "Bitcoin" in a P2P manner. The underlying technology for Bitcoin outlined in Nakamoto's paper was termed Blockchain, which refers to a particular way of organizing and storing information and transactions. Subsequently, other ways of organizing information and transactions for asset transfers in a P2P manner were devised—leading to the term DLT to refer to the broader category of technologies.

DLT facilitates the recording and sharing of data across multiple data stores (ledgers), which each have the exact same data records and are collectively maintained and controlled by a distributed network of computer servers, called nodes [11].

Blockchain is a particular type of DLT, uses cryptographic and algorithmic methods to create and verify a continuously growing, append-only data structure that takes the form of a chain of so-called 'transaction blocks'—the blockchain—which serves the function of a ledger.

One of the members (nodes) initiates a new addition to the database. A new "block" of data may contain several transaction records. Information about this new data block is then shared across the entire network, containing encrypted data so transaction details are not made public, and all network participants collectively determine the block's validity according to a pre-defined algorithmic validation method termed as consensus mechanism. Only after validation, all participants add the new block to their respective ledgers (Fig. 2). Through this mechanism each change to the ledger is replicated across the entire network and each network member has a full, identical copy of the entire ledger at any point in time. This approach can be used to record transactions on any asset, which can be represented in a digital form. The transaction could be a change in the attribute of the asset or a transfer of ownership.

Blockchain this becomes a vehicle for trust, through the transparency of the public record and the validation of inputs from unconnected parties along the supply chain.

Distributed ledger supply chains are being developed and tested around the globe on different types of application domains. Directly linked to supply chains is the example of the IBM Food Trust blockchain, which went live as a commercial

Fig. 2 Block creation and validation

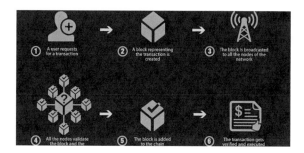

product in 2018. During the proof of concept phase, IBM worked with Walmart, who challenged them to trace mangos from farm to store.

Using Walmart's existing systems, this process took almost a week to run, while the blockchain-based system completed the task in 2.2 s [1].

4 An Intelligent Digital Supply Chain

In essence, blockchain is a data structure that makes it possible to create a digital ledger of transactions and share it among a distributed network of computers. After a block of data is recorded on the blockchain ledger, it is computationally infeasible to change or remove it.

When someone wants to add to the ledger, participants in the network, all of which have copies of the existing blockchain, run algorithms to validate the proposed transaction. If a majority of nodes agree that the transaction looks valid—that is, identifying information matches the history of a blockchain—then the new transaction will be approved and a new block added to the chain. The transaction is fulfilled or executed only when it has been approved for addition to the blockchain. Each block is connected to the previous block via a hash (tamper-proof digital fingerprint). On the blockchain, users can observe transactions that have occurred, so they know which outputs are available for spending and which ones have been consumed. Each block in the blockchain represents, in effect, the claim by someone on the network that the transactions contained inside the block are the first ones to spend the inputs involved, and therefore any transaction in the future that attempts to spend the same inputs should be rejected as invalid [9].

Thus a blockchain offers transparency and "democracy" in the handling of transactions submitted to it, developing and strengthening the bonds within the value added community it serves.

Based on such properties the authors have proposed the evolution of value added communities along the structure and concept of a blockchain. The objective is to create a closely-knit community of suppliers that serve the needs of a main manufacturer, organisation, or service provider.

- The community could comprise a group of companies, from similar industries or not, that will utilise the regular services of the supply chain.
- The members of the supply chain will be companies offering complementary services and/or the same services with different levels of capacity and the ability to serve the main core of the community with greater flexibility and meeting their changing needs at short notice.
- Different suppliers can offer their capacity and availability details to satisfy the requirements of each job, in the form of a transaction.
- As each job or part of are assigned to a member of the community the details are shared with the rest across the Blockchain, so each other member is notified of what jobs or parts of remain unfulfilled and can make offers, as each transaction is received and approved by the members
- No financial details are shared for each agreement reached, thus confidentiality is not breached, but transparency as to which member is assigned a specific job is maintained.

4.1 A Block Chain Prototype

Figure 3 shows the diagrammatic representation of an Intelligent Agent enhanced Blockchain that is currently under development at the University of the West of England. A cluster of Raspberry Pi computers (Fig. 4) is utilised to build a prototype supply chain and subsequently a full value-added community as shown in Fig. 3. The first stage of implementation is complete at the time of writing with successful testing of all functionality.

- Intelligent Agents systems can be utilised at each company to negotiate the details of each agreement. Once an agreement is reached, the members of the community are notified through the Blockchain.

Fig. 3 Blockchain based value-added community

Fig. 4 A blockchain prototype

Depending on the level of importance simple agreements can fully automated and negotiated and agreed by AI systems, or the AI negotiators requiring approval by a human decision maker.

Regardless, the utilisation of AI will minimise the complexity of decision making and it will speed up the negotiation with very high levels of accuracy, consistency and integrity.

- The Blockchain will ensure that the information shared will stay secure and only within the community subscribed to it. Only the required level of details will be shared.

  The information available at any one time will current and it will allow AI systems on draw on it plan the sharing of jobs, negotiate with suppliers and process jobs in an orderly, timely, secure fashion at very fast speeds.

  This ability will allow companies to revise plans speedily and be able to respond to changing circumstances with high levels of flexibility.

  Supply chain will become an enabler to flexible operations at times of high volatility.

- A diagrammatic representation of the resulting structure is shown in Fig. 3, with Intelligent agents supporting transaction analysis and response to each transactions submitted on the block chain be members of the value-added community.

4.2   The Case of Chain of Custody—Future Work

Upon completion of the above value-added community as a working prototype of Intelligent Agent supported blockchain, the authors intend to complete a proof of concept project applied on a chain of custody system that monitors and audits the required quality standards in an industry focused on sustainable products in the print sector. The project will involve a full supply chain with manufacturers, suppliers, retailers, and the chain custodian. The aim of the project is to establish the potential of the enhanced value-added community concept to apply and enhance every type of supply chain, whether manufacturing or service oriented.

5   Conclusion

Supply chains need to become flexible, responsive, transparent, intelligent and secure. Major events like the current Covid-19 induced crisis can put any such systems at risk. The proposal put forward here is for a resilient system based on a value-added community that is supported by blockchain technology and artificial intelligence in the form of intelligent agents to support supply chains. The authors believe that the proposed system can provide the qualities that will allow a harmoniously functioning supply chain to be capable of responding quickly to any disturbance, and to be able to create a sustainable local/regional ecosystem of interrelated companies.

References

1. Baucherel K (2018) Blockchain from Hype to Help, ITNOW 60(4):4–7. https://doi.org/10.1093/itnow/bwy087
2. De Lara S, Grech C (2018) Blockchain and transaction regulation. ITNOW 60(4):24–25. https://doi.org/10.1093/itnow/bwy094
3. Han C, Zhang Q (2020) Optimization of supply chain efficiency management based on machine learning and neural network. Neural Comput Appl. https://doi.org/10.1007/s00521-020-05023-1
4. McManus J (2020) Competing in the digital age. ITNOW 62(3):52–53. https://doi.org/10.1093/itnow/bwaa084
5. Patsavellas J (2019) Stand by for… disruption. In: Developing an eco-system for supply chain success. In search of certainty, IET
6. Pimenidis E, Miller CJ, MacEachen CF (2002) A new look at project estimation for E-business. In: Proceedings of the international conference on project management "Breakthrough with Project Management—In the Era of Global Revolution by IT" (ProMAC2002), 31 July–2 August 2002, Singapore, pp 353–358
7. Pimenidis E, Iliadis L (2003) E-collaboration in the rural Areas, in proceedings of the itafe'03. In: International congress on information technology in agriculture, food and environment, 7–10 Oct 2003. Ege University, Izmir, Turkey, pp 80–84
8. Pimenidis E, Bolissian J (2004) Value added communities and their impact on rural areas in Greece. In: Proceedings of the international conference on information systems & innovative

technologies in agriculture, food and environment (HAICTA2004), 18–20 March 2004, Co-organised by the Hellenic Association for ICT in Agriculture Food and Environment, vol 2. Aristotle University and University of Macedonia, Thessaloniki, Greece, pp 119–125

9. Stallings W (2017) A blockchain tutorial. Internet Protocol J 20(3):2–24
10. The Economist Intelligence Unit (2020) The great unwinding, Covid-19 and the regionalisation of global supply chains. The EIU Ltd
11. World Bank (2017) Worldbank.org—distributed ledger technology and Blockchain Fintech notes. https://bit.ly/2Ew6IpH. Accessed 12 Sept 2020

Does the GDPR Protect UK Consumers from Third Parties Processing Their Personal Data for Secondary Purposes? A Systematic Literature Review

David Sinclair and Arshad Jamal

Abstract Consumers control over their personal data is something the GDPR is meant to protect but there seems to be a gap in that protection when secondary processing is undertaken by data brokers. An assessment of this protection was undertaken using a systematic review of the available literature. a systematic review of 20 scholarly papers was conducted using the established guidelines and steps including undertaking a CIMO-Logic exercise, developing research objectives, undertaking a literature search, selecting study materials and undertaking a quality assessment. Consumers are being manipulated by primary collectors to provide personal data that is sold to brokers for secondary processing. This results in them losing control over that data, which the GDPR should protect. There appears therefore to be a gap in the protection afforded to consumers by the GDPR, which requires further research. This review is to the best of my knowledge the first on this specific topic and in identifying further areas for research it is hoped that this study will add value to academic knowledge. There were significant limitations in undertaking the study due to extenuating technical issues and the results of this study should be treated with caution and if possible, re-run at a later date. The study makes five recommendations for further research.

Keywords Consent · Consumer · Data broker · GDPR/general data protection regulation

1 Introduction

The internet is an essential requirement for most people, at home and at work. Being connected to the internet has completely changed communication, shopping and work (Jay 2019, p. 113). UK online shopping, is increasing at 129% a week [16]

D. Sinclair · A. Jamal (✉)
Northumbria University London, London, UK
e-mail: arshad.jamal@northumbria.ac.uk

D. Sinclair
e-mail: david.sinclair@northumbria.ac.uk

© The Author(s), under exclusive license to Springer Nature Switzerland AG 2021
H. Jahankhani et al. (eds.), *Cybersecurity, Privacy and Freedom Protection in the Connected World*, Advanced Sciences and Technologies for Security Applications, https://doi.org/10.1007/978-3-030-68534-8_24

and this generates unprecedented volumes of data including personal and GDPR[1] special categories of personal data (Jay 2019, p. 113).[2]

An industry of intermediaries, known as data brokers ('Brokers) has emerged to buy personal data ('Data'), process and manipulate that Data into saleable products, which they sell to a range third parties, for marketing and other purposes. There is little transparency between primary collectors, consumers and Brokers, who are considered untrustworthy. There is however also a willingness by consumers to sell/trade their Data for benefits [14].

In its raw form this data has little value but once it is processed and refined it gains a significant monetary value as a commodity [1]. Primary collectors are keen to collect Data either by consumers sharing the data in return for benefits[3] or without consumers knowledge [18].

Once Brokers obtain information, consumers lose control over that Data, which can be processed and resold or rented without their knowledge [15]. It is the data controllers (primary collectors and Brokers) who decide what happens to consumer's Data [22].

Brokers compile and aggregate Data from a variety of sources and these practices take place in the shadows without consumers knowledge or consent, compromising consumers right to privacy [12].

The Brokering industry is unregulated and Brokers do not want attention as this could draw consumer attention to their activities, which could result in consumer access to the data they hold [2].

The proliferation of online channels and increased internet access via mobile devices has increased the quantity and quality of data available to Brokers but they are looking for the right quality of Data from primary collectors. These collectors therefore manipulate consumers into providing them with more Data than they require for their purposes in order to benefit from the additional income. At all times, power lies with the Brokers, who decide how, when and by whom consumers Data is processed for secondary purposes [13].

Because Data is collected by primary collectors, who are generally big name businesses such as supermarkets, clothing brands and social media platforms, they are trusted by consumers, which leads to consumers having a 'perception of privacy' and they disclose more data than is necessary for, e.g. their purchase [13].

---

[1]Regulation (EU) 2016/679 of the European Parliament and of the Council of 27 April 2016 on the protection of natural persons with regard to the processing of personal data and on the free movement of such data, and repealing Directive 95/46/EC (General Data Protection Regulation) [4].

[2]Defined in GDPR Article 9(1) as personal data revealing 'racial or ethnic origin, political opinions, religious or philosophical beliefs, or trade union membership and the processing of genetic data, biometric data for the purpose of uniquely identifying a natural person, data concerning health or data concerning a natural person's sex life or sexual orientation.

[3]Such as discounts on shopping or access to online services.

1.1 Purpose Statement

Secondary data processing risks consumer privacy and the GDPR was enacted to give them greater control over the processing of their Data and thereby protect their privacy. A systematic literature review ('SLR') was undertaken of the available literature, to investigate whether (and if so, to what extent) the GDPR protects UK consumers from third party (Broker) secondary processing their Data.

An evaluation of the literature was undertaken to examine consent, Broker processing and the requirements of the GDPR, to determine whether the GDPR is effective in protect consumers.

The study has three objectives that are set out under 'Research Objectives' below.

Having drawn conclusions, this article will identify areas where further academic and/or legal research is required.

2 Research Methodology

This research uses systematic literature review methodology and follows the established guidelines published by Kitchenham and Charters [11] and Hoda et al. [8]. An initial review of the 'grey' literature was undertaken in order to obtain an understanding of the practitioner's view of Brokers secondary processing of consumers Data and to identify the key terminology used. CIMO-Logic was used to develop the research question and objectives [5].

1. **Context**—EU and UK law, data brokers and those that collect data directly from data subjects. The relationships to be studied are those between the data subject and the primary processor and those between the primary and secondary (data broker) processors.
2. **Intervention**—The event to be investigated is the primary processor obtaining GDPR compliant consent to the secondary processing of personal data by data brokers.
3. **Mechanism**—A data subject is purported to have given consent the secondary processing of her or his personal data in return for some benefit, i.e. the free use of a search engine or for points on a store card that can lead to discounts on goods or service.
4. **Outcome**—The effects of the intervention are that peoples' personal data is being processed by data brokers for purposes never envisage by the data subject, who has not consented to that processing, or whose consent to that processing is not GDPR compliant. While data subjects may have considered that secondary processing of their data would lead to, e.g. targeted advertising, which they may find beneficial, they do not realise that the same processing is being used to build (an often inaccurate) profile of them that could have significant, adverse life consequences for them.

2.1   Research Objectives

The CIMO review identified key themes for the study that enabled a search against key words. In addition, EDPB[4] and ICO[5] guidance on the GDPR and consent were reviewed to identify key legal issues. The following three study objectives were developed:

1. To identify and describe the GDPR factors required for a third-party organisation to be able to rely on an individual's consent to the secondary processing of personal data and special categories of personal data (consent to processing).
2. To understand the GDPR methods that primary data collectors use to obtain valid consumer consent to the use of their personal data and special categories of personal data (lawfulness of processing).
3. To identify if, having given consent, it is possible for a consumer to use the GDPR to control the use of their personal data processed by third parties (consumer control).

2.2   Search Process

The aim was to include between 20 and 30 documents in the study and the inclusion criteria were that the title had to include two of the search terms and the abstract had to include a discussion of Data processing in relation to issues that would affect consumers. A summary of the search process is shown in Fig. 1.

The search and document selection process used followed guidance provided by Hoda et al.[6] This involved searching standard online databases that were recommended by Northumbria University for information security research, i.e. Science Direct, IEEE Xplore, ACM and Springer, together with legal databases Practical Law and Lexis PSL that were used for legal texts and commentary.

However, shortly after the search process started, a significant cyber-attack on Northumbria University denied accessing to either the University library or any of the required databases. This attack stopped the search and created a significant time constraint on the study.

In order to continue, a Google Scholar search was undertaken using the same criteria and filters. A significant drawback with Google Scholar is that it only shows abstracts and not full documents, which delayed the study until the University came back on-line.

When the University's systems were restored the search was completed and additional documents were located on Science Direct.

---

[4]European Data Protection Board (formally the Article 29 Data Protection Working Party), which is make-up of the data protection regulator from each of the EU Member States.

[5]Information Commissioner's Office.

[6]Hoda et al. [8] Systematic literature reviews in agile software development: A tertiary study [8].

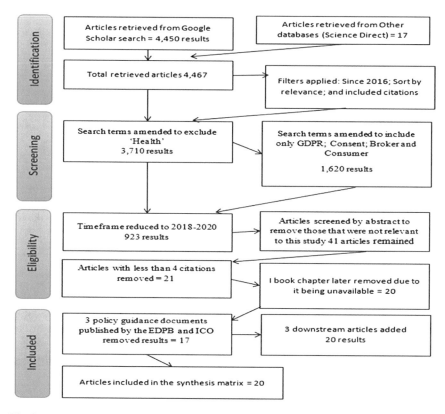

Fig. 1 Search process adapted from https://tamu.libguides.com/systematicreviews

A search was undertaken for all relevant papers published between May 2018 (i.e. when GDPR was enacted) and the time of the search (i.e. August 2020). IEEE and ACM returned no results. The search of ACM had to be modified to fit the search criteria options available in the advance search feature, which did not include all Boolean options.

The search criteria used were:

"All Metadata":GDPR) OR "All Metadata":General Data Protection Regulation) AND "All Metadata":Consent) AND "All Metadata":data broker) OR "All Metadata":information broker) AND "All Metadata":consumer.

2.3 Study Selection

A final inclusion criteria that was applied to the remaining 42 documents was that each of the articles had to have been cited at least three times and a book chapter had to be excluded due to the book being unavailable.

Number	Document Title
A1	Personal data trading scheme for data brokers in IoT data marketplaces
A2	The impact of user location on cookie notices (inside and outside of the European Union)
A3	Never mind the data: The legal quest over control of information & the networked self
A4	Pursuing consumer empowerment in the age of big data: a comprehensive regulatory framework for data brokers
A5	Data analytics in a privacy-concerned world
A6	Privacy and personal data collection with information externalities
A7	Pricing privacy - the right to know the value of your personal data
A8	Corporate digital responsibility
A9	AI and Big data: A blueprint for a human rights, docial and ethical impact assessment
A10	Security towards the edge: Sticky policy enforcement for networked smart objects
A11	Visions of Technology: Big Data Lessons Understood by EU Policy Makers in Their Review of the Legal Frameworks on Intellectual Property Rights...
A12	Big Data and discrimination: perils, promises and solutions. A systematic review
S13	Power to the people? The evolving recognition of human aspects of security
S14	What if you ask and they say yes? Connsumers' willingness to disclose personal data is stronger than you think
S15	The Invisile Middlemen: Acritique and Call for Reform of the Data Broker Industry
S16	Data Brokers and Data Services
S17	Does the GDPR Enhance Consumer's Control over Personal Data? An Analysis from a Behvioual Perspective
S18	EU General Data Protection Regulation: Changes and implications for personal data collecting companies
S19	Forgetting personal data and revoking consent under te GDPR: Challenges and proposed solutions
S20	Information asymmetries: recognizing the limits of the GDPR on the data-driven market

Fig. 2 SLR search final documents

Final exclusion criteria were applied in that abstracts that discuss data processing in relation to health data, blockchain, transport, financial and tax, vendor apps, or ownership of data and those that discussed processing related to non-EU countries, smart cities, or autonomous vehicles were excluded. A further three downstream articles were added.

A total of 20 documents remained, the full text of which were checked for duplicates and those not relevant to the topic and these 20 documents were discussed and agreed by the review panel (shown at Fig. 2).

2.4 Quality Assessment

The quality of the documents was evaluated using criteria developed for this study and shown in Fig. 3.

3 Findings

Information was extracted from the 20 articles reviewed using a structured extraction form, this information was then put into a synthesis matrix (see Fig. 4).

This allowed the development of a SLR Summary of Review form to be completed. The extracted information did not align exactly with that required to meet the three objectives set for this study but instead fell into three main themes of Consent; GDPR; and Consumers.

1. Did the article state its purposes?
2. Did the article set out a context?
3. Did the article discuss threats, consumers, brokers, or the GDPR?
4. Did the article come to any Conclusions?
5. Did the article discuss future research?

6. Did the article suggest any academic or practitioner change?

A summary of the results is shown below opposite:

Study	Total
A1	6
A2	6
A3	4
A4	8
A5	9
A6	8
A7	6
A8	5
A9	5
A10	6
A11	7
A12	6
S13	4
S14	5
S15	6
S16	5
S17	7
S18	8
S19	7
S20	6

Fig. 3 Quality assessment results

Article	Main themes identified in articles	
A1	The widespread use of the internet and data services and big data continues. Data brokers have emerged to buy and sell data about individuals to third parties. There is little transparency between primary providers, consumers and brokers who are considered untrustworthy. There is however a willingness to sell. A model is proposed to provide a better deal for consumers.	• Data brokers have emerged to buy and sell personal data; • Lack of transparency in this secondary processing; • Consumers willing to sell/give away their data.
A12	This SLR article considers the risk of consumer discrimination in Big Data analytics and identifies shortcomings in the law and highlights 'obstacles to fair data mining'. Highlights the lack of legal and social sciences research in this area. However, the article did not consider the legal position or GDPR in any detail.	• Risk of consumer discrimination from Big Data analytics; • Lack of legal and social sciences research in this area; • No GDPR or general legal position considered.

Fig. 4 Synthesis matrix

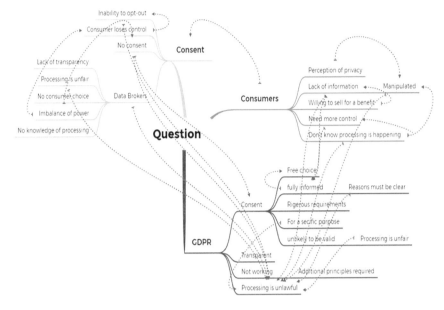

Fig. 5 Thematic map

A thematic map (Fig. 5) was developed for this study to identify sub-themes and the relationships between those themes.[7]

3.1 *Consumers*

Consumers are generally unaware that primary collectors collect far more of their Data than is needed for primary purposes. This Data is used for secondary processing by intermediaries such as Brokers [17].

This is because primary collectors fail to provide consumers with all of the relevant information that they need to make decisions. Where information is provided, it is hidden in lengthy privacy statements and policies on websites and/or shrouded in large amounts of highly technical text. Were sufficient information to be provided to consumers, they would not comprehend what they are being told or understand the logic behind the secondary processing of their Data.

Consequently, primary collectors are not complying with the GDPR's provisions. Despite this, primary and secondary process consumers data, which is unlawful continues and is increasing at a significant rate [17].

---

[7]As set out in 'Research Objectives' above.

Even if primary collectors and Brokers obtained consumer consent and that consent covered all of the intended processing, this would not diminish their obligations as data controllers to observe the GDPR's processing principles, in particular, the 'necessity' of collection for a 'specified purpose' to be 'fair' [6].

The GDPR of itself, is unable to mitigate Brokering practices, nor can it provide sufficient transparency to enable consumers to make informed choices and thereby give GDPR valid consent. The lack of transparency means that consumers don't have control over what happens to their Data. Discounting consumers being unaware that secondary processing of their Data takes place, it is unlikely that they would in any event, consent to the open-ended secondary processing of their Data [17].

Providing consumers with free choice (and thereby control over their Data) is not realistic when that control comes through consumers being provided by information, which is not, in fact, provided [19].

There is an acknowledgement that there is an excessive over-use of Data and a GDPR and regulatory failure to prevent Data sharing between primary collectors and Brokers. The data collection and processing that takes place is such that it now enables Brokers to infer information about non service users from information they have collected about service users, so called profiling, that the GDPR unable to prevent [3].

Primary Data collectors and Brokers are failing, almost universally, to provide sufficient and/or adequate information to consumers in breach of the GDPR. Added to which, the Regulation place a significant level of responsibility on consumers to inform themselves before giving consent, by making them (and not data controllers) responsibility for reading and understanding all of the information provided to them [17].

It is a fundamental tenet of the GDPR that a consumer can withdraw consent to Data processing at any time and that withdrawing consent should be as easy as giving it. However, controllers often make opt-outs invisible and imperfect and consumers are rarely provided with adequate, visible and understandable information to enable them to make an informed choice [17].

Opt-outs are confusing or non-existent because the Broker is generally invisible and the website does not express whether individuals can opt out. Where they do discover the broker and opt out, they may still never know whether their choice has been implemented. In short, brokers and primary collectors are failing to comply with GDPR provisions on consent, transparency and the provision of information and on data subject access rights. Consumers therefore have no real choice [2].

3.2 GDPR

Brokers have emerged to buy and sell data about individuals to third parties, with little or no transparency over their operations, which has led to them being considered untrustworthy. However, there is wide acceptance that Brokers rarely ever steal Data but instead purchase it from consumers who are willing to sell that data in return for benefits, or from primary collectors who also operate outside the law in collecting that data [14].

Existing safeguards, including those imposed by GDPR relating to the Brokering industry are poor. However, a more sophisticated, model-based approach to data protection, giving consumers better control over their Data could be developed but this would involve Brokers voluntarily accepting that approach and this is unlikely to happen [21].

It is accepted that technology alone cannot deliver a complete security solution and consumers must understand the threats they face and be able to protect themselves. However, the GDPR does not give these human aspects the attention they merit, instead it focuses on technical security and less on policy, training and education [7].

It is the 'technical complexities and multiple data-exploiting practices primary collectors and Brokers that make it hard for consumers to gain control over their Data.' The GDPR addresses the need for more consumer control but the lack of enforcement means that its new Data processing principles, which are designed to empower consumers are ineffective [19].

Brokers compilation, aggregation of individuals Data and sale of that information takes place in the shadows without consumers knowledge or consent compromises consumers rights to privacy and leaving them vulnerable to 'predatory and unsavoury marketing practices'. EU data protection provides the right framework to protect consumers, but there is little or no enforcement of that legislation [12].

A study of cookie notices found that 65% of site operators did not comply with legal requirements and Brokers regularly collect consumers internet browsing behaviour without consent, both of which is in breach of the GDPR [18].

Consent is one of the GDPR six lawful grounds[8] for processing[9] Data. However, in addition to obtaining consent to lawfully process Data, controllers must comply with the GDPR Article 5 data processing principles[10] both in obtaining consent and in processing the Data. To do otherwise makes any consent and/or processing unlawful [6].

Article 5 requires secondary processing to be fair, lawful, transparent and meet the purpose limitation principle, which neither primary collectors and/or Brokers can achieve because it is impossible for them to determine, at the time of collection, what future processing will take place. Consumer GDPR rights are therefore, extinguished [15].

---

[8]Set out in Article 6(1) as: consent; the performance of a contract; legal obligation; vital interests; public interest/exercise of official authority; and legitimate interest.

[9]Defined by Article 4(2) as: 'Any operation or set of operations which is performed on personal data or on sets of personal data... such as collection, recording, organisation, structuring, storage, adaptation or alteration, retrieval, consultation, use, disclosure by transmission, dissemination or otherwise making available, alignment or combination, restriction, erasure or destruction'.

[10]Processing is lawful, fair and transparent; it is limited to a specified, explicit and legitimate purpose and not further processed for an incompatible purpose; it remains accurate and up to date; it is subject to storage limitation; it remains secure; and the controller shall be able to demonstrate compliance with the lawful, fair and transparent processing.

Guidance suggests that the 'imbalance of power' that exists between Brokers and consumers means mean that in any event, consent cannot be regarded as 'freely given' as consumers are unable to exercise real choice over what happens to their Data. This is because the consumer is unaware that processing is taking place. Consent cannot therefore be valid [20].

Karanasiou and Douilhet argue that the GDPR increases consumer rights over their Data, giving them greater control. They also suggest that this could also limit Brokers secondary processing, a view supported by the EDPB. Countering this argument is that there is a significant imbalance of power between consumers and Brokers, who are generally unknown to the consumer whose data they are processing. Therefore, in order for consumer rights to be effective, the GDPR would have to be enforced.

The GDPR is seen by authors as a potentially effective means of regulating Brokers but as consumers are unaware of Brokers activities, they are unable to exercise their rights. Karanasiou and Douilhet [10] argues that for the GDPR to be effective in protecting consumers, Brokers would have to agree to be bound by the GDPR's requirements. Karanasiou believes that the GDPR, of itself, is unable to protect consumers, nor can it require the provision of sufficient transparency by Brokers, to enable consumers to make informed choices.

3.3 Consent

The GDPR[11] provides the conditions for consent to be valid and the EDPB sets out the elements of, and conditions for obtaining consent, which are expansive [6]. The ICO provides that the standard for obtaining consent is a high one [9] and that the GDPR imposes rigorous requirements on those seeking consent [6].

Consumers must be given an option to express consent but as they have insufficient information to consider the consequences of providing that consent, they simply consent when they are confronted with a consent request. Added to which, consumers often simply consent whenever they are confronted with a request to do so [19].

Even if primary data collectors have consent to process consumers' Data, they are unlikely to have obtained valid consent to sell/pass Data to Brokers unless the consumer has been provided with all relevant information about the specific processing to be undertaken, by whom and for what purposes [20].

Once used for secondary processing the consumer loses all control over what arguably ceases to be their Data [19]. The current consent model is therefore, not effective (Jay 2019) and consumer rights need to be increased to protect consumers [22].

However, if the GDPR's provisions are evaluated from a behavioural perspective, it is possible to predict the Regulation's effectiveness in providing increased consumer control rather than assuming that consumers have better control. The study found

---

[11] Article 4(11) provides that consent must be freely given, specific, informed and clear affirmative action unambiguously indicating the data subject's wishes.

that consumers do not have control due to a lack of information and a lack of the implementation of data protection by design and by default [19].[12]

4 Discussion of Findings

The first objective in this study is to understand the methods that primary data collectors use to obtain valid consumer consent to the use of their Data in order to establish the lawfulness of processing of that Data for secondary purposes.

Consent is generally the only GDPR lawful ground for the secondary processing of Data, but the conditions for GDPR valid consent are rigorous according to the EDPB [6] and set an extremely high hurdle for primary collectors and Brokers to overcome according to the ICO [9].

Brokers compiling and aggregating Data, which takes place in the shadows without consumers knowledge or consent, which according to Kuempel [12], compromises consumers rights to privacy and leaving them vulnerable to 'predatory and unsavoury marketing practices' and breaches both the requirements for consent and the Article 5 principles. Consent, if obtained, will therefore, be invalid.

Oh et al. [14] argue that the lack of transparency by primary collectors in informing consumers about Brokers activities and processing cannot meet the GDPR principles and unless the Article 5 principles are met, consent will be invalid and processing will be unlawful.

It is, according to van Ooijen and Vrabec [19] hard for consumers to gain control over their Data and while the GDPR addresses the need for more consumer control, the lack of enforcement makes the Regulation ineffective.

Karanasiou and Douilhet [10] also argue that the GDPR increases consumer rights over their Data and that the GDPR does give consumers greater control, which could limit Brokers secondary processing, but again this would require the GDPR to be enforced.

The second objective is to understand the methods that primary data collectors use to obtain valid consumer consent to the use of their Data and special categories of Data (lawfulness of processing).

The study identified that there are three actors involved in the transfer of Data. The GDPR imposes duties on primary collectors (as data controllers), who collect Data from consumers for an initial purpose and on Brokers, who in receiving that Data also become controllers. The GDPR was enacted to protect consumers, who do not have GDPR duties.

For processing to be lawful, consumers must give consent for a specified purpose or purposes that are not incompatible with each other. However, Politou et al. [15] have identified that this is generally not possible for primary collectors or Brokers, at the time Data is collected from consumers, to envisage all secondary processing and

---

[12]Required by GDPR Article 25.

processors, breaching the specified purpose and transparency principles and making any consent invalid.

van Ooijen and Vrabec [19] state that it is hard for consumers to gain control over their Data because of the 'technical complexities and multiple data-exploiting practices' of primary collectors and Brokers and while the GDPR addresses the issue of consumer control through the principles, the lack of enforcement makes it ineffective.

Kuempel [12] argues that EU data protection provides the right framework to protect consumers. Karanasiou and Douilhet consider the GDPR to be a potentially effective means of regulating primary collectors and Brokers but they argue that consumers need to be aware of them and their activities.

According to Wieringaa et al. [21], existing safeguards provided by the GDPR are poor and consumers need to be provided with greater control by the Regulation. Until this happens the GDPR does not protect consumers.

While some of the authors allude to the lack of enforcement of GDPR provisions against primary collectors or Brokers being a key factor in not providing consumers with more control over their Data. This issues is not however, discussed in any of the articles reviewed.

This may, in part, be that poorly drafted provisions and a general lack of enforcement make GDPR duties difficult to enforce.

The third objective of the study is to identify if, having given consent, whether consumer can use the GDPR to control the use of their Data processed by third parties for secondary purposes.

The growth of the internet has changed the way people communicate, shop and work has according to Jay (2019) generated Data and special categories of Data at an unprecedented rate.

Adesina [1] found that the right quality of Data has a significant monetary value and so Brokers are keen to purchase that Data from primary collectors. van Eijk et al. [18] found that this leads those collectors to manipulate consumers into providing them with more of the Data that is required by Brokers, which is beyond that required for their specific purposes. Primary collectors do not therefore provide consumers with all relevant GDPR required information.

Oh et al. however, found that consumers too easily consent to the processing of their Data in return for benefits such as discounts on shopping, without as van de Waerdt [17] identified, taking the time to read and understand any information provided to them by controllers. van Ooijen and Vrabec [19] found, consumers simply consent when they are confronted with a request to do so [19].

Consumers can, only be expected to read and understand what they are consenting to where they are fully informed and that information is transparent, which Politou et al. [15] found does not happen. This according to Mazurek and Malagocka [13] is because consumers are manipulated by primary collectors into disclosing more Data than they need to meet their legitimate requirements.

Van de Waerdt [17] argues that consumers are generally unaware that their Data is sold to Brokers or what secondary processing is undertaken on their Data and by whom, which according to Politou et al. [15] means consumers lose all control over their Data.

As van de Waerdt [17] says, even discounting that consumers would not have taken the time read and understood relevant information provided to them and are therefore unaware that their Data is being processed for secondary purposes, or by whom, it is unlikely that they would consent to the open-ended processing of their Data.

5   Conclusions and Further Research

The GDPR should protect consumers personal data but there appears to be a gap in that protection when Data is collected and used by Brokers for secondary processing.

An assessment was undertaken using a systematic review of 20 scholarly papers using established SLR guidelines to develop assessment criteria and determine whether the GDPR protects consumers. CIMO-Logic was used to identify three objectives and develop a research question. Quality criteria were developed and recorded and the search results were summarised with 20 articles being selected for the study. Information was extracted from the 20 articles reviewed using a structured extraction form and the information was then put into a synthesis matrix to identify key themes. The author developed a thematic map to provide an insight into the key themes and sub-categories of those themes, which were found to be Consumers, GDPR and Consent.

While these objectives were not directly met, the study identified the three related areas of consent to processing, GDPR provisions for that consent and consumer control as being important.

The key finding from the study was that the GDPR of itself does not effectively protect consumers, although it was clear that authors believed the GDPR should give consumers greater control over their Data and that it provided a good framework for the regulation of the Brokering industry. There is, however, a lack of primary research in this area.

While not discussed to any degree in the literature, the key issue is not a lack of GDPR provisions but of enforcement of those provisions by the regulatory authorities and if verified by further research, this would constitute a significant gap in the GDPR's ability to protect consumers.

This review is to the best of the author's knowledge the first research into this specific topic and in identifying further areas for research it is hoped that this study will add value to academic knowledge.

There were significant limitations in undertaking the study due to extenuating technical issues and the results of this study should be treated with caution and if possible, re-run at a later date. The study makes five recommendations for further research.

The recommendations are that further research is required:

1. To undertake study which uses larger data set and multiple reviewers to evaluate and verify the findings of this research.
2. To determine the extent to which the GDPR could be made effective in protecting consumers Data processed for secondary purposes.
3. To look into the enforcement of the GDPR to protect consumers Data.
4. To explore the brokering industry and its operations and GDPR compliance.
5. To extend this research to evaluate the actions (or otherwise) consumers take to protect their data.

References

1. Adesina A (2018) Data is the new oil [Online]. Available at: https://medium.com/@adeolaadesina/data-is-the-new-oil-2947ed8804f6. Accessed 23 Aug 2020
2. Alowairdhi A, Ma X (2019) Data Brokers Data Serv. https://doi.org/10.1007/978-3-319-32001-4_298-1
3. Choi JJ, Joen D-S, Kim B-C (2019) Privacy and personal data collection with information externalities. J Public Econ 173:113–124
4. Commission E (2016) Regulation (EU) 2016/679 of the European parliament and of the council. s.l.:s.n
5. Denyer D, Tranfield D (2009) Producing a literature review, in Buchanan and Bryman (2009), SAGE Handbook of Organizational Research Methods (Chapter 39). SAGE Publications Ltd, London, England
6. EDPB (2020) Guidelines 05/2020 on consent under Regulation 2016/679. European Data Protection Board, Brussels
7. Furnell S, Clarke N (2012) Power to the people? The evolving recognition of human aspects of security. Comput Secur 31:983–988
8. Hoda R, Sallehb N, Grundy J, Teea HM (2017) Systematic literature reviews in agile software development: a tertiary study. Inf Softw Technol 85:60–70
9. ICO (2018) The general data protection regulation lawful basis for processing: consent [Online]. Available at: Available at: https://ico.org.uk/for-organisations/guide-to-data-protection/guide-to-the-general-data-protection-regulation-gdpr/lawful-basis-for-processing/. Accessed 20 May 2018
10. Karanasiou AP, Douilhet E (2016) Never mind the data: the legal quest over control of information & the networked self. In: IEEE international conference on cloud engineering workshop (IC2EW), Berlin, pp 100–105
11. Kitchenham B, Charters S (2007) Guidelines for performing systematic literature reviews in software engineering (version 2.3). Technical report, Keele University and University of Durham
12. Kuempel A (2016) The invisible middlemen: a critique and call for reform of the data broker industry. Northwestern J Int Law Business 36(1):207–234
13. Mazurek G, Malagocka K (2019) What if you ask and they say yes? Consumers' willingness to disclose personal data is stronger than you think. Bus Horiz 62(1):751–759
14. Oh H, Park S, Lee GH, Heo H, Choi JK (2019) Personal data trading scheme for data brokers in IoT data marketplaces. IEEE Access 7:40120–40132
15. Politou E, Alepis E, Patsakis C (2018) Forgetting personal data and revoking consent under te GDPR: challenges and proposed solutions. J Cybersecur 4(1):1–20
16. Skelton P (2020) Internetretailing.net [Online]. Available at: https://internetretailing.net/covid-19/covid-19/online-shopping-surges-by-129-across-uk-and-europe-and-ushers-in-new-customer-expectations-of-etail-21286#:~:text=Themes%20%3E%20COVID%2D19-,Online%

20shopping%20surges%20by%20129%25%20across%20UK%20and%20Eur. Accessed 23 Aug 2020
17. van de Waerdt P (2020) Information asymmetries: recognizing the limits of the GDPR on the data-driven market. Comput Law Secur Rev 38:1–18
18. van Eijk R, Asghari H, Winter P, Narayanan A (2019) The impact of user location on cookie notices inside and outside of the European Union. In: Workshop on technology and consumer protection (ConPro'19)
19. van Ooijen I, Vrabec HU (2019) Does the GDPR enhance consumers' control over personal data? An analysis from a behavioural perspective. J Consum Policy 42:91–107
20. WP 29 (2018) Article 29 working party guidelines on consent under regulation 2016/679. Article 29 Data Protection Working Party, Brussels
21. Wieringaa J, Kannan PK, Ma X, Reutterer T, Risselada H, Skiera B (2019) Data analytics in a privacy-concerned world. J Business Res 5:1–11
22. Yeh C-L (2018) Pursuing consumer empowerment in the age of big data: A comprehensive regulatory framework for data brokers. Telecommun Policy 42:282–292

Bibliography: List of Grey Literature

23. Hickey A (2019) Report: GDPR regulators digging into data brokers [Online] Available at: https://www.ciodive.com/news/report-gdpr-regulators-digging-into-data-brokers/545682/. Accessed 6 Sept 2020
24. Hintze MLG (2017) Meeting upcoming GDPR requirements while maximizing the full value of data analytics [Online]. Available at: https://papers.ssrn.com/sol3/papers.cfm?abstract_id= 2927540. Accessed 23 Aug 2020
25. Katwala A (2018) Forget Facebook, mysterious data brokers are facing GDPR trouble. Wired [Online]. Available at: https://www.wired.co.uk/article/gdpr-acxiom-experian-privacy-intern ational-data-brokers. Accessed 5 Sept 2020
26. Lemarchand L (2017) Why you should not by from a Data Broker [Online]. Available at: https://www.mediadev.com/gdpr-data-broker/. Accessed 5 Sept 2020
27. Ram AMM (2019) Data brokers: regulators try to rein in the 'privacy deathstars' [Online]. Available at: https://www.ft.com/content/f1590694-fe68-11e8-aebf-99e208d3e521. Accessed 23 Aug 2020
28. Shah SMKA (2020) Secondary use of electronic health record: opportunities and challenges [Online]. Available at: https://ieeexplore.ieee.org/document/9146114. Accessed 19 Sept 2020
29. Wlosik M (2019) What is a data broker and how does it work? [Online]. Available at: https:// clearcode.cc/blog/what-is-data-broker/. Accessed 18 Aug 2020

Identification of Critical Business Processes: A Proposed Novel Approach

Yousuf Alblooshi⊙**, Amin Hosseinian-Far**⊙**, and Dilshad Sarwar**⊙

Abstract Critical Business Processes (CBPs) are processes that are crucial to the financial stability and operations of an organisation. This paper focuses on surveying the literature, while presenting a critical synthesis of the findings of previous studies on CBPs. The paper seeks to extensively and critically review the current literature to understand state-of-the-art methods and key research gap for CBP identification. While this paper targets the process of identifying the gap in literature, it helps in finding out what is needed for mitigating it, motivating the future researches in this area, and pushing the boundary between human and machine interaction in key strategic decisions for organisations along with security implications.

Keywords Critical Business Processes (CBPs) · Strategic decisions making · Security

1 Introduction and Background

Business processes are critical to the operation of any business as they involve determining the overall functionality and pertinent issues, such as the risks that the organisation may be exposed to when undertaking a certain activity [1, 2]. Every company is involved in the identification of critical business processes. Chang [3], affirmed that how well these processes were identified and organised determining whether an organisation succeeded or failed. For instance, appropriate advertising process gave a company adequate potential clients and appropriate manufacturing process ensured that the product was of high quality and priced reasonably, which improved customer

Y. Alblooshi · A. Hosseinian-Far · D. Sarwar (✉)
University of Northampton, Northampton NN1 5PH, UK
e-mail: Dilshad.Sarwar@Northampton.ac.uk

Y. Alblooshi
e-mail: Yousuf.Alblooshi@Northampton.ac.uk

A. Hosseinian-Far
e-mail: Amin.Hosseinian-Far@Northampton.ac.uk

© The Author(s), under exclusive license to Springer Nature Switzerland AG 2021
H. Jahankhani et al. (eds.), *Cybersecurity, Privacy and Freedom Protection in the Connected World*, Advanced Sciences and Technologies for Security Applications, https://doi.org/10.1007/978-3-030-68534-8_25

experience. Through the literature, business processes were considered a set of inter-connected organisational tasks that were designed to take inputs and change them into desired outputs [4].

Critical Business processes (CBPs) were considered vital to the stability of the organisations in terms of finance and operations [5]. Therefore, identifying CBPs was key for business continuity of the organisation, while at the same time being an essential step to comply with national and international regulations and standards (e.g., ISO/IEC 27,001, NCEMA). Consequently, the aim of identifying CBPs was to achieve the objectives of organisations by aligning businesses processes with organisational objectives, which required continuous enhancement of the CBPs [4]. Nevertheless, more often, identification of CBPs has been proven to be challenging, given that the majority of business founders/owners did not have a comprehensive understanding of the CBPs.

Current standards only provided high-level guidelines to companies and directly relied on input from independent business units. Anand et al. [6] asserted that, the input is usually subjective. Thus, business units might show various sets of priorities and objectives that had incompatible and incomplete comprehension of their activities and goals of their peers. In addition, Chang [3] affirmed that different business units might show conflicting specifications as a result of unpredicted inter-relation between their activities due to lack of a systematic framework that can be used to tackling these discrepancies within the existing mechanisms, which rendered the final CBP self-contradictory. Therefore, this created an urgent need to devise a framework that can help organisations to identify CBPs and conduct process improvement. The study explored the literature concerning this fact, while finding out that, many previous researchers attempted creating a novel framework for CBP identification appropriate for all business organisations regardless of their size. Surveying the literature review will provide a value for this research, while assisting in specifying the gaps existed.

2 The "Business Process"—Background

A glance on the business process historical evolvement provides rich information on this valuable technique. "Adam Smith" referred to the term "processes" in his famous example of the pin factory, while stating the way by which a pin was made [7]. In the early 1980s, "August-Wilhelm Scheer" founded his company, IDS Scheer, in Germany, with the entry of Phase 3. Later on, he introduced the concept of "Architecture of Integrated Information Systems" (ARIS) in 1991, as a system concept that would allow company data to be connected to information flows, work and control [8]. While simultaneously supervising diverse management viewpoints for the sake of business clarification and also for the better control of circumstances and execution of processes within the organisation. Following the argument of von Rosing et al. [7], "John Zachman", American business and IT consultant, and an Enterprise Architecture pioneer, published his initial structure entitled "*A Structure for the Architecture of Information Systems*", in a report in the Journal IBM Systems.

Such published report has resulted in a continued growth, which is known recently as the Zachman Enterprise Architecture System-Conceptual Development of a System. While has proposed for defining the related standards and finding out the way by which an organisation is built and structured using accountabilities, roles, interrogations, tasks, Power, information and process of gathering information. von Rosing, et al. [7] referred too to "Taiichi Ohno and Shingeo Shingo" who has developed the "*Just in Time*" since World War II, Toyota's, and "*Pull Process*" principles. Nevertheless, the TPS has experienced industrial growth and has grown substantially since its establishment in the early 1970s while being continuously developed both by business practitioners and academic researchers [7]. In 1990, at the time when Japanese experience was spreading to the West, "James Womack" summarised the TPS concepts to establish Lean Manufacturing, whereas one cannot ignore the success achieved by those companies that adopted such techniques [7]. In 2005, "James P. Womack and Daniel T. Jones" published an article in the Harvard Business Review outlining the latest philosophy known as "Lean Consumption" [7, 9]. By the experimental design of the production process, the correct level of task division was specified.

Contrary to the view of "Smith" which was restricted to the same functional domain and included operations in the manufacturing phase in direct sequence, the new phase model incorporated cross-functionality as an essential aspect. The division of labour was generally adopted according to his views, although the incorporation of tasks into a formal or cross-functional framework was not recognised as an alternative solution until very further. The appropriate intensity of the division of tasks was calculated by the supply chain experimental setup. As "Smith" has said that the division of labour was generally introduced according to his ideas but, until much later, the integration of tasks into an organised or cross-functional system was not seen as an option.

Indeed, previous researchers have defined the term "business process" differently. Andersson et al. [10], referred to the historical evolvement of the term, "process" that was derived from the Latin word, "processus" or "processoat", while indicating to a performed action of something, along with the way by which it was done. Accordingly, the "process" is a series of interlinked activities and tasks that are carried out in response to a given situation that seeks to achieve a particular outcome from a used method. Processes are continuously taking places and are considered as the pillars of all actions, including significant concepts, such as time, space, movement, and these forms adhere to nature.

While referring to the term, "The business process" or "the business operation", Andersson et al. [10] denoted it to a set of individuals or related equipment, coordinated activities or even tasks in which a specific sequence for a particular client or customer produced a service or product (served a particular business purpose). One can understand that, the "business procedures" are carried out at all levels of an organisation, considering its obviousness or ambiguousness to the clients. A business process may also be represented (modelled) as a flowchart of a set of operations with interleaving decision points, or as a process matrix of a sequence of action with related laws accumulating data in the process [11]. The advantages of using business processes include increased customer satisfaction and versatility in adapting to the market's rapid changes. The Process-oriented organisations can decompose institutional departmental barriers and work at eliminating functional silos.

Therefore, a "process" is a series of interrelated processes and actions conducted in the sense of an occurrence to accomplish a common end result for the user of the product. Processes are the basis of all acts concerning principles like time, space and motion and they form and conform to nature itself [7, 9]. Understanding the meaning of the term, "process" requires an in-depth clarification of the "business process" itself.

3 Business Process

The word 'process' was derived from the Latin word processus or processoat, that gave the meaning of "a performed action of something that is done, and the way it is done" [10]. Therefore, a process is a series of interlinked activities and tasks that are carried out in response to a situation that seeks to achieve a particular outcome for the method used. Processes are continuously taking place and happening all around us, with all that is done during the day. These are the pillars of all actions that include concepts such as time, space, and movement, and these forms and adhere to nature. A business process or business operation is a set of individuals or equipment related, coordinated activities or tasks in which a specific sequence for a particular client or customer produces a service or product (serves a particular business purpose).

Anand et al. [6] defined a business process as a collection of different activities that involved a combination of one or more inputs resulting in an output that was of value to the customer. A slightly generalised definition, according to the authors, is the specific ordering of activities across time and place, having a beginning and end with clear input and output, that collectively generate a valuable output to stakeholders.

Therefore, in literature, a process involved a series of steps aimed at achieving a task, and successful implementation of these steps during crises translated to business continuity. To ensure business continuity, there is a need to undertake proper business continuity planning [12]. Some of the experts [13] posited that, the definitions of business planning were generated from several disciplines, such as risk management and facilities management (also see [4] and [14] for other applications). According to the experts [13], organisations ought to make sure that their CBPs were simple to be taken advantage if a crisis occurred. The authors further suggested that, organisations should establish a systematic process of identifying, managing and monitoring risks in line with the process of continuity of a business. While most of the studies advocated for this systematic process, the main problem that most business owners face recently, is the challenges in the identification of such risks in a business process. As Gates [15] noted, the ability to identify critical business processes would allow the organisation to operate as efficiently as possible and to be fully aligned to its strategic goals.

4 Properties of Business Processes

Paying attention to the properties of business processes is an essential aspect of organisational process management. As long as there is a possibility of displaying an asset at the business model level of the process, all process instances based on the business process model must have this property revealed. Although structural process dependence is essential, dependence is the key that should be taken with data generated during business processes. Data dependency between activities is studied in business process models. Attention lies at the structural properties of process models; Properties are neither unique to application nor exclusive to a domain. In definition, the condition is close to that of database theory standardisation. With accordance to the explanation of an author [16], when all tables are found in, for example, a relational database schema is in third normal form. Then any irregularities will no longer occur during database runtime submissions. In the sense of Petri nets, the structural properties of the business processes were studied. The initial soundness criterion implemented in the sense of workflow networks was based on structural soundness. Although soundness is an essential criterion for particular situations it appears to be too solid. It has developed lazy soundness by providing sophisticated monitoring of flux structures for business process patterns, like that of the discriminator patterns [16].

Classification of business processes is vital for the development of effective business process frameworks [17, 16]. Some of the previous researchers [18] encouraged a meaningful way of classifying the business process is by identifying and defining business processes ranging from core to management processes. The current study emphasises on the identification of the core or critical processes, which are an end-to-end and cross-functional process that delivers value directly. As mentioned, critical business processes are the primary process representing essential activities performed by an organisation to attain its short-and long-term objectives. According to Rahimi et al. [19], the business processes make up the value chain, which is a set of a high level of critical processes which are interconnected that adds to the product and service. The value chain ensures the creation and delivery of quality products and services that ultimately deliver values to consumers.

Similarly, a research conducted by some authors [20] shows that critical processes can exist within organisational functions but typically move across different functions and departments, and even across and between different organisations. In their study, Cici and D'Isanto [21] found that the number of critical processes in a company, irrespective of its size, range from 4–8; most of these processes are involved in the development and creation of products and services, the promotion and delivery to customers, and consumer feedback. Overall, critical processes are critical to the value chain and work flawlessly together to attain real customer value.

Research shows that the capacity of an organisation to recognise, identify, and control its critical processes falls under the strategic capability for that particular organisation [22]. It is appeared that, critical processes are of strategic significance, as they have a great impact in different organisations and play a vital role in their success. It is noteworthy that if critical processes are performed good, world-class service delivery can be provided; however, if they are inefficient or not managed well, they could pose a major strategic weakness. Maniora [23] states that critical processes are functional and facilitate distribution by offering the outcomes and value in the form of goods, services or knowledge directly to customers. The recognition and comprehension of important business processes is thus a key starting point for a positive approach to business process management [24].

Four core business management functions represent the business cycle phases. These are: Development, Marketing, Production and Administration (Management) this simple distinction is also rather informative, as these specific duties cover all types of corporations and all aspects of business units. The development involves product and business creation, as well as that of the company and its employees. Development requires responding to the needs of all business activities and is necessary. Marketing states that the task is market formation. The company cannot exist without the demand from customers who need structures. The selling also requires the word, marketing. Production is the entire process of manufacturing products and fulfilment services that customers need and in the last Administration (management) includes all behaviour needed to monitor the resources. The definition includes all essential functions of management support to the operation in a business unit [25].

All the organisation's activities will strive to maximise the business processes. For a company with a diverse range of goods and services, it is likely that extras that do not fall within your core business can drain resources without adding value to customers. Helping tasks that made some sense could have missed their importance in older organisations at the time. For example, one may have adopted compliance requirements to collect data some years ago. He used the info, that does not need it, but his staff still faithfully produces the reports. If any feature does not bring value to the business process or require it, consider flowering it away. That is one of the reasons why mapping the core business processes are so critical.

Most of the major companies consider outsourcing services to companies whose primary activity is support functions. Most of them are considering cost-savings possibilities in this way. In the end, turning to process-oriented performance improvement management, there will be a need for a baseline. The best initiative for analysing the business process architecture is with the main functions. After all, they are the tasks which depend on everything else in the company. Business process integration involves the incorporation of both structure and actions of affected data objects. Research in federated information technology has so far primarily addressed the integration of object model or presumed the identification of the lower-level operations to be carried out if they concentrate on the representation of activities. Based on earlier work showing how to use a key distinction between possible memantine communication to improve the classification process, this examines business processes from the perspective of a prospective integration method or designer and also provides a description of various types of similarities between activities in the process [16, 17].

From this definition, one can establish a collection of options for business process integration and, moving to a more comprehensive stage, present the set of options available based on different forms of interaction between the activities of the processes to be incorporated [26].

Given that, the models of business processes [27] can be defined via many methods. They include BPMN model OMG [28] and the Trend Initiative [29], among others. Börger [6] experiments have demonstrated that the methods do not have the appropriate means to catch situations for businesses beneficial to evaluate, control and communicate the resulting models [27]. However, for different purposes there is still little agreement about how to better characterise market processes. Both businesses are subject to threats regardless of scale [30]. Therefore, it is important to define key business processes for business continuity. The information system (IS) is utilised by companies to boost their key operations' productivity and quality [31]. The combined initiative of information systems, their job processes, their staff, and deployment methodologies will achieve total productivity and performance. In this segment organisations are observed to be sensitive to threats regardless of size; hence it is important to recognise core business processes that promote an organisation's continuity and sustainability. Hence, figuring out the differences in business processes across domains has many advantages.

5   Business Process Modelling

Previously, a great deal of focus has been given to the terms "company operations" and "workflow control." These are used for the development of specific office or manufacturing operations. As people rely on process analyses and templates, "workflow administration" is used because people often advocate the implementation of previously evaluated systems by computer technologies [32]. The term "market system modelling" is utilised in general.

A modern and streamlined solution is introduced in this study to the topic of process management. To date, current frameworks either concentrated on one aspect of workflow management (modelling or implementing) either or several different vocabulary constructs have been implemented for business process modelling. This allows the incorporation and application of modelling workflow definitions of "legacy" enterprise processes into a Workflow Management Model in a collaborative environment. This innovative method is a crucial factor in the efficient usage of process control systems. The framework for the definition of the SAP Rl3 method was used again in the EPC business process modelling vocabulary. There is also a lot of awareness and explanations of business processes in this organisation.

Business process modelling is used when people focus on process design and development, whereas "workflow administration" is commonly used when people often seek by information technology to facilitate the execution of the previously studied processes. A novel and streamlined approach to the process management problem is suggested in this paper. To date, current frameworks have either focused solely on one aspect of workflow management (modelling or execution) or implemented several new vocabulary constructs for business process modelling. This new method allows reusing of an existing language for business process modelling. It blends it with the validation and verification of the modelled processes and the efficient execution of those processes on database-centric systems.

The value chains have a high-level framework of the tasks the company carries out. To offer a more detailed view of this, these business functions at the top are broken down into smaller functions granularity and, essentially, efficient market processing operations. The technique of choice is practical decomposition. Business processes consist of a number of associated operations whose structured execution leads to the realisation of a corporate intent in a technical and organisational context, which can define the business processes by business process templates. Because this section focuses on the arrangement of tasks to be done, disregarding business processes' technological and organisational climate, the term "process model" is used. A notation is required to describe the process models and provide notational elements for the conceptual elements of process met models. For example, if the process met model has a concept called the activity model then a notational element for expressing the activity models is needed.

Coordination is an essential feature of a business process management program of work among the company staff. To accomplish this program information must be given about the organisational structures under which conduct the company method. The met model stage, as in process modelling and data modelling provides how models can be represented, in this case, organisation. At this point, the definitions are places, responsibilities, teams, and relationships supervisor's roles. There are a few systematic examples in organisational modelling rules on how to express organisational structures, and notes to be presented [22].

The space for modelling business processes is organised conceptual models used. Although the terms given have the concepts behind those terms and their relationships used informally in previous chapters, conceptual models will now be discussed in greater detail. These models are presented in the "Unified Modelling Language", which is object-oriented modelling and design language. Business processes are operations the organised execution of which takes place some aim for the company. Those can be device processes, user contact activities, or physical labour. Neither is manual operations assisted. Intelligence networks an example of a manual activity is sending a parcel to a business partner. Activities of user interaction go one step further; these are activities in which awareness workers are doing the work, using information systems. Physical exercise isn't necessarily contributing. An example of human communication behaviour is the input of data about an insurance argument in the sense of a call centre. Since humans use data systems for conducting these activities includes applications with acceptable user interfaces to allow successful work. Such applications must be linked to back-end applications that store the information input and make it available for future use. Many tasks that are carried out during the establishment of a business process are manual, but state changes are reached by user engagement practices in the business process management system. For example, there is an information system that can monitor the delivery of a parcel. Usually, the receiver accepts the actual delivery of a package. The actual delivery is important logistical information business processes that need information systems to reflect properly. A logistics process requires many types of activities. Those happenings are also used as monitoring information for the customer. Computer operations do not require a human user; information systems do them. An example of device operation is to collect stock details from a stockbroker application or to check a bank's balance account. The specific parameters needed for the invocation are presumed ready. If a human user makes this information accessible, then contact is user activity. Both kinds of activities include access to the software systems concerned [22]. This will lead to classifying the business process is critical.

6   Classification of Business Process

Classification of business processes is vital for the development of effective business process frameworks. Researchers [18] have encouraged a meaningful way of classifying business process by identifying and defining business processes ranging from core to management processes. The current study emphasises on the identification of the core or critical processes, which are an end-to-end and cross-functional process that delivers value directly. As mentioned, critical business processes are the primary process representing essential activities performed by an organisation to attain its short-and long-term objectives. According to Rahimi et al. [19], the business processes made up the value chain, which was a set of high level interconnected critical processes that added to the product and service. The value chain ensures the creation and delivery of quality products and services that ultimately deliver values to consumers. Similarly, research by Varbanov et al. [20] showed that critical processes could exist within organisational functions but typically moved across different functions and departments, and even across and between different organisations. In their study, Cici and D'Isanto [21] found that, the number of critical processes in a company, irrespective of its size, ranged from 4–8; most of these processes were involved in the development and creation of products and services, the promotion and delivery to customers, and consumer feedback. Overall, critical processes are critical to the value chain and work flawlessly together to attain real customer value.

Research showed that, the capacity of an organisation to recognise, identify, and control its critical processes fell under the strategic capability for that particular organisation as they had a major impact in different organisations and were vital to their success. It is noteworthy that if critical processes are performed well, world-class service delivery can be provided; however, if they are inefficient or not managed well, they could pose a major strategic weakness. Maniora [23] stated that, critical processes were operational and enabled delivery by directly providing the outputs and value to clients in the form of products, services, or information. Therefore, the identification and understanding of critical business processes is a vital starting point for a successful business process management approach [24].

While surveying the literature, the researcher found out that, several approaches were used to describe models of business processes [27]. They included the OMG standard BPMN [7], and Workflow Pattern Initiative among others [29]. A study by Börger [6] affirmed that the approaches did not provide practitioners with the best ways of capturing business scenarios that were useful in analysing, managing and communicating the resulting models [27]. Instead, the lack of consensus regarding how to best describe the business processes for various reasons remains to be the case. All businesses, irrespective of their sizes, were subject to risks [30]. Thus, identifying core business processes is essential in facilitating business continuity while considering Information System (IS) as what is utilised by organisations to improve the efficiency and effectiveness of their core activities [31]. Overall, effectiveness and efficiency can be achieved through the integrated effort of information systems, its work systems, its employees, and implementation methodologies.

Many of the previous researchers stated that, Design Science Research (DSR) was a research methodology for information systems that offered guidelines for evaluation and iteration in research projects [33]. Such design-science paradigm had been utilised in areas, such as engineering [34], digital forensics [2], education [35], and Information Systems (IS). The application of DSR to information systems was attributed to Hevner [33], "by engaging the complementary research cycle between design-science and behavioural-science to address fundamental problems faced in the productive application of information technology". The behavioural-science, which was also called kernel theories, involved the design and verification of theories that explained human, organisation dynamics, or theories from other fields [33]. The design-science, also called design theory, extended organisation and human capabilities by creating novel artefacts [33].

There were different DSR approaches, though Peffers et al. [36] restructured prior research in DSR into six activities and established a concrete guideline for researchers in this field. These activities included "problem identification and motivation, the definition of the objectives for a solution, design and development, demonstration, evaluation, and communication." (see Fig. 1 for a pictorial illustration).

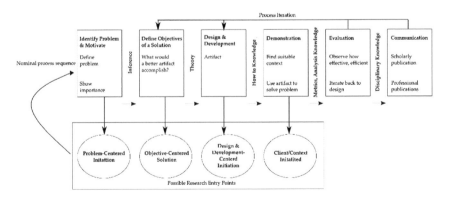

Fig. 1 Design science research [35]

7 Conflicts Between Critical Business Processes (CBPs)

No doubt that, conflicts are unavoidable in organisations, but when there are conflicts between the critical business processes, the process of ending such conflict will vary. Some authoris [8] argued that, conflicts between CBPs could range from small to major encounters that might be highly visible and perhaps led to business destruction; therefore, organisations should find out swift, effective, and low-cost methods of resolving them. Conflicts between CBPs cause major inconveniences to all stakeholders, including customers, employees, and investors, whereas organisations must be proactive in developing ways of reducing or managing them, while finding out innovative ways to address potential conflicts before even they occur [37–39].

Allowing conflicts between business processes to fester and leaving them unresolved could be detrimental, as the conflicts could lead to significant damage as a result of highly problematic litigious situations. Therefore, the desire for conflict resolution between CBPs creates the foundation for attaining a good understanding of all CBPs [8, 40–42].

However, confrontation involves minimising, removing, or resolving certain types of dispute and all kinds of confrontation. Thomas and Kilmann [37] defined five models for dispute management; fight, negotiate, collaborate, avoid, and satisfy. Conflicts resulting from concurrent workflow processes should be carefully addressed when describing simultaneous workflow processes. By analysing the conflicts that are immanent in the concurrent workflow definition before runtime, business process designers and many other workflow management system owners would find it very helpful. Businesses should take advantage of acceptable dispute styles and rates, considering that, it is the purpose of dispute mediation and not the goal of dispute settlement. This while considering the fact that, business processes have been developed in recent years with increased protection and considering compliance issues. For example, recognising process-related security properties is critical because a conflict of interest may emerge from the simultaneous assigning of decision-making and control tasks to the same subject area [8, 39, 40].

In this context, process-related access control systems are typically used to establish constraints on authorisation, such as duty separation (SOD) and duty binding (BOD), to decide the subject is authorised (or obliged) to perform a specific function. SOD restrictions impose conflict of interest policies in a workflow system by specifying that two or more activities have to be carried out by separate persons. Conflict of interest results from the reciprocal distribution of two mutually exclusive entities to the same subject (e.g. permits or tasks) [41, 42].

Accordingly, tasks may be described as dynamically mutually exclusive (on the level of process type or dynamically reciprocal exclusive (on the instance level of the process). A static, then the restriction of mutual exclusion (SME) for all process instances can be seen universal in the Info System. Two roles for SMEs can therefore never be delegated to the same topic or position and two functionally mutually exclusive tasks that be delegated to the same subject. Still, they must not be carried out in the same method instance by the same subject [42].

While paying consideration to the theme that, process management system adaptability (PMS) is essential to agile enterprise intelligence networks, generally the changes in process-oriented applications may occur at two levels; the process form or the level of the process case. A type of process reflects a specific business activity (e.g., managing a patient's purchase order or treatment). At the same time, a System Schema explains it which determines the activity gathering and sets the control and the data flow between them [38]. Accordingly, new process instances should be created and executed based on such process schema with a focus on the given process logic. For example, changes to the process type are necessary to adapt the process-oriented information system to changed business operations or new legislation. They are handled by (structural) modification of the respective process scheme which leads to the respective form of news schema version. The propagation of a process type change to process instances already running is also needed particularly for long-running processes. Function instances for which this is possible to support the new schema and can therefore be moved to it [38].

On the other hand, concomitance to the workflow process is recognised as one of the major sources that trigger such an incorrect representation of the process. Therefore, conflicts generated by simultaneous workflow processes should be treated carefully when defining simultaneous workflow processes [41, 42].

Researchers suggest a set-based constraint method to evaluate potential read–write conflicts and write-write conflicts between activities that read and write to the shared variables. The system consists of two phases. Within the first step, a formal workflow description creates set constraints. The minimum solution of the set constraints is found in the second phase [41]. However, in some cases, conflicts may arise between business processes and critical business processes either due to resource requirements or dependency conflicts. In such cases, organisations have to identify which processes to keep, delay, and stop.

Once again and as it was mentioned in this paper; the critical business processes are those directly affecting an organisation's ability to protect its assets, meet its critical needs, and satisfy mandatory regulations and requirements and must be restored immediately after disruption to maintain business continuity. Therefore, by using the process classification frameworks discussed in this paper, an organisation's assessment team can identify critical processes, their dependencies, and outcomes in addition to the conflicting business processes, considering that, the CBP has priority at all times with the execution of the conflicting BP depending on the availability of resources and the expected impact on the organisation for non-execution [16, 17].

Since non-technical businesspeople began using IT, there has been a conflict between The Organisations and the businesses they are charged with helping. The IT Company is a profoundly complicated area that requires both different skills to produce and sustain. The Company's vocabulary is fundamentally different from the market vocabulary. IT talks about servers, equipment, licensing software, infrastructure, technology, and power. The Company uses terms such as forecasts, margins, and time-to-market, transparency for costs, end-user experience, flexibility, compliance, benefit, monitoring, and agility [39, 42].

Owing to these competing considerations, the company frequently fails to understand the significance of a seemingly straightforward request. These requests may include the priority of having extra fund, stating the project's costs, the availability of human resources, further explanation of the project's details, resisting an innovation or validity of new software, and other conflicted requests and aims. This may consequently, end up with a conflict [8, 39–42].

Typically, organisations resolve such conflicts by continuing execution of the critical business process while the business process can be delayed or performed concurrently with another activity. This allows companies to fulfil their core objectives and maintain regulatory compliance while deferring non-critical processes to appropriate times. The organisation will measure the impact of disturbances and place them in readiness stages to help identify the organisational continuity and disaster recovery criteria. Focusing on clear and accurate essential business processes has the potential to dramatically reduce operational costs and promote a disruptive product and marketing technologies that can make a massive impact in the industry [40].

However, where a business process conflicts against another non-critical process, then the one with the highest number of dependencies takes precedence as non-execution would create issues in multiple parts of the system [40]. The organisations which are facing such conflicts should perform a comprehensive analysis of their processes to identify critical business processes, their dependencies, and outcomes to help inappropriate resource allocation policies. Not only did the business obligate to identify the critical business functions but it also developed possible responses that aligned with the level of risk significance [40]. The business processes could then be rescheduled appropriately to minimise duplication of effort and other resource usage inefficiencies.

Workflows coexistence is regarded as one of the most relevant examples of this incorrect description of the workflow method. The contradictions created by parallel workflow processes will also be closely addressed in describing concomitant workflow processes. But whether a workflow process is non-conflict or not, without experimental execution, is very difficult to determine; it would be a very tedious and time-consuming task for process designers [37]. When identifying contradictions in the concomitant workflow concept prior to runtime, then business process managers and several other workflow management systems consumers will find it extremely beneficial.

8 Contribution to Knowledge

While surveying the literature, the matter was clear that, the development of a novel framework for identification of critical business process is crucial to improving decision-making processes, which can enhance the sustainability of businesses. Thus, while finding out the gap in the literature concerning the absence of a framework, the structure and procedure and organisational and control system should be designed in line with the objectives to improve transparency and accountability in decision-making processes involving the identification of critical business processes. The framework must be tested and implemented with the highest quality and management teams must be thoroughly trained to handle and utilise the framework to reap maximum benefits as much as possible. Any imperfect testing and implementation could have detrimental effects on the overall organisations.

With the findings of this study, organisations that are interested in improving their performance and sustainability can utilise the framework to help them in identifying critical business processes. The major contribution of this study is to establish a novel framework that can be used to identify critical business processes in any business organisations regardless of their sizes. The framework can be utilised in obtaining conflict-free CBPs for any business organisation that is willing to improve their business processes, as the framework provides clear guidelines for the implementation of business continuity. Overall, the application of the framework can make a huge difference in improving business decisions and achieving sustainability goals.

Also, the study findings would be crucial in uncovering problems faced by business owners that identify critical business processes. The implication of this is that engaging in the activity, business organisations will be well informed of the problems that lie ahead. This is vital in establishing a way of managing any challenge that might adversely affect the CBPs identification process. In addition, the findings will inform policymakers of areas to concentrate on how to improve current techniques used to identify business processes.

Furthermore, this research has a high potential to address the challenges encountered by business organisations and management teams while identifying critical business processes. Identification of these processes is crucial in assisting organisations in understanding potential risks that can affect business continuity. Policymakers can use this information to make amendments on the current tools.

9 Conclusion and Further Work

The framework will be utilised in obtaining conflict-free CBPs for XXX, as a case study, as well as any organisation regardless of their sizes willing to improve their business processes. This will assist professionals in development for business continuity (which is also a necessary step for certification, e.g., ISO/IEC 22,301, 27,001) with much less time and effort. In addition to that, the results will push forward the frontier in Information System research in optimising CBP identification. The framework will provide a clear guideline to the responsible managers for implementing business continuity (which is entirely dependent on CBP identification), without seeking external support. Nevertheless, it will provide a mechanism for conflict resolution in CBP, reported by different business units, with minimum effort.

The development of a novel framework for identification of CBPs is vital to improving decision-making processes, which can enhance the sustainability of businesses. With the findings of this study, organisations that are interested in improving their performance and sustainability can utilise the framework to help them to determine their core competencies that require improvement to run the critical business functions.

The organisation will be aware of their risks ahead as they identified CBPs which are within these process, which can be mitigated proactively and efficiently. Besides, the findings will support the governance policymakers to enhance and expand the scope of their policy by enforcing the new framework of CBPs identification techniques to all business units. A further work for this research is an investigation into automatic CBP identification using probabilistic techniques, and development of a functional automatic tool which could reliably identify CBPs with a reasonable computational time.

References

1. AlBlooshi Y, Hosseinian-Far A (2019) A novel framework for identification of critical business processes. In: ICGS3 2019: 12th international conference on global security, safety & sustainability
2. Montasari (2018) Testing the comprehensive digital forensic investigation process model (the CDFIPM). In: Technology for smart futures. Springer, Berlin, pp 303– 327
3. Chang JF (2016) Business process management systems: strategy and implementation. CRC Press
4. Miller HE, Engemann KJ (2019) Business continuity management in data center environments. Int J Inf Technol Syst Approach (IJITSA) 12(1):52–72
5. Trkman P (2010) The critical success factors of business process management. Int J Inf Manage 30(2):125–134
6. Anand A, Wamba SF, Gnanzou D (2013) A literature review on business process management, business process reengineering, and business process innovation. In: Workshop on enterprise and organizational modeling and simulation, pp 1–23, 2013.
7. von Rosing M, Foldager U, Hove M, von Scheel J, Bøgebjerg AF (2015) Working with the business process management (BPM) life cycle, pp 265–341
8. Prasad R, Sundaram G (2018) Resolving conflicts between multiple software and hardware processes. USA Patent U.S. Patent 9,940,188
9. Rosing MV, Scheer AW, Scheel HV (2017) The complete business process handbook: body of knowledge from process modeling to BPM. Morgan Kaufmann
10. Bider I, Johannesson P, Andersson B, Perjons E (2005) Towards a formal definition of goal-oriented business process patterns. Bus Process Manage J
11. Greasley A (2003) Using business-process simulation within a business-process reengineering approach. Bus Process Manage J 9(4):408–420
12. Chapman CL (2017) The influence of leadership on business continuity planning: a qualitative phenomenological study, Doctoral dissertation, University of Phoenix
13. Zhang X, McMurray A (2013) Embedding business continuity and disaster recovery within risk management. World 3(3)
14. Schätter F, Hansen O, Wiens M, Schultmann F (2019) A decision support methodology for a disaster-caused business continuity management. Decis Support Syst 118:10–20
15. Gates LP (2010) Strategic planning with critical success factors and future scenarios: an integrated strategic planning framework. Carnegie-Mellon Univ Pittsburgh Pa Software Engineering Inst, Pittsburgh
16. Rábová I (2009) Business rules specification and business processes modelling. Agri Econ (Zemědělská Ekonomika) 55(1):20–24

17. Aitken C, Stephenson C, Brinkworth R (2010) Process classification frameworks. in Handbook on business process management, vol 2. Springer, Berlin, pp 73–92
18. vom Brocke J, Zelt S, Schmiedel T (2016) On the role of context in business process management. Int J Inf Manage 36(3):486–495
19. Rahimi F, Møller C, Hvam L (2016) Business process management and IT management: the missing integration. Int J Inf Manage 36(1):142–154
20. Varbanov PS, Sikdar S, Lee CT (2018) Contributing to sustainability: addressing the core problems, pp 1121–1122
21. Cici C, D'Isanto D (2017) Integrating sustainability into core business. Symph Emerg Issues Manage 1:50–65
22. Weske M (2012) Business process management architectures. In: Business process management. Springer, Berlin, pp 333–371
23. Maniora J (2017) Is integrated reporting really the superior mechanism for the integration of ethics into the core business model? an empirical analysis. J Bus Ethics 140(4):755–786
24. Weilkiens T, Weiss C, Andrea G, Duggen KN (2016) OCEP certification guide, business process management—fundamental level. Morgan Kaufmann
25. Bicevskis J, Cerina-Berzina J, Karnitis G, Lace L, Medvedis I, Nesterovs S (2020) Domain specific business process modeling in practice. https://www.lu.lv/fileadmin/user_upload/lu_portal/projekti/datorzinatnes_pielietojumi/publikacijas/Publik_kol_3_1.pdf. Accessed 29 Aug 2020
26. Olshanskiy O (2018) Developing the structure and classification of trade enterprise business processes. https://economyandsociety.in.ua/eng/. Accessed 29 Aug 2020
27. Börger E (2012) Approaches to modeling business processes: a critical analysis of BPMN, workflow patterns and YAWL. Softw Syst Model 11(3):305–318
28. BPMN (2011) Business process model and notation. OMG. https://www.omg.org/spec/BPMN/2.0
29. Russell N, van der Aalst WM, ter Hofstede AH (2016) Workflow patterns: the definitive guide. MIT Press
30. Hopkin P (2018) Fundamentals of risk management: understanding, evaluating and implementing effective risk management. Kogan Page Publishers, London
31. Kasemsap K (2015) The role of information system within enterprise architecture and their impact on business performance. in: Technology, innovation, and enterprise transformation. IGI Global, pp 078–1102
32. Xu X (2014) Research on task-based usage control core models in workflow for manufacturing environment. Adv Mater Res 886:378–381
33. Hevner AR (2004) Design science in information systems research. MIS Q 28(1):75–105
34. Simon HA (1996) The sciences of the artificial. MIT Press
35. Heathcote D, Savage S, Hosseinian A (2020) Factors affecting university choice behaviour in the UK higher education. Educ Sci 10:199
36. Peffers K, Tuunanen T, Rothenberger MA, Chatterjee S (2007) A design science research methodology for information systems research. J Manage Inf Syst 24(3):45–77
37. Thomas KW, Kilmann R (1974) Thomas-kilmann conflict mode instrument, Texedo. Xicom, NY
38. Rinderle S, Reichert M, Dadam P (2004) ON dealing with structural conflicts between process type and instance changes. In: Business process management. BPM. Springer, Berlin, pp 274–289
39. Kuester JM, Gerth C (2018) Computing dependent and conflicting changes of business process models. USA Patent U.S. Patent 9,959,509
40. Ramadan Q, Strüber D, Salnitri N, Riediger V, Jürjens J (2018) Detecting conflicts between data-minimisation and security requirements in business process models. In: ECMFA, 2018 modelling foundations and applications. Springer, pp 179–198
41. Lee M, Han D, Shim J (2001) Set-based access conflicts analysis of concurrent workflow definition. In: Proceedings of the third international symposium on cooperative database systems for advanced applications. CODAS. IEEE, pp 172–176

42. Schefer S, Strembeck M, Mendling J, Baumgrass A (2011) Detecting and resolving conflicts of mutual exclusion and binding constraints in a business process context. In: OTM 2011 on the move to meaningful internet systems: OTM 2011. Springer, pp 329–346

A Critical Overview of Food Supply Chain Risk Management

Maryam Azizsafaei, Dilshad Sarwar, Liam Fassam⊕, Rasoul Khandan⊕, and Amin Hosseinian-Far⊕

Abstract Due to the increasing occurrence of disruptive events caused by both human and also natural disasters, supply chain risk management has become an emerging research field in recent years, aiming to protect supply chains from various disruptions and deliver sustainable and long-term benefits to stakeholders across the value chain. Implementing optimum designed risk-oriented supply chain management can provide a privileged position for various businesses to extend their global reach. In addition, using a proactive supply chain risk management system, enterprises can predict their potential risk factors in their supply chains, and achieve the best early warning time, which leads to higher firms' performance. However, relatively little is known about sustainable risks in food supply chains. In order to manage the ever-growing challenges of food supply chains effectively, a deeper insight regarding the complex food systems is required. Supply chain risk management embraces broad strategies to address, identify, evaluate, monitor, and control unpredictable risks or events with direct and indirect effect, mostly negative, on food supply chain processes. To fill this gap, in this paper we have critically discussed the related supply chain risk management literature. Finally, we propose a number of significant directions for future research.

Keywords Supply chain risk · Food supply chain risk · Sustainable development · Risk assessment

M. Azizsafaei · D. Sarwar (✉) · L. Fassam · R. Khandan · A. Hosseinian-Far
University of Northampton, Northampton NN1 5PH, UK
e-mail: Dilshad.Sarwar@Northampton.ac.uk

M. Azizsafaei
e-mail: Maryam.Azizsafaei@Northampton.ac.uk

L. Fassam
e-mail: Liam.Fassam@Northampton.ac.uk

R. Khandan
e-mail: Rasoul.Khandan@Northampton.ac.uk

A. Hosseinian-Far
e-mail: Amin.Hosseinian-Far@Northampton.ac.uk

© The Author(s), under exclusive license to Springer Nature Switzerland AG 2021 413
H. Jahankhani et al. (eds.), *Cybersecurity, Privacy and Freedom Protection in the Connected World*, Advanced Sciences and Technologies for Security Applications, https://doi.org/10.1007/978-3-030-68534-8_26

1   Introduction

The global population has increased rapidly in the past few decades, which can have a direct effect on increasing demands for food products. Population growth can exert tremendous pressure on natural resources that contribute to global climate change and global warming [1, 2]. It also can reduce the level of sustainable development in different countries. Achieving the UN Sustainable Development Goals (SDGs) is crucial in order to harmonize key dimensions of industrial growth, economic growth, social involvement, and environmental protection [3]. One of the major sustainability challenges is food security noted as one of the Sustainable Development Goals (SDG 2) [4].

Considering specific characteristics of food commodities such as perishability and its dynamic system, food supply chains (FSCs) are far more complex than other industries such as manufacturing/service [5, 6]. Food is considered as the vital and most basic human need for survival. Through the years, FSCs have had to deal with massive challenges such as food price fluctuation, climate change, food wastage, food and nutrition security, governance problems, and value-distribution across FSCs [7, 8]. On the other hand, the food network is characterized by a dynamic environment with customers who have increasing demands for food safety and sustainable food commodities. Food consumers also have an intense concern regarding how food products are supplied [9]. The environmental performance of food supply chains is immensely affected by the downstream processes that frequently include elements such as distribution of food products through various channels or drop-off points. One of the critical success elements for improving the food distribution system is selecting efficient logistics strategies and adopting appropriate technologies [10].

In this paper, we examine food supply chain risks and security within the context of the UK food supply chain. The rest of the paper is structured as follows: Sect. 2 outlines a brief overview of Food Supply Chain Management. Section 3 discusses sustainable supply chains in the UK context. Sections 4 and 5 outline risk management within supply chains in general and food supply chains respectively. Section 6 provides a detailed review of literature on supply risk management. The paper is concluded in Sect. 7.

2   Food Supply Chain Management (FSCM)

The food industry is characterized by a dynamic environment due to the changing demands of its customers [11–13]. Based on this characteristic of the food industry, companies should have the flexibility to promptly adjust their strategies and redesign their resources [14–17]. Adopting mass production is another characteristic in food industries. In addition, the whole processes across the supply chains such as purchasing, manufacturing, financing, and sales and marketing have been affected by globalization conditions, and integrated in order to generate global chains [13].

To improve efficiency of operation and to provide necessities, FSC needs to recognize and identify characteristics in this industry. Moreover, Rivera et al. (2014) argue that the characteristics between two main categories of fresh food and long-life products are different [18]. Therefore, each of the categories requires its own specific strategy. Due to low-profit margins in food supply chains, product differentiation is the most common strategy adopted in the food industry [19]. One of the important factors deliberated in food markets as the differentiation strategy is product freshness [20].

Risk assessment is an essential method for minimizing waste in FSC and preventing food and resource wastage. It also supports organizations to establish resilient strategies to achieve food security. Christopher and Peck have argued that the contemporary challenges in the food businesses are about handling and mitigating risks using resilient supply chain principles in various enterprises [21]. Evaluating risks in FSCs can lead to the improvement of other key aspects such as sustainability and performance [22].

In pursuance of supplying safe and reliable commodities, entire supply chain roles should be aware of different potential risks either within and outside their systems. Existing literature highlight that enterprises are required to follow an explicit design to recognize and evaluate risks within their supply chain. The risk evaluation ultimately supports organizations to implement a preventive and sometimes a reactive plan to turn strategies to actions and appropriately manage potential risks [23]. Reducing uncertainties and liability across supply chains are the expected results that aim to achieve a high level of supply chain performance [24].

3   Sustainable Food Supply Chain Management in the United Kingdom

In line with the growing concerns, sustainability has been examined by consumers in recent decades. Particularly the concept of organic food as well as fair trade are under the scrutiny of consumers. Due to the dynamic environment in the food supply chains, food safety and sustainable production, and distribution are considered as the most significant customers' expectations [9]. The UK Sustainable Development Commission [25], according to its strategy and various stakeholder's perspectives, has constructed an applicable international framework for high priority principles of sustainable food and farming industries (Table 1).

4   Sustainable Chain Risk Management

George presented an alternative definition for risk to avoid basic faults in the previous definition which was offered by Insurance Terminology of the American Risk and Insurance Association in 1966 [26]. He defined risk. "not as uncertainty, but as the

Table 1 UK Sustainable Development Commission Priorities [25]

UK Sustainable Development Commission Priorities	
1	Safe, healthy products, nutrition and information for consumer
2	Rural and urban economies and communities
3	Viable livelihoods from sustainable land management
4	Operate within biological limits of natural resources
5	Reduce energy consumption, minimize inputs, renewable energy (environmental performance)
6	Worker welfare, training, safety and hygiene
7	High standards of animal health and welfare
8	Sustaining resources for food production

objective probability that the actual outcome of an event will differ significantly from the expected outcome." As noted in Table 2 there are other definitions that can characterize the following formulas for risk.

The coherent and consistent services from suppliers are expected by customers throughout the world. However, due to the increasing complexity in the current competitive global market, it is challenging to guarantee seamless supply chains [30]. Given the existing supply chains susceptibility and disruptions' intensity, operations management practitioners and researchers have concentrated on investigating the phases of supply chain risk management (SCRM). According to the Business Continuity Institute survey in 2016, 73% of organizations noted that they intend to include risk management approaches over their supply chain processes [31]. Such a high rate indicates that there is significant amount of resources required in organizations to mitigate supply chain risks. The main reasons for supply chain disruptions are categorized into two specific internal and external groups. Some examples of external events include natural catastrophes, changes in legislation and regulation, and market development. Instances of internal events are fraud, accidents, theft, epidemic disease, and sabotage [32]. Financial stability, organization's reputation, and customers' desires are certainly affected by these disruptions [33]. The growing probability of disruptive events with significant impacts has directed organizations to employ diverse proactive and reactive strategies for risk mitigation. Figure 1 illustrates the most commonly adopted supply chain risk strategies for mitigating risks.

Table 2 Risk definitions

Citation	Formula
[27]	Risk $=$ Probability $\times$ Impact
[28]	Risk $=$ Hazard $+$ State of the system $+$ Consequences
[29]	Risk Source $=$ Hazard $+$ Vulnerability $\rightarrow$ Disruption $\rightarrow$ Consequences

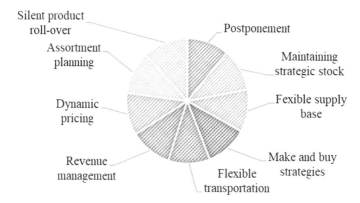

Fig. 1 Most commonly adopted supply chain risk strategies, adapted from [34]

Supply chain vulnerability is defined by [21] as disclosure to shocks emerging from within and outside of the supply chains. The vulnerability has been determined from three important characteristics including the tendency to risk, the ability to resist the shock, and strength-building [21]. Contrary to the vulnerability concept, resilience in supply chains is the ability to recover and resist disturbances (e.g., supplier failure, inadequate demand prediction, etc.) and has a positive implication [35]. Resilience as a proactive approach aims to improve an organisation's ability to mitigate various risks [28].

Kumar et al. define supply chain risks as the possible deviations from the primary objective that, ultimately, target the reduction of value-added processes at various stages [36]. Zsidisin define risk in supply chains as "the probability of an incident associated with inbound supply from individual supplier failure or the supply market occurring, in which its outcomes result in the inability of the purchasing firm to meet customer demand or cause threats to customer life and safety" [37].

Kern et al. have categorized risk into two main groups; Disruptive risks, and operational risks [38]. An example of an operational risk is the inappropriate or failed activities that can cause a supply-demand inconformity [39]. Equipment failure, supply failure, and strategy failure are other instances of operational risks. On the other hand, disruptive risks occur as a result of human-made or the natural catastrophe; such as terrorist attacks, natural disaster, and economic dilemma [40]. Disruptive risks are less controllable than operational risks. Another classification that has proposed by [41] is (a) internal risk events occurring within a firm's supply chain and (b) external risks arising from the environment surrounding a supply chain system.

5   Food Supply Chain Risk Management

Due to the dynamic market environment, the importance of achieving competitive advantages in the global trade, and complicate relationships among supply chain network actors (i.e. suppliers, producers, logistics providers, service providers, customers, etc.), FSCs are susceptible to various types of risks. Manuj and Mentzer argue that a risk management method incorporates three essential phases that are: risk identification, risk evaluation, and risk mitigation [42].

The empirical examination of supply chain risk management is broadly detailed in various literature [43, 44]. To provide a better understanding of risk in the context of supply chains, Rao and Goldsby offered a systematic classification of risks [45]. Tummala and Schoenherr developed a broad method to govern potential risks in supply chains by adopting a risk management procedure [46]. According to this management approach, risk identification is considered as the first stage in the risk analysis and is followed by risk assessment, and risk monitoring stages. The process of risk assessment is about identifying the most appropriate mitigation and proactive strategy based on the identified risks. The risk impacts on supply chain and their measurement techniques hinge on the architectural assessment "impact area" of various risks [47]. Supply chain risk management (SCRM) encompasses processes of risk recognition, risk measurement, risk handling, risk analysis, risk monitoring across the risk management framework [48].

The most cited and adopted supply chain risk classification in different research studies is conducted by [49]. He analyzed more than 200 quantitative articles between 1964 and 2005 and classified supply chain risks into two major risk types that include disruption and operational risks. Disruption risks are affected by man-made and natural failure (i.e. terrorist violations, hurricanes, earthquakes, storm, economic disaster [40]. Operational risks emerge during the business procedure execution or different supply chain practices [39]. Heckmann et al. argued that operational risks in FSCs include supply failure, demand fluctuation and uncertainty, price variance in the market, and cost growth due to machine/equipment failure or management failure [50].

Risk classifications in the supply chain are also provided in SCRM literature such as [21, 45]. Olson and Dash suggested that for simplifying supply chain risks, such risks can be classified into three main groups [51]. These groups include: internal to firm, external to firm but internal to supply chain network. and external to network. Totally, with their sub-categories, five categories that can be generated include the internal process, internal control, demand and supply in supply chain network, and environmental risks. Christopher and Peck defined processes as sequences of value-adding activities adopted by various firm; they also argued that the internal process risks can disrupt these processes in focal firms [21]. They also stated that internal control risks are arising from misapplication of policies, rules, and procedures for controlling processes in firms. In terms of demand risks in supply chains, they argue that it is related to potential disruptions that have negative effects on the downstream flows in supply chains such as materials, cash, and information. On the other hand, supply risks have adverse impacts on upstream flows of supply chains. Various focal firms, upstream and downstream supply chains, and even market places are affected by the final category which is the environmental risks. According to a research review by Goh et al., most risks are categorized based on their source, which are typically within supply networks or their external environment [52]. Many recent studies are focusing on supply chain risks due to the number of related occurrences causing disruption and lowering organizational performances [22, 53].

According to the research undertaken by several researchers (e.g. [54, 55]), there is another classification for risks in food supply chains. It is believed that risks are mainly emerging from sources such as weather, natural disaster, biological and environmental-elements, market-related elements, logistical and infrastructure factors, political factors, public policy and institutional elements, and management and operational influences (Table 3).

Performance measurement has many overlaps with the risk management field, and many scholars consider risks as major sources for compromising performance in supply chains (e.g., [55, 57]). In order to provide further insights and offer integration among supply chain actors as well as to generate useful information for ideal decision making, performance management is considered as an effective suite of techniques and tools [58]. The main performance measurements suggested in previous studies include financial, especially total cost, level of responsiveness to customers, flexibility, food safety, and quality time, particularly lead time, and processes [57].

5.1 Sustainable Supply Chain Risk Management

Due to severe pressure from various stakeholders, organizations around the world are concentrating on the sustainability of their product/service and their operation, and the triple bottom line framework (i.e. environmental, social, and economic performance) has become a focal point [59]. Many current analytical and empirical studies are now focused on sustainable operations. Sustainable operations are related to concepts such as innovation and adopting new technology, remanufacturing, supply chain analysis and design, product development, reverse logistics, and applying appropriate inventory management methods to minimize waste. Nevertheless, a few research works [16, 60] have attempted to evaluate the linkage between sustainable practices adoption and supply chain risks with a view to assessing those risk by different methods. According to Carter and Rogers, sustainability is one of the capabilities in organizations to identify and mitigate social, economic, and environmental risks across the SC [61]. Other studies in finance and strategy have also investigated the linkage between Social Responsibility (SR) and risk management in firms (i.e. evaluate concerning stock market efficiency and performance) [62]. Taylor and Vachon argued that the value and importance of sustainability can inform supply chain risk mitigation approaches [60].

Table 3 Main classification of risks emerging in agri-food supply chains, adapted from [56]

Risk	Definition
Weather related risk	Result of hail and wind catastrophe and to immense humidity or extreme rain that can increase the possibility of pests and diseases
Natural disaster risks	Extensive typhoons, droughts, cyclones, hurricanes, earthquakes, floods, and volcanic activity
Biological and environmental related risks	The biological risk can be from various sources such as bacteria, plants, in-sects, viruses, birds, animals, and humans. Some of these risks frequently have negative impacts on the quantity of production and postharvest, but some of these may have an effect on the quality of products as well. Environmental-related risks are caused by environmental degradation such as soil erosion or factory pesticide or sewerage flow into water sources
Market-related risks	Mainly, market risks are caused by reasons such as demand fluctuation, price change, change in quality standards, short in supply and access to various desirable products and services
Logistical and infra-structure risks	Lack of reliable and affordable transport, inappropriate communication management and information sharing, high energy consumption due to improper route planning and transportation mood selection can cause logistics and infra-structure risks
Political risks	Political risks are related to politico-social vulnerability inside or outside of a country, trade disruptions due to contention with other neighboring countries or traders, seizure of the asset due to dispute or regulation changes by foreign countries and investors
Public policy and institutional risks	Changing monetary, uncertain financial policies (e.g., credit, savings, insurance) and tax policies; changing regulatory and legal procedures are major causes of public policy and institutional risks
Management and operational risks	Weak system management regarding making decisions about capital and as-set allocation, sources selection, quality control, planning, and forecasting, using the high capacity of machines and equipment and maintaining those, and communication and leading labor and employees are the main sources of management and operational risks

6   Review of Supply Chain Risk Appraisal Approach

Risk assessment is defined as an explicit, systematic process that is both complicated and evolving. In line with this evolvement, adopting comprehensive quantitative risk assessments is more common in recent literature. However, various firms, specifically small and medium-size organizations, encounter many difficulties within their quantitative risk assessment implementations. The main reasons for these difficulties include lack of proficiency, knowledge, scheduling and time management, motivation, engagement, and capital. In addition, due to the lack of access to quantitative data and an applicable model with appropriate parameters, quantitative risk assessments are not always usable [63].

According to [64], when risk managers are struggling with the aforementioned problems, they can adopt qualitative risk assessment for prioritizing risks, setting appropriate strategies and policies, and risk resource allocation. In order bridge the gap between the two different approaches (i.e. qualitative and fully quantitative), various semi-quantitative scoring systems and other techniques such as decision trees have also been introduced e.g. by [65, 66].

In the past decade, there have been growth in studies concentrated on supply chain risk assessment. Gaudenzi and Borghesi suggested an AHP-based framework for examining supply chain risks [67]. Chang et al. introduced an exploratory technique to develop optimum decisions for minimizing risk in FSCs [68]. In order to present a comprehensive system thinking approach in the SCRM field, Ghadge et al. conducted a systematic literature review [69]. There are a few significant contributions to the field of SCRM. Hossein Nikou and Selamat presented a literature review on supply chain risk management to evaluate the potential risks across the Malaysian FSCs [70]. Manning and Soon, in order to drive SC agility and stability in various organizations, designed a resilience model for FSCs [71]. Fearne et al. focused on the mitigation approaches for the risks related to fresh beef supply chains [72]. There are other studies such as [73] that evaluated the relationship between potential risks and organizational performance in food supply chains, particularly for fresh food retailer networks. Ding et al. measured indicators of quality performance in the FSCs in the Australian beef processing sector [74]. Various risk impacts on food processing performance are also highlighted in [75]. Dani and Deep, conducted a research review on various risk response development approaches [76]. Wang et al. established a new risk assessment methodology for studying aggregative food safety risks in the food supply chain using fuzzy set theory and AHP [77]. The main important literature focusing on food supply chain risk assessment is provided and examined in Table 4.

Table 4 Summary of the SCRM assessment literature

Source	Aim	Risks involve	Method
[78]	To establish a ranking for suppliers based on aspects determined by micro/macroeconomic features	1. Food quality 2. Corruption 3. Environmental sustainability 4. Logistics 5. Price 6. Production volume 7. Economic growth	Technique for Order Preference by Similarity to the Ideal Solution (TOPSIS), Elimination et Choix Traduisant la Realité (ELECTRE), Cross-Efficiency (CE)
[79]	To develop sustainable framework to minimize food waste.	1. Lack of skilled personnel 2. Poor leadership 3. Failure within the IT system 4. Capacity 5. Poor customer relationship	Pareto analysis Decision-Making, Trial and Evaluation Laboratory (DEMATEL)
[53]	To evaluate the impact of possible demand disruptions in FSCs	1. Demand disruption	Game theory
[80]	Review the mathematical models generated in agricultural business	1. Seasonality 2. Supply 3. Lead-times 4. Perishability	Review Paper
[40]	Risk assessment with two different approach and creating novel approach for assessment	1. Macro level risks 2. Operational risks external to the firm 3. Internal risks	Hierarchical holographic modelling and FL
[81]	To model a government-manufacturer-farmer game for FSCs risk management	1. Society health risks from chemical additive	Game theory
[39]	Reduce the occurrence of the food safety issues and ensure the quality of the people's life	1. Safety risk	Fuzzy AHP
[82]	Develop a model by adopting AHP approach for supply chain risk assessment	1. Earthquake 2. Financial Crisis 3. Supply interruptions 4. Inaccurate demand forecasts 5. Technology upgrades 6. Machine breakdowns	Orders-of magnitude and AHP
[50]	A critical review of supply chain risk	1. Network risk 2. Process risk	Review Paper

(continued)

Table 4 (continued)

Source	Aim	Risks involve	Method
[34]	A literature review regarding supply chain risk management	1. Macro risk factors 2. Micro risk factors 3. Demand risk factors 4. Manufacturing risk 5. Supply risk factors	Review Paper
[83]	Examine the research literature related to food supply chain risk assessment for realizing progress in this area	1. Planning 2. Quality of raw materials 3. Resource allocation 4. Production 5. Specification change 6. Delay 7. Defects 8. Reputation 9. Contract risks 10. Supply	Review Paper and survey
[84]	Propose an incentive scheme include two contracts (i.e. wholesale-market-clearance and wholesale-price-discount sharing) for eliminating "double marginalization" in three-tier supply chain	1. Poor logistics contracts	SIM
[85]	Managing and mitigating risks in food supply chain	1. Macro level risks 2. Demand management risks 3. Supply management risks 4. Product/service management risks	ISM Modelling
[86]	Qualitatively examine the various types of uncertainty effecting on transport operations instead of evaluating the each involve risk	1. Delays 2. Delivery constraints 3. Lack of coordination 4. Variable demand 5. Poor information	Review Paper
[87]	To examine risks in FSCs	1. The quality risks 2. The logistics and inventory control risks 3. The structural risks 4. The information risks 5. The cooperation risks 6. The market risks 7. The environmental risks	System dynamics

(continued)

Table 4 (continued)

Source	Aim	Risks involve	Method
[45]	SCRM review	1. Environmental factors 2. Industry factors 3. Organisational factors 4. Problem-specific factors and 5. Decision-maker related factors	Review Paper
[88]	Identify the relationship between cold chain and developing economies in India	1. Information 2. Communications technology	Fuzzy Interpretive Structure Modelling (FISM) approach
[49]	Perspectives in supply chain risk management	1. Operational risk 2. Uncertain cost 3. Disruption risk 4. Natural and man-made disasters 5. Economic crises	Review Paper
[89]	Model for inbound supply risk	1. Internal risk 2. Quality risk 3. External risk 4. Demand risk 5. Natural or man-made disaster 6. Security	AHP
[67]	Proposed a method to assess supply chain risks according to supply chain objectives	1. Transport/distribution 2. Manufacturing 3. Order cycle 4. Warehousing 5. Procurement	AHP
[41]	To understand the business needs for (SCRM) from a practitioner overview.	1. Loss of IT 2. Fire 3. Loss of site 4. Employee health and safety 5. Customer health and product safety 6. Industrial action 7. Loss of suppliers 8. Terrorist damage 9. Pressure group	Exploratory quantitative survey and qualitative focus group discussions
[47]	Managing risk to avoid supply chain breakdown	1. Supply risk 2. Strategic risk 3. Regulatory risk 4. Customer risk 5. Operations risk 6. Impairment asset risk 7. Competitive risk 8. Financial risk 9. Reputation risk	Supply chain risk tool

(continued)

Table 4 (continued)

Source	Aim	Risks involve	Method
[90]	Managing complex problems associating with both operational and supply chain risk for minimising the costs	1. Length of harvest season 2. Crop size under climatic variations	SP

7 Conclusion and Future Research

We have provided a detailed narrative extraction on key literature related to food supply chain risk management. We believe that the future direction of research for food supply chain risk management embrace the capabilities offered by technologies such as Artificial Intelligence (AI). Breakthroughs in advance digitization, information systems, robotics, technological development, and Artificial Intelligence (AI) will be the driving force of the "fourth industrial revolution" [91]. AI for instance, can provide a distinctive ability in which machines obtain intelligence for making decisions through minimizing human intervention. Machine Learning (ML) is one of these methods, which is the key to unlocking meaning from the dataset through learning from experience [92]. It has revealed in the literature that the machines could possess higher level accuracy in final results compared by human being's outputs in many fields throughout the decision-making processes [93], for instance prediction of cancer [94], drug discovery and development [95], big data [96], and genomics [97].

Despite enthusiasm regarding AI in recent years, there a few vendors in the food industry that apply machine learning in their system. Most early AI applications are mainly adopted by industries such as pharmaceutical, healthcare, cosmetics and, retail. One future research direction in the field is to investigate the possibility of applying ML methods to develop predictive analytics for sustainable supply chain risk management. The research in this spectrum is still at its initial stages, providing purely theoretical schemes that have not been thoroughly tested or applied in real-world contexts.

Through applying machine learning approaches, we could pave the way for automating prediction by training datasets for such predictive analytics tools. The automation will enable organizations to predict supply chain risks, mitigate those, and put measures in place to develop resilience. Subsequently, the negative impacts of events such as unprecedented weather and supply chain shock (e.g. COVID-19) will be greatly reduced. It also can support organizations to provide sustainability to the sector and other intangible benefits (i.e., social value).

References

1. Hosseinian-Far A, Pimenidis E, Jahankhani H, Wijeyesekera DC (2011) Financial assessment of London Plan Policy 4A.2 by probabilistic inference and influence diagrams. Artif Intell Appl Innov 51–60
2. Hosseinian-Far A, Jahankhani H (2015) Quantitative and systemic methods for modeling sustainability. In: Green information technology. Morgan Kaufmann, pp 83–92
3. Desa UN (2016) Transforming our world: the 2030 agenda for sustainable development. United Nations, New York, USA
4. Abdella MG, Kucukvar M, Cihat Onat N, Al-Yafay HM, Bulak ME (2020) Sustainability assessment and modeling based on supervised machine learning techniques: the case for food consumption. J Clean Prod 119661:251
5. Ali SM, Nakade K (2017) Optimal ordering policies in a multi-sourcing supply chain with supply and demand disruptions-a CVaR approach. Int J Logist Syst Manag 28(2):180–199
6. Singh A, Shukla N, Mishra N (2018) Social media data analytics to improve supply chain management in food industries. Transp Res Part E: Logist Transp Rev 114:398–415
7. Gokarn S, Kuthambalayan TS (2017) Analysis of challenges inhibiting the reduction of waste in food supply chain. J Clean Prod 168:595–604
8. Fredriksson A, Liljestrand K (2015) Capturing food logistics: a literature review and research agenda. Int J Logist Res Appl 18(1):16–34
9. Beske P, Land A, Seuring S (2014) Sustainable supply chain management practices and dynamic capabilities in the food industry: a critical analysis of the literature. Int J Prod Econ 152:131–143
10. Tarantilis CD, Ioannou G, Prastacos G (2005) Advanced vehicle routing algorithms for complex operations management problems. J Food Eng 70(3):455–471
11. van der Vorst J, Beulens A (2002) Identifying sources of uncertainty to generate supply chain redesign strategies. Int J Phys Distrib Logist Manag 32(6):409–430
12. Wiengarten F, Pagell M, Fynes B (2012) Supply chain environmental investments in dynamic industries: comparing investment and performance differences with static industries. Int J Prod Econ 135(2):541–551
13. Trienekens J, Wognum P, Beulens A (2012) Transparency in complex dynamic food supply chains. Adv Eng Inform 26(1):55–65
14. Teece DJ, Pisano G, Shuen A (2009) Dynamic capabilities and strategic management. Knowl Strategy 18(7):77–116
15. Barreto I (2010) Dynamic capabilities: a review of past research and an agenda for the future. J Manag 36(1):256–280
16. Foerstl K, Reuter C, Hartmann E, Blome C (2010) Managing supplier sustainability risks in a dynamically changing environment-sustainable supplier management in the chemical industry. J Purch Supply Manag 16(2):118–130
17. Zhu Q, Sarkis J, Lai K (2013) Institutional-based antecedents and performance outcomes of internal and external green supply chain management practices. J Purch Supply Manag 19(2):106–117
18. Rivera X, Orias N, Azapagic A (2014) Life cycle environmental impacts of convenience food: comparison of ready and home-made meals. J Clean Prod 54(4):1513–1520
19. Ahumada O, Villalobos JR (2009) Application of planning models in the agri-food supply chain: a review. Eur J Oper Res 196(1):1–20
20. Lütke Entrup M (2005) Advanced planning in fresh food industries: integrating shelf life into production planning. Physica, Heidelberg
21. Christopher M, Peck H (2004) Building the resilient supply chain. Int J Logist Manag 15(2):1–14
22. Govindan K (2018) Sustainable consumption and production in the food supply chain: a conceptual framework. Int J Prod Econ 195:419–431
23. de Oliveira UR, Marins F, Rocha H (2017) The ISO 31000 standard in supply chain risk management. J Clean Prod 151:616–633

24. Mangla SK, Kumar P, Barua MK (2016) An integrated methodology of FTA and fuzzy AHP for risk assessment in green supply chain. Int J Oper Res 25(1):77–99
25. DEFRA (2002) The strategy for sustainable farming and food: facing the future. DEFRA Publications, London
26. George LH (1967) An alternative to defining risk as uncertainty. J Risk Insurance 34(2):205–214
27. ISO (2002) ISO/IEC GUIDE 73:2002; risk management—vocabulary—guidelines for use in standards. ISO [Online]. https://www.iso.org/standard/34998.html. Accessed 2020
28. Elleuch H, Dafaoui E, Elmhamedi A, Chabchoub H (2016) Resilience and vulnerability in supply chain: literature review. IFAC-PapersOnLine 49(12):1448–1453
29. Gourc D (2006) Vers un modèle général du risque pour le pilotage et la conduite des activités de biens et de services: Propositions pour une conduite des projets et une gestion des risques intégrées. Institut National Polytechnique de Toulouse, Toulouse
30. Bode C, Wagner SM, Petersen KJ, Ellram L (2011) Understanding responses to supply chain disruptions: insights from information processing and resource dependence perspectives. Acad Manag J 54(4):833–856
31. Alcantara P, Riglietti G (2016) Supply chain resilience report 2019. Business Continuity Institute (BCI), UK
32. Ivanov D, Pavlov A, Pavlov D, Sokolov B (2017) Minimization of disruption-related return flows in the supply chain. Int J Prod Econ 183:503–513
33. Hendricks KB, Singhal VR (2003) The effect of supply chain glitches on shareholder wealth. J Oper Manag 21(5):501–522
34. Ho W, Zheng T, Yildiz H, Talluri S (2015) Supply chain risk management: a literature review. Int J Prod Res 53(16):5031–5069
35. Leat P, Revoredo-Giha C (2013) Risk and resilience in agri-food supply chains: the case of the ASDA pork link supply chain in Scotland. Supply Chain Manag Int J 18(2):219–231
36. Kumar SK, Tiwari MK, Babiceanu RF (2010) Minimisation of supply chain cost with embedded risk using computational intelligence approaches. Int J Prod Res 48(13):3717–3739
37. Zsidisin G (2003) A grounded definition of supply risk. J Purch Supply Manag 9(5–6):217–224
38. Kern D, Moser R, Hartmann E, Moder M (2012) Supply risk management: model development and empirical analysis. Int J Phys Distrib Logist Manag 42(1):60–82
39. Xiaoping W (2016) Food supply chain safety risk evaluation based on AHP fuzzy integrated evaluation method. Int J Secur Appl 10(3):233–244
40. Nakandala D, Lau H, Zhao L (2017) Development of a hybrid fresh food supply chain risk assessment model. Int J Prod Res 55(14):4180–4195
41. Jüttner U, Peck H, Christopher M (2003) Supply chain risk management: outlining an agenda for future research. Int J Logist Res Appl 6(4):197–210
42. Manuj I, Mentzer JT (2008) Global supply chain risk management strategies. Int J Phys Distrib Logist Manag 38(3):192–223
43. Manhart P, Summers JK, Blackhurst J (2020) A meta-analytic review of supply chain risk management: assessing buffering and bridging strategies and firm performance. J Supply Chain Manag
44. Ghadge A, Wurtmann H, Seuring S (2020) Managing climate change risks in global supply chains: a review and research agenda. Int J Prod Res 23(3):313–339
45. Rao S, Goldsby TJ (2009) Supply chain risks: a review and typology. Int J Logist Manag 20(1):97–123
46. Tummala R, Schoenherr T (2011) Assessing and managing risks using the Supply Chain Risk Management Process (SCRMP). Supply Chain Manag 16(6):474–483
47. Chopra S, Sodhi MS (2004) Managing risk to avoid: supply-chain breakdown. MIT Sloan Manag Rev 46(1)
48. Neiger D, Rotaru K, Churilov L (2009) Supply chain risk identification with value-focused process engineering. J Oper Manag 27(2):154–168
49. Tang CS (2006) Perspectives in supply chain risk management. Int J Prod Econ 70(3):455–471
50. Heckmann I, Comes T, Nickel S (2015) A critical review on supply chain risk—definition, measure and modeling. Omega (United Kingdom) 52:119–132

51. Olson DL, Dash D (2010) A review of enterprise risk management in supply chain. Kybernetes 39(5):694–706
52. Goh M, Lim JY, Meng F (2007) A stochastic model for risk management in global supply chain networks. Eur J Oper Res 182(1):164–173
53. Ali SM, Rahman MH, Tumpa TJ, Moghul Rifat AA, Paul SK (2018) Examining price and service competition among retailers in a supply chain under potential demand disruption. J Retail Consum Serv 40:40–47
54. Fitzgerald KR (2005) Big savings, but lots of risk. Supply Chain Manag Rev 9(9):16–20
55. Yeboah NE, Feng Y, Daniel OS, Joseph NB (2014) Agricultural supply chain risk identification—a case finding from Ghana. J Manag Strategy 5(2):31
56. Jaffee S, Siegel P, Andrews C (2010) Rapid agricultural supply chain risk assessment: a conceptual framework. In: Agriculture and rural development discussion paper, vol 47, no 1, pp 1–64
57. Aramyan LH, Lansink A, van der Vorst J, van Kooten O (2007) Performance measurement in agri-food supply chains: a case study. Supply Chain Manag 12(4):304–315
58. Chan FT, Qi HJ (2003) An innovative performance measurement method for supply chain management. Supply Chain Manag 8(3):209–223
59. Hosseinian-Far A, Pimenidis E, Jahankhani H, Wijeyesekera DC (2010) A review on sustainability models. In: International conference on global security, safety, and sustainability, pp 216–222
60. Taylor KM, Vachon S (2018) Empirical research on sustainable supply chains: IJPR's contribution and research avenues. Int J Prod Res 56(1–2):950–959
61. Carter CR, Rogers DS (2008) A framework of sustainable supply chain management: moving toward new theory. Int J Phys Distrib Logist Manag 38(5):360–387
62. Jo H, Na H (2012) Does CSR reduce firm risk? Evidence from controversial industry sectors. J Bus Ethics 110(4):441–456
63. Prakash S, Soni G, Rathore A, Singh S (2017) Risk analysis and mitigation for perishable food supply chain: a case of dairy industry. Benchmarking 24(1):2–23
64. Coleman ME, Marks HM (1999) Qualitative and quantitative risk assessment. Food Control 10(4–5):289–297
65. Ross T, Sumner J (2002) A simple, spreadsheet-based, food safety risk assessment tool. Int J Food Microbiol 77(1–2):39–53
66. Davidson VJ, Ryks J, Fazil A (2006) Fuzzy risk assessment tool for microbial hazards in food systems. Fuzzy Sets Syst 157(9):1201–1210
67. Gaudenzi B, Borghesi A (2006) Managing risks in the supply chain using the AHP method. Int J Logist Manag 17(1):114–136
68. Chang W, Ellinger AE, Blackhurst J (2015) A contextual approach to supply chain risk mitigation. Int J Logist Manag 26(3):642–656
69. Ghadge A, Dani S, Kalawsky R (2012) Supply chain risk management: present and future scope. Int J Logist Manag 23(3):313–339
70. Hossein Nikou S, Selamat H (2013) Risk management capability within Malaysian food supply chains. Int J Agric Econ Dev 1(1):37–54
71. Manning L, Soon JM (2016) Building strategic resilience in the food supply chain. British Food J 118(6):1477–1493
72. Fearne A, Hornibrook S, Dedman S (2001) The management of perceived risk in the food supply chain: a comparative study of retailer-led beef quality assurance schemes in Germany and Italy. Int Food Agribus Manag Rev 4(1):19–36
73. Srivastava SK, Chaudhuri A, Srivastava RK (2015) Propagation of risks and their impact on performance in fresh food retail. Int J Logist Manag 26(3):568–602
74. Ding MJ, Jie F, Parton KA, Matanda MJ (2014) Relationships between quality of information sharing and supply chain food quality in the Australian beef processing industry. Int J Logist Manag 25(1):85–108
75. Chaudhuri A, Srivastava SK, Srivastava RK (2016) Risk propagation and its impact on performance in food processing supply chain: a fuzzy interpretive structural modeling based approach. J Model Manag 11(2):660–693

76. Dani S, Deep A (2010) Fragile food supply chains: reacting to risks. Int J Logist Res Appl 13(5):395–410
77. Wang X, Chan HK, Yee R, Diaz-Rainey I (2012) A two-stage fuzzy-AHP model for risk assessment of implementing green initiatives in the fashion supply chain. Int J Prod Econ 135(2):595–606
78. Puertas R, Marti L, Garcia-Alvarez-Coque J (2020) Food supply without risk: multicriteria analysis of institutional conditions of exporters. Int J Environ Res Public Health 17(10):3432
79. Mithun Ali S, Moktadir MA, Kabir G, Chakma J, Rumi M, Islam MT (2019) Framework for evaluating risks in food supply chain: implications in food wastage reduction. J Clean Prod 228:786–800
80. Behzadi G, O'Sullivan MJ, Olsen TL, Zhang A (2018) Agribusiness supply chain risk management: a review of quantitative decision models. Omega (United Kingdom) 79:21–42
81. Song C, Zhuang J (2017) Modeling a Government-Manufacturer-Farmer game for food supply chain risk management. Food Control 78:443–455
82. Dong Q, Cooper O (2016) An orders-of-magnitude AHP supply chain risk assessment framework. Int J Prod Econ 182:144–156
83. Sun Q, Tang Y (2014) The literature review of food supply chain risk assessment. Int J Bus Soc Sci 5(5):198–202
84. Cai X, Chen J, Xiao Y, Xu X, Yu G (2013) Fresh-product supply chain management with logistics outsourcing. Omega (United Kingdom) 41(4):752–765
85. Diabat A, Govindan K, Panicker VV (2012) Supply chain risk management and its mitigation in a food industry. Int J Prod Res 50(11):3039–3050
86. Sanchez-Rodrigues V, Potter A, Naim MM (2010) Evaluating the causes of uncertainty in logistics operations. Int J Logist Manag 21(1):45–64
87. Liu M, Fan H (2011) Food supply chain risk assessment based on the theory of system dynamics. In: 2nd international conference on artificial intelligence, management science and electronic commerce, AIMSEC 2011—proceedings, pp 5035–5037
88. Joshi R, Banwet DK, Shankar R (2009) Indian cold chain: modeling the inhibitors. Br Food J 111(11):1260–1283
89. Wu T, Blackhurst J, Chidambaram V (2006) A model for inbound supply risk analysis. Comput Ind 57(4):350–365
90. Allen SJ, Schuster EW (2004) Controlling the risk for an agricultural harvest. Manuf Serv Oper Manag 6(3):225–236
91. Dogru AK, Keskin BB (2020) AI in operations management: applications, challenges and opportunities. J Data Inf Manag 1–8
92. Farsi M, Daneshkhah A, Hosseinian-Far A, Chatrabgoun O, Montasari R (2018) Crime data mining, threat analysis and prediction. In: Cyber criminology. Springer, Cham, London, pp 183–202
93. Mohri M, Rostamizadeh A, Talwalkar A (2018) Foundations of machine learning, 2nd ed. In: Adaptive computation and machine learning. The MIT Press, Massachusetts
94. Cruz J, Wishart D (2006) Applications of machine learning in cancer prediction and prognosis. Cancer Inform 2:59–77
95. Lo D, Wu F, Chan M, Chu R, Li D (2018) A systematic review of burnout among doctors in China: a cultural perspective. Asia Pac Family Med 17(3):1–13
96. Esmaeilbeigi M, Chatrabgoun O, Hosseinian-Far A, Montasari R, Daneshkhah A (2020) A low cost and highly accurate technique for big data spatial-temporal interpolation. Appl Numer Math 153:492–502
97. Chatrabgoun O, Hosseinian-Far A, Daneshkhah A (2020) Constructing gene regulatory networks from microarray data using non-Gaussian pair-copula Bayesian networks. J Bioinform Comput Biol 2050023

Knowledge Sharing and Internal Social Marketing in Improving Cyber Security Practice

Hiep Cong Pham, Mathews Nkhoma, and Minh Nhat Nguyen

Abstract This paper presents two new ways to establish effective cyber security practice among employee users. Peer knowledge sharing has been used widely in organizations to promote innovation and efficiency, hence its applicability to encourage safe cyber security practice can be fruitful. Internal social marketing is a marketing technique that aims to promote social responsibility to achieve social objectives such as sharing responsibility in ensuring cyber security effectiveness. Using in-depth interviews with employees in organizations located in Ho Chi Minh City, Vietnam, our study explores effective methods of promoting knowledge sharing among users and how it impacts security practice. Similarly, 7Ps in a mixed social marketing approach are evaluated to capture comprehensive security social space that users normally interact with in their quest for cyber security compliant practice. Initial findings of our studies provide practical implications to security professionals to create more supporting and enabling communication infrastructure that serves sustained behavioral changes in complying and co-creating cyber security practice among users.

Keywords Security compliance · Security practice · Social marketing

1 Research Background

The security risks to an organisation's sensitive information are constantly growing and both external and internal attackers are becoming more sophisticated and persistent [1]. Juniper Research predicts data breaches will cost $8 trillion globally by 2022

H. C. Pham · M. Nkhoma (✉) · M. N. Nguyen
RMIT University Vietnam, Ho Chi Minh City, Vietnam
e-mail: mathews.nkhoma@rmit.edu.vn

H. C. Pham
e-mail: hiep.pham@rmit.edu.vn

M. N. Nguyen
e-mail: s3576018@rmit.edu.vn

© The Author(s), under exclusive license to Springer Nature Switzerland AG 2021
H. Jahankhani et al. (eds.), *Cybersecurity, Privacy and Freedom Protection in the Connected World*, Advanced Sciences and Technologies for Security Applications, https://doi.org/10.1007/978-3-030-68534-8_27

[2]. Technical measures have been effective and robust in preventing cyber risks from information security breaches [3]. However, research also shows that a majority of organisational security incidents are directly or indirectly caused by employees who violate or neglect the information policies of their organisations [1, 4] thus, employee compliance choices are critical to organisational security [5]. Certainly, this challenge requires more than technical measurements to overcome. Acknolwledging this challenge, this paper presents the authors' latest findings on employing new methods to encourage safe and compliant security practice from employees.

2 Important of Knoweldge Sharing on Enhancing Cyber Security Compliance

Prior studies have established that users' personal factors such as attitude, self-efficacy and perceived response costs associated with security tasks can affect their intention to comply with information security policies and practices [6]. Additionally, knowledge about cyber security and motivation to protect from cyber risks, are necessary to enhance cyber security practice. Security self-efficacy, which is the combination of the individual's security knowledge, skills and expertise, enables the individual to perform security tasks, as well as cope with changing security requirements. A lack of knowledge about information security leads to low levels of engagement of employees in cyber security practice, thus jeopardising overall organisational information security [7].

Although knowledge could be learned through documented procedures, there are certain tasks which require more practical instructions and certain expertise of information security. An employee can unintentionally violate the security policy of the organisation which could put them and organisation into a security risky situation. Researchers in this regard, place a high stake on knowledge sharing in the workplace through means of social exchanges and informal peer discussions. Safa and Von Solms [8] discussed the potential benefits of an information sharing culture and how it could improve staff efficacy regarding security awareness, promoting sharing knowledge on policies and procedures as well as compliance improvements in the security programs. Furthermore, knowledge sharing enables information workers to develop ideas, share information security concerns and collaborate within the workplace environment regarding solving cyber information threats [3]. Previous research has shown that knowledge sharing among users within organisations is an effective way to increase awareness and compliance with information security policies [8, 9]. Given the increasing number of cyber risks, effective 'real-time' knowledge sharing could help employees protect against potential risks [10]. However, few research studies have examined how employees practice knowledge sharing in the context of cyber security [3]. Furthermore, it is not known whether such practices affect security behaviour in the workplace.

3 Internal Social Marketing and Cyber Security Compliance Behaviour

Besides encouraging cyber security knowledge sharing among employees, another method has been proposed to enhance cyber security practice through internal social marketing (ISM). ISM combines social and internal marketing, applying internal marketing to influence employees' attitude and behavior towards organizational changes, but to aiming to achieve social, rather than commercial objectives [11].

ISM is concerned with employees' perspectives and actions that management can undertake to develop the requisite behavioral outcomes. In order to encourage and lead employees to regularly practice cyber security compliance, an organization needs an approach that helps to build understanding of employees' motivations and behaviors within the social system [11]. ISM is premised on the tools and techniques of commercial marketing applied to the organizational context.

ISM often uses a 7Ps framework to design a servicescape that meets the needs of the employee and achieves the organization's goals [11]. In cyber security, these elements could be designed so as to enhance employee engagement with the product (i.e. participating in the creation of a secure and safe cyber-environment). Furthermore, when it comes to cyber security, there is a more macro-social goal—one of creating a safer cyberworld for all. As such, ISM can more effectively motivate employees and support organizational security compliance objectives. However, even the effect of applying internal social marketing on fostering intrinsic motivations and enhancing cyber security compliance from employees has been acknowledged, this impact still needs further investigation and to be supported by further research evidences.

4 Findings and Discussions

The following findings were established based on our recent in-depth interviews with 30 participants in several organisations in Vietnam. The participants were asked to provide opinions and share experience in their daily cyber security practice with regard to knowledge sharing and the impacts of social marketing elements on their practice. Full descriptions of the data collection and analysis procedure can be found here [12].

4.1 Knowledge Sharing Techniques

* Social media for urgent security incident updates.

In order to enhance the cyber security knowledge sharing among employees, our study pointed out two critical methods that encourage the willing and intention to share knowledge from peer-to-peer. The first method is the application of social media as an alternative cyber security communicating channel. Since many employees use social media applications for exchanging work-related information such as Facebook Messenger, Skype and Zalo (a local Vietnam developed chat application) due to their popularity, instant responses, rich multimedia contents including visual and audio file sharing. The application of social media tools has been recorded as very useful for the organisation to communicate security topics, especially in urgent situations since all staffs can join and discuss about the problems or shared information easily and immediately.

Our findings show that organisations can adopt social media tools to communicate security issues due to their high availability and accessibility on mobile devices [13]. As people carry their mobile devices with them most of the time, consequently disseminating urgent security messages can reach most people almost instantly.

Furthermore, participants expressed the need that social media security messages should be brief (due to mobile device small screens) and visual to depict the contents more clearly to avoid TL; DR (too long; didn't read) responses. Additionally, social media information should be framed to directly relate to each group's interest to avoid flooding irrelevant updates on their professional social media channels. Too much irrelevant information leads to ignorance/avoidance of the messages. Only significant and urgent notifications should be sent to these channels. Otherwise, when too much irrelevant information has been sent to people daily via such social media channels, employees can feel bored, tired and raise the intention to block or ignore future messages from IT department.

In addition, social media applications support group discussion, where people can comment and contribute to the information security problems in a normative setting where modelling of appropriate behaviour is shared amongst peers.

However, it is important to acknowledge the inherent risks from sharing confidential information on social media which can be used for unintended purposes due to its widespread distribution nature. Social media can provide an efficient solution to information sharing. It is important to develop secure and private social media tools restricted to authorized users in order to protect the organisation. Topics of a commercially sensitive nature will still require specialist expertise. Social media brings a new challenge to organisations to manage its appropriate use among employees, but social media is accessible and useful as a community building tool and enables diffused leadership for cyber security.

* Local security experts as a source for domain-specific security knowledge.

Our participants acknowledge that advice from local experts is important and useful to enhance security practice. We found that that local security experts can be an alternative source for seeking security advice or solutions rather than relying solely on the IT department. Moreover, because a local security expert is perceived as part of the employees' peer group, they can sometimes provide better and more relevant advice to their colleagues (e.g. based on their experience about the job requirements) than IT staff. Moreover, trust and respect between employees and the local experts are important, according to some participants. A high level of trust among colleagues and local security experts could initiate open discussions on the security issues, both in regular as well as in serious incidents. Because each department has a different policy—for example a finance department cares more about personal trading policy than the marketing department—having an expert who has experience and knowledge about cyber security in the department is good idea, since they know what problems that the department usually faces during work and employee within a department can trust them to ask.

Sourcing security information from a departmental expert can be viewed as an effective way of improving security knowledge and gaining timely advice. Most participants emphasized the importance of having a designated colleague in providing job-specific security requirements. A local security expert who was experienced and possessed job-related knowledge of information can be timelier and more useful than IT staff. Furthermore, since some organisations did not conduct a formal staff orientation for new staff members, who normally do not have the domain knowledge to respond to security issues alone. Therefore, direct senior staff member's advice can provide new users with specific and uniquely relevant job-related security knowledge and requirements. Sharing knowledge between a designated local security expert to other colleagues is, therefore, a supplementary approach to enhance security knowledge of the employees.

Given the complexity of security requirements and the apparent lack of timely security training, local experts can provide contextual and well-timed advice in response to an immediate threat [14, 15]. As a point of close contact, local experts can facilitate regular exchanges of best practices on similar issues and develop a communal approach to mitigate security challenges. The findings support the work of Lave and Wenger [16] whereby learning is situated in everyday practice and people develop their skills by learning from senior staff and mentors.

4.2 ISM's Impacts on Employees' Cyber Security Compliance

A marketing mix in social marketing can be conceptualized as "the 4Ps": product, price, promotion, and place. In services, this is expanded to "the 7Ps", adding physical evidence (or environment), processes, and people. Our study [12] applied and examined how 7Ps can provide a systematic review of cyber security activities and their impact on user security practice. In ISM, the 7Ps provides a framework of activities that can be deployed to motivate employees to behave in the way that benefits both the organization and employees.

The product is secure and reliable information and the ability to do a job; consistently and when you want to (i.e. no downtime). Despite a relatively high awareness of cyber security, there is a considerable gap between users' descriptions of what is necessary and their reported behaviors. Many participants also had varying perceptions of cyber security needs from. Most participants looked at security compliance as bureaucratic and time-wasting tasks, with little benefit to users. Thus, achieving this goal requires flexibility and autonomy and consideration for the goals of the employee. For example, a virus scanning should not severely slow down a computer, otherwise the users may skip it to resume their work.

The need to communicate and promote the idea of cyber security is essential to any compliance program. However, the promotion of cyber security needs to be user orientated; designed for different skills sets, roles and communication needs. Importantly, it has to be both personally relevant and interesting in order to be motivating. Just because someone 'should' comply does not mean that they will. Promoting personal responsibility (autonomy and competence) without overt fear messaging will assist in increasing engagement. Effective promotion should allow users to engage with security requirements at the right time and in the right place, as most participants only undertook recommended security measures if such measures directly influenced their job (i.e. if they didn't need to engage with the policy to do their job, they didn't use it at all). Thus, more creative promotion such as the use of visual and interactive content, and potentially using social media on both PC and mobile platforms could be used to reinforce policy requirements.

Written policies—promotions—should be short and simple and should deliver clear guidance. The main reasons many users skip reading policy documents are due to the complexity of information, the use of jargon, and the seemingly huge amount of technical knowledge required to understand it. These findings are consistent with Brennan et al.'s [11] recommendation that policy should have a clear statement, which is transparent and articulated in terms that an individual can engage with. Another strategy to engage users is to make communication accessible and enjoyable such as gamifying security training [17]. This might also build on the social (people) aspects of compliance.

Compliance comes with costs both for the organisation and individual. Thus, there is a price element that applies to cyber security compliance. The price of cyber security is considered as the time employees have to spend time on training programs and day-to-day security tasks, such as scanning for viruses and reading repeated notifications. If participants are unfamiliar with the tasks, they might take significantly greater amounts of time and effort to complete them, thereby shifting the burden of associated compliance costs to those least likely to be able to "afford" them. Users easily voice that costs are time and effort, as well as loss of productivity. But other potentially more extreme costs to the organization are not well articulated by and indeed are probably unknown to users. This represents another opportunity to explore whether users can be engaged in cyber security issues sufficiently to become motivated to act, either by learning more about managing cyber security.

Rewards can be used to enhance compliance. However, introjecting rewards can decrease self-determination, so these have to be used with caution. Therefore, intrinsically motivating rewards such as training and development, smoothing workflows, providing autonomy, and recognizing competency are more likely to result in enhanced compliance. Further, decreasing job stress by saving time and effort will also be self-motivating and incentivizing.

The place where people are being asked to respond is also important to security so that supporting measures and locations should encourage safe practice rather causing risks and mistakes from users. We found that most participants reported the 'place' of security (non) compliance behavior occurred mainly through digital channels, including accessing emails and websites, and downloading software. Managing the place element of security behaviour can help security practitioners to consider different usage contexts that users can be exposed to security risks and develop necessary counter measures and awareness training. Further, compliance behaviour often takes place in an uncontrolled environment and ensuring that support is available when and where it is needed, as well as in the form that is most useful at the time, was found to be a critical element in enhancing a cyber security system.

Processes and procedures must support user's cyber security. For example, streamlined and integrated procedures are more conducive to compliance, as they decrease the cognitive load without being repetitive or boring. User involvement should be decreased where possible to increase salience of personal engagement when necessary. Use humans to do human things, leave virus checkers to run in the background. Processes can also be both a barrier and a facilitator to effective cyber security. If processes are too complicated and the user is not motivated or able to improve their knowledge in order to comply, then security will be at risk. Additionally, if processes are too simple, security measures may also be seen as ineffective. Co-creation of processes is one way of ensuring the balance between organizational and user needs.

Furthermore, to support users' active participation in security processes, companies could provide online training, handbooks, IT support, online systems, and virtual helpdesks, to help competent users navigate cyber security processes by themselves.

People factors are important to cyber security: people both make and break the security systems that organisations rely upon to profitably operate. In this sense, understanding the human factors that facilitate and moderate the system enhances cyber security. For example, Pham, El-den [18] demonstrated that it is only when they find IT staff competent and effective in managing IT systems, people will be more engaged in security activities themselves. Effective IT support processes should provide a responsive and effective personal help desk to reduce work interruption, offset the effects of decreased productivity, and increase employee satisfaction [19].

Moreover, as the results show, people do not want to be disempowered widgets in a computerized system. They do want to be involved and engaged and seen as participating in creating a secure environment. organizations can try to improve self-efficacy of employees to enhance their confidence and motivation to comply [20], but they also need to have a chance to use those skills or they may get bored and find something new to do that may damage the security of the environment [21].

The physical evidence of security measures reminds users of the reality of cyber-attacks and their consequences. It does so without the need for personal experience and risk taking. Understanding the difference between front (visible) and backstage (invisible) elements can be helpful in developing interventions that are user friendly (or merely invisible) and where the user has the 'script' to enable them to participate in securing the environment. Each piece of evidence serves as a reminder of the requirement for personal involvement and responsibility for cyber security. Especially, physical evidence such as office design plays an important role in developing and maintaining security awareness, which may positively impact on employees' attitudes on perceiving and evaluating cyber-threats and its consequences [6, 22].

5   Conclusion

Effective security program relies greatly on whether users exercise safe security practice without strict and expensive monitoring. Users should be regarded as important assets not just internal threats to the protection of organisational information assets. This paper highlights that distinguishing the types of cyber security information (e.g. policy and procedure, security risk warning, job-related advice) and using alternative knowledge disseminating methods (i.e. social media or local experts), employees can be more paying attention to security communication, hence be more consciously taking care of organisational security requirements. The use of social media and local experts as alternative channels would ensure employees to access security information and maintain their awareness, in addition to formal periodical trainings. Our use of internal social marketing approach moves away from traditional approaches which are mostly based on the use of formal sanctions and security fear-based communication, by creating a behavioural infrastructure through the deployment of multi-level resources that aim to achieve intrinsic motivation towards compliance.

References

1. Sindiren E, Ciylan B (2018) Privileged account management approach for preventing insider attacks. Int J Comput Sci Net-Work Secur 18(1):33–42
2. Juniper Research (2017) Cybercrime & the internet of threats. https://www.juniperresearch.com/document-library/white-papers/cybercrime-the-internet-of-threats-2017. 30 May 2018
3. Rocha FW, Antonsen E, Ekstedt M (2014) Information security knowledge sharing in organizations: investigating the effect of behavioral information security governance and national culture. Comput Secur 43:90–110
4. Ashenden D (2008) Information security management: a human challenge? Inf Secur Tech Rep 13(4):195–201

5. Sommestad T, Hallberg J, Lundholm K, Bengtsson J (2014) Variables influencing information security policy compliance: a systematic review of quantitative studies. Inf Manag Comput Secur 22(1):42–75

6. Sommestad T, Karlzén H, Hallberg J (2015) The sufficiency of the theory of planned behavior for explaining information security policy compliance. Inf Comput Secur 23(2):200–217

7. Pham CH, El-den J, Richardson J (2016) Stress-based security compliance model-an exploratory study. J Inf Comput Secur 24(3):326–347

8. Safa NS, Von Solms R (2016) An information security knowledge sharing model in organizations. Comput Hum Behav 57:442–451

9. Mallinder J, Drabwell P (2013) Cyber security: a critical examination of information sharing versus data sensitivity issues for organisations at risk of cyber attack. J Bus Contin Emerg Plan 7(2):103

10. Torres HG, Gupta S (2018) The misunderstood link: information security training strategy

11. Brennan L, Binney W, Hall J (2015) Internal social marketing, servicescapes and sustainability: a behavioural infrastructure approach. Innovations in social marketing and public health communication. Springer, New York, pp 87–105

12. Pham HC, Brennan L, Parker L, Phan TN, Ulhaq I, Nkhoma MZ, Nguyen MN (2019) Enhancing cyber security behavior: an internal social marketing approach. Inf Comput Secur 28(2):133–159

13. Kwahk K-Y, Park D-H (2016) The effects of network sharing on knowledge-sharing activities and job performance in enterprise social media environments. Comput Hum Behav 55:826–839

14. Shafiq M, Zia-ur-Rehman DM, Rashid M (2013) Impact of compensation, training and development and supervisory support on organizational commitment. Compens Benefits Rev 45(5):278–285

15. Raineri N, Paillé P (2016) Linking corporate policy and supervisory support with environmental citizenship behaviors: the role of employee environmental beliefs and commitment. J Bus Ethics 137(1):129–148

16. Lave J, Wenger E (1991) Situated learning: legitimate peripheral participation. Cambridge university press

17. Burke B (2016) Gamify: how gamification motivates people to do extraordinary things. Routledge

18. Pham CH, El-den J, Richardson J (2015) Influence of security compliance demands and resources on security compliance-an exploratory study in Vietnam. In: Pacific Asia conference on information systems (PACIS 2015). Singapore

19. Salanova M, Llorens S, Cifre E (2013) The dark side of technologies: technostress among users of information and communication technologies. Int J Psychol 48(3):422–436

20. Rhee HS, Kim C, Ryu YU (2009) Self-efficacy in information security: its influence on end users' information security practice behavior. Comput Secur 28:816–826

21. Johnston TM, Brezina T, Crank BR (2019) Agency, self-efficacy, and desistance from crime: an application of social cognitive theory. J Dev Life-Course Criminol 5(1):60–85

22. Kokolakis S (2017) Privacy attitudes and privacy behaviour: a review of current research on the privacy paradox phenomenon. Comput Secur 64:122–134

Transformation of Cybersecurity Posture in IT Telecommunication: A Case Study of a Telecom Operator

Ahmed Adel, Dilshad Sarwar, and Amin Hosseinian-Far

Abstract Organisations are facing sophisticated and advanced persistent threats (APT) that are targeting sensitive information assets. Any form of cyber-presence can be typically attacked by adversaries, and the motives of such attacks are context dependent. Besides, users and organisations are prone to software vulnerabilities, misconfigurations, outdated systems and several other systemic deficiencies which can be leveraged to compromise enterprise assets and gain an initial foothold within an organisation network. The aim of the paper is to develop a flexible and generally comprehensive organisational strategy to defend against the massive increase in cyberattacks, in order to protect the strategic business objectives of an organisation and keep an alignment between business objectives and security. Moreover, this paper reflects on the work undertaken by multiple teams within the chosen case study organisation to enhance the cybersecurity.

Keywords Cybersecurity · Security operation centre · Cyberculture · IT telecommunication · Cyber resilience

1 Introduction

Cyber Security, a term that is used a lot nowadays. In an era of information and communication technology (ICT), the value of the data is very crucial, especially when it comes to personal information, credit cards information, or financial information. Today everything is moving through the Internet, all aspects of the business

A. Adel (✉)
University of Northampton and Ericsson, Northampton, England
e-mail: ahmed.adel@ericsson.com

D. Sarwar · A. Hosseinian-Far
University of Northampton, Northampton, England
e-mail: dilshad.sarwar@northampton.ac.uk

A. Hosseinian-Far
e-mail: amin.hosseinian-far@northampton.ac.uk

© The Author(s), under exclusive license to Springer Nature Switzerland AG 2021
H. Jahankhani et al. (eds.), *Cybersecurity, Privacy and Freedom Protection in the Connected World*, Advanced Sciences and Technologies for Security Applications, https://doi.org/10.1007/978-3-030-68534-8_28

operations pass through the Internet. On this basis it is therefore imperative that the IT infrastructures are secure.

Cybersecurity is a culture that must be introduced to everyone to increase awareness and reduce negative impacts on organisational information and data security. No hacker will spend time and effort to hack any system that contains no information or invaluable information. Therefore, understandably sectors such as telecommunications and banking are under attack by cyber criminals on an ongoing basis. Recent reports indicate 43% of telecommunication organisations suffered a DNS-malware attack [15]. Moreover, the top five cybercrime attacks were Phishing scams, Identity Theft scams, Online Harassment, Cyberstalking and Invasion of privacy [31]. This paper will talk about cybersecurity in the area of telecommunications, and look at the importance of the security department within organisations, focusing on how to spread the value of cybersecurity to reach a state where all employees in the organisation support cybersecurity. A case study will be conducted in the area of telecommunications, in the region of the Middle East and Africa. It was challenging to know where the organisation under the study is on the cybersecurity posture spectrum. Unfortunately, after severe incidents and significant compromise attempts, the organisation obtained management commitment and support for the new security strategy and program. Security can't start from the middle of the organisation. Ultimately, to achieve significant improvement in information security, senior management and the board of directors must be held accountable for information security governance. They must provide the necessary leadership, organisational structures, oversight, resources and processes to ensure that information security governance is an integral and transparent part of enterprise governance.

2   Literature Review

Cyber-Security became an important area to consider for any organisation running its business on networks or the Internet. So, security awareness became essential for all employees working in all areas. The process of saving valuable information became the responsibility for everyone to undertake. The security department has the upper hand to make sure the organisations processes are in place, regarding tools, and procedures, but the implementation and respect to the process will remain the success factor. For any organisation to move from the ignorance to the awareness, the organisations are required to follow specific steps for that. [12, 23, 24] seek to disseminate information security awareness process that aims to cultivate positive security behaviours. To reach that they found that using either behavioural intention model based on the Theory of Reasoned Action, or the Protection Motivation Theory and the Behaviourism Theory are imperative. They refined both the process and the model. This was then tested through action research and it was found that whether the organisation have or even implement an information security policy, this does not guarantee employees will understand their role in the organisation using security processes and save information assets. They also found that it is critical to design an information security awareness campaign to ensure objectives and requirements.

Bada et al. [4] reviewed current information on security- awareness campaigns and the effectiveness of these campaigns on employees. They then examined the factors responsible for the change in online behaviour, such as personal, social and environmental factors. And, they finally summarised the most critical components for a successful cybersecurity awareness campaign, also, furthermore factors were also examined which could lead to a campaign's failure.

de Bruijn and Janssen [10] indicates that society is turning into a cyber-physical community entirely depending on Information and Communication Technology (ICT) due to the rapid change in the digital life we are living. In return, this makes the need for cybersecurity is a must.

Limba et al. [19] elaborate that for critical infrastructure that uses technologies based on communication and information technology, it depends on cybersecurity. Organisations are trying to make themselves safer from vulnerabilities. They provided theoretical aspects that can be used to ensure security on the critical infrastructure. The cybersecurity model is analysed from management perspectives and is not concerned with technological issues. They also explained that the private sector is much less inclined to share information about specific attacks, although such information could suggestively contribute to the field of cybersecurity. The model consists of six core sections (Fig. 1).

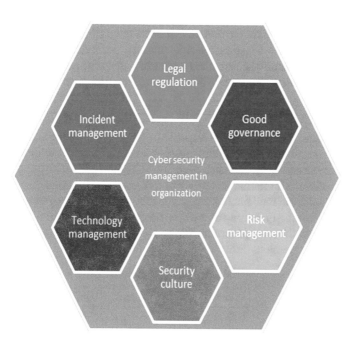

Fig. 1 Cyber Security Model *Source* Limba et al. [19]

Limba et al. [19] have defined the security levels from initial, medium and the highest level of the cybersecurity management model, which they call it interoperability level, which characterised by the full interconnection of all management model dimensions. They indicate that on this level, the organisation is operating as a vast army of soldiers and the cybersecurity model is an inherent part of the organisation which is in line with De Bruijn and Janssen [10].

Nowadays, when the threats are rapidly disseminating everywhere, Organisations needs to evolve solutions that have more complex measures. As cybersecurity management model considering all strategic aspects.

Sallos et al. [25] Agree on the last point of Limba et al. [19] that Organisations must consider cybersecurity through a strategic lens. It must be as a function that must be adopted by everyone. They find that Cybersecurity management is not straight forward, as it is a culture and not a task. It also requires focused consideration in terms of strategies, structures and practices. Sallos et al. have set out the basis holistic view of knowledge which focus on action-results. Sallos et al. also highlighted the importance of a knowledge-based approach to be taught as a concept for the employees.

Von Solms and von Solms [34] highlighted that higher management must clearly understand the cybersecurity, and it's the security department task to make it clear for them to ensure the more senior management buy-in. They succeeded to define the relationship between cybersecurity and information security, concerning the governance perspective. By this, they were able to ensure the board of directors by in, for investment. Moreover, understanding of what cybersecurity cause to the business can if it is absent. They succeeded to make a clearly state that the Cybersecurity target is to protect the organisation against the risks that may harm the business as a whole. The more the organisation is dependent on the Internet, the higher the cyber threat.

Paul et al. [21] state that to enhance cybersecurity, the organisation must study first what is cybersecurity and what are the risks that may occur. They suggest that every organisation should do the following.

- Learning about the basics of Computer and Cyber related terms and concepts.
- Learning about the basics of Security related concepts such as (Computer Security, Network Security, Database, Web Security, IT Security).
- Learning about Information Security and Information Assurance.
- Learning about the fundamental characteristics of Information Assurance.
- Learning about the Function and Role of Information Assurance in general.
- Learning about the laws governing IT and Cyberworld.
- Learn about expressions Table 1.

Table 1 Cybersecurity Expressions adapted from Paul et al. [21]

Cyber café	Cybercrime	Cybernetics	Cyberspace
Cyber hygiene	Cyberwarfare	Cyber organism	Cyberlaw
Cyberattack	Cyberculture	Cyberage	Cyber forensic

Table 2 Cyber Security Functions adapted from Paul et al. [21]	Cybersecurity analyst	Cybersecurity expert	IT Manager	IT security analyst
	Cyber forensic expert	Ethical hacker	Data security analyst	Web security analyst

Paul et al. [21] also recommend that each organisation at least must have one of the following functions if not all when it comes to huge organisations with valuable information system in place (Table 2).

Al-Mohannadi et al. [1] clarify the threat of cybersecurity threats among IT employees. within this, it has been highlighted that a Cyber-attack is one of the critical issues for most of the organisations. Organisations and Governments are doing their best to protect valuable data from being stolen. There are many systems such as Intrusion Detection System (IDS), Intrusion Prevention System (IPS), firewall, packet shaping devices which are available to protect networks. There are also attack modelling techniques that organisation can perform patterns from them to understand the nature of the attack. Thus today, most of the organisations have a security operation centre (SoC) to be the first and most solid layer to protest the organisation from cyber-attacks. They concluded that employees must have awareness sessions and learn about cybersecurity, and this is in line with Bada et al. 4. They also highlighted that IT employees need to improve their knowledge about cyber threat. It was also found that there is a gap of understanding between Security operation team and other IT experts. SoC team are generally capability of safeguarding from cybersecurity threats if they can identify it. Demertzis et al. [9] highlighted the importance of Security operation center as a central level responsible for monitoring, analysing, assessing and defending the organisation asset. This is the most important department under any IT infrastructure domain.

Vukašinovi [32] highlighted that each organisation must have cybersecurity measures. In the field of telecommunication, he found that according to research, mobile phone users are increasingly exposed to cyber-attacks. After analysing more than 400 000 applications available in the most popular apps and Google applications, it was found that 14 000, or 3% have security vulnerabilities, including sensitive information such as location, text messages and contacts. He discovered that Cyber-security attacks are divided into two groups; the first group is the passive and the second group involves active attacks. In the passive aggression, the attacker gets rights without changing the content of messages. In the active attacks, the attacker can modify, delete, copy the contents of files, set himself as an authorised user, disable functions and do whatever he wants. The protection of any network systems should follow the following:

- Confidentiality
- Integrity
- Availability.

Table 3 Cybersecurity areas adapted from Vähäkainu and Lehto [33]

Infrastructure security	Endpoint security	Application security	IoT-security
Web-security	Security operations and incident response	Threat intelligence	Mobile security
Cloud security	Identity and access management	Network security	Human security

Vuka agrees with most of the above articles that there is no fully protected computer network. The most secure system is one that is not connected to the Internet at all. The protection given to the network systems is a must nowadays. And, by ensuring protection, Organisations will enable preventing unauthorised intrusion. Monitoring systems need to be used to reduce the security risks of intervention into systems.

Vähäkainu and Lehto [33] highlighted the importance of artificial intelligence (AI) to help cybersecurity management. They indicate that organisations benefit from the ability of (AI) systems to improve their expertise quickly and from sharing it to all those who need it. They discussed the following cybersecurity areas (Table 3).

Their study highlighted that information on 11 artificial intelligence solutions were gathered. These perspectives were divided into the following areas:

- infrastructure security
- endpoint security
- web security
- security operations and incident response
- threat intelligence, mobile security and human security.

Vähäkainu and Lehto [33] have concluded that the (AI) system should detect and quickly react to any attack, such as an abnormal login, and/or suspicious usage of cloud services. There are many ways to detect threats. But, as an organisation may face up to 200,000 information security events per day the investigation of the threats by using human information security specialists is expensive and is time-consuming, and therefore (AI) is a must nowadays.

3   Methodology

The methodology used in this paper is the Design Science Research Methodology. Bisandu [6] explained Design Science as it is concerned with knowledge acquisition that relates to designs and activity which offer a specific guide to for evaluation and iteration within a project. While research methodology is also seen as an action plan, strategy, process, behind the choice of and methods and linking the choice of methods use [2, 6]. Design Science Research Methodology (DSR) is considered to be the other side of Information System research that evaluates information Technology artefacts needed to solve problems identified in an organisation.

Design science research contributes highly in the field of Information Systems as it measures the way it is applied to business needs. It solves an existing problem. Therefore, there is considered to be one of the most useful methodologies in these fields [6, 14, 22].

This methodology is that it is used when there is crucial dependence upon human cognitive abilities to produce effective solutions. Or personal social skills are a critical dependence upon to deliver effective solutions which is the case in this paper and the case study too [6, 7, 13].

The Design Science Research Methodology has found to be an excellent method in the Information Science and Computer Science because it is a method that works with human, organisational social kind of problem-solving through artefact development [6, 14].

4   Case Study

4.1   Understanding the Cybersecurity Posture

The security status of your enterprise's software and hardware, networks, services, and information; your ability to manage your defences; and your ability to react to and recover from security events are collectively referred to as your cybersecurity posture. Understanding and defining the full scope of your cybersecurity posture is essential to protecting your business against breaches [5].

To understand and optimise your cybersecurity posture, you need to:

- Analyse what it currently looks like
- Identify the possible gaps
- Then take action to eliminate those gaps.

The thing that kept me awake at night (as NATO military commander) was cybersecurity. Cybersecurity proceeds from the highest levels of our national interest-through our medical, our educational, to our personal finance (systems). Admiral James Stavridis, Ret. Former-NATO Commander [11].

Unfortunately, most people do not understand the gravity of the problem until it personally affects them through identity theft or other malicious activity. Unsurprisingly, however, the rate of cybersecurity-related crime is exploding, and a recent study claims that there is a new victim of identity theft every 2 s in the United States alone [28].

On top of that, Half of the cyberattacks are targeting small businesses that usually don't have sufficient cybersecurity to protect themselves from such threats. In a Statista report in 2018, the most notorious cyber-attacks experienced by companies of all sizes include phishing (37%), network intrusion (30%), inadvertent disclosure (12%), stolen/lost device or records (10%), and system misconfiguration (4%) [27] (Fig. 2).

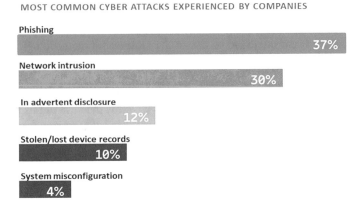

MOST COMMON CYBER ATTACKS EXPERIENCED BY COMPANIES

Phishing — 37%

Network intrusion — 30%

In advertent disclosure — 12%

Stolen/lost device records — 10%

System misconfiguration — 4%

Fig. 2 Most common cyber-attacks *Source* [20]

4.2 The Turning Point

Understanding the current security posture and identifying gaps in existing organisational security systems is very important. It will require skilled resources and tools, audits- both internal and external are one of the main processes used to determine information security deficiencies from control and compliance standpoint and are one of the essential resources in strategy development. Early detection of security problems and solving it will be a cost-effective solution for developing secure systems [36].

The turning point had started when there was a successful attempt to attack one of the mission-critical systems and investigations found that the system was compromised. The attacker targeted confidential information assets. Fortunately, the effort failed due to unexpected server behaviour, and the attacker zipped the theft data in the same server, causing service interruption due to file system utilisation. On top of that, this incident reported a service availability issue. This is considered as cybercrime [16].

4.3 Gap Analysis is the Base for Strategy Development

In addition to the costs that companies face to deal with the immediate effects of an incident, security incidents can cause more costly, long-term harm such as damage to reputation and brand. Beyond the impact to market capitalisation, if the issue threatens the public good, regulators may intervene, enacting stricter requirements to govern future business practices. Data breaches stain the reputations of companies both big and small, damaging the brand and reducing consumer trust, and sometimes the consequences can affect the company for years to [30] notes Paul Bischoff, researcher and privacy advocate at Comparitech.

Muddy Water – global attack geography 2018

Countries targeted by the Muddy Water spear-phishing campaign in 2018, according to Kaspersky Lab detection data

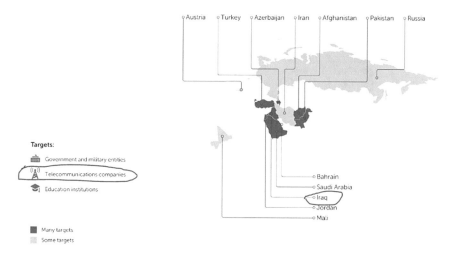

Targets:

🏛 Government and military entities

📡 Telecommunications companies

🎓 Education institutions

■ Many targets
▨ Some targets

Fig. 3 *Source* Kaspersky [17]

In our case and after the security incident; the response and actions of management were hugely influential; their support helps to develop the implementation of the remediation strategy. According to Kaspersky Lap detection data; we found that a sizeable Iraqi telecom provider was deeply compromised with 10% of their endpoints infected with POWERSTATS (Fig. 3).

As with any complex problem, a lot of questions need to be answered to identify where you are:

- Do we have a real-time inventory of all our assets, asset calcification?
- Are we able to continuously monitor our assets?
- Do we have security incident management in place?
- Do we have a backup process, DR, and BCP?
- Do we have security metrics?

Gap analysis [8] identified 39 Security gaps and deviations in security tools, processes, execution and management system. End of service, End of life (EOS/EOL) operating system are still in the operation state, hardening, patch management and user access management are not defined. Security policy, regulatory standard and governance, risk management are not defined; some systems were compromised already.

The gaps between the current stat and desired stat paved the way to develop the security strategy and developing the road map to achieve the objectives. The desired defined based on the outcomes set by management, regulatory requirement and a variety of frameworks to ensure defence in depth (Fig. 4).

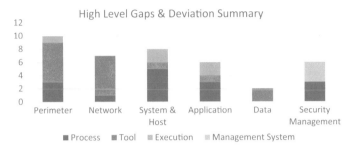

Fig. 4 Gaps and deviation summary (graph developed by Authors from real organisation data)

The full security audit was finalised after the major security breach, below are the summary of the gaps identified:

- VPN Users gain access directly, Multi-factor authentication is missing, VPN user time out is not in place.
- Anti-Spam and Anti-Virus implemented; the effectiveness of the tool usages not measured.
- The absence of log management and the centralised log management solution available for the network.
- No real-time security attributes monitoring of the devices and nodes except bandwidth monitoring.
- No distributed denial-of-service (DDOS) protection implemented.
- No Advance Threat protection.
- Terminal Access Controller Access Control System (TACACS) is not implemented for all network devices.
- Zone segregation is not defined.
- No Next-Gen firewall available, current firewall policy is based access-list only. Ext to int for any to an internal subnet is allowed, many test rules also presented, No FW rule validation in place.
- Antivirus management servers are exposed with full Internet access, same control to be validated on other servers.
- Network Segmentation for Management & data is not implemented for all.
- Vendor allowed to connect the corporate network with their device. No control over for third party / non-standard systems to connect thus causing vulnerable.
- Asset management tracker in place, however no EOS/EOL details for the devices and tools are highlighted. The tracker is not updated regularly.
- No defined hardening processes. Antivirus & Vulnerability status are verified during the commissioning process.
- Undefined patch management process in place. Patch management is followed on reactive/OnDemand manner and no zero-day patch management process, still many critical patches are in an open state, and many EOS/EOL systems are running in the infrastructure.

- Vulnerability scan performed for critical system only, the mitigation process is not well defined and not aligned with GRC; even some critical vulnerabilities are still open.
- No IDM tool in place, No ID & access revalidation in place and No password age and password reuse policy implemented.
- All the clients not configured to route through a proxy.
- Clear text services are like telnet, FTP, HTTP is allowed from outside.
- No defined policy in place.
- Only a few components have the process documents and no procedure or SOP available.
- No defined metrics for all the Security measures.
- There is limited security governance in place; Security governance is not aligned with the Org level. Elevation/escalation of compliance & security weakness requires leadership decision.
- Business continuity plan (BCP) and Disaster Recovery (DR) process are not in place, No BCP/DR test executed. Only critical systems data backup is happening and stored in DR locations.

4.4   Road Map for Remediation

An implementation plan was formalised as a road map to close all tactical and strategic gaps [18]. The Implementation plan started immediately after the assessment conclusion. The business case developed for investment in security tools and controls. Prioritising action items and addressing the most critical vulnerabilities and issues first guide the roadmap of our entire defence strategy and influence our security spending.

To improve and raise your Cybersecurity posture and awareness, you don't need to invest endlessly in new security tools. The truth is that 80% of data breaches can be prevented with necessary actions, such as vulnerability assessments, patching, and proper configurations. An example is Phishing attacks are the most common cybersecurity attack. This type of attacks is a big part of why there are so many compromised passwords. In the last year, 76% of businesses reported that they had been a victim of a phishing attack (Info Security 2017), security awareness is the most effective control to mitigate the risk of phishing attaches [26].

In addition to the immediate response to isolate the compromised system and the eradication effort, the Tactical (6 months) and strategic (3 years) action plan formalised, and our security program started immediately with ultimate commitment from C-level management to ensure the required resources.

4.5 Actions Were Taken to Eliminate the Gaps

The steps taken consist of controls, processes, and practices to increase the resilience of the computing environment and ensure that risks are known and handled effectively. These activities dealt with by an internal team supported by external vendors as needed.

4.5.1 Security Enhancements in the Security Tools

Although cybersecurity spans technical, operational and managerial domains, a significant portion of the actual implementation of the information security program is likely to be technical [35] below are the summary of the security enhancements in the security tools;

- SEIM tool is currently used by the security team and SOC for log monitoring and security management.
- Multi-factor authentication is currently used for VPN, Webmail and servers access.
- Network (LAN and Wireless) security controls have been implemented based on least privilege using Cisco ISE.
- Enterprise password management tool used for local admin control.
- Windows and Linux security patching tools are currently used for centralised patch management.
- Advanced Threat Protection ATP is used.
- Next-generation firewalls are used in the network.
- Next-generation Antivirus is used in all clients and servers.
- New proxy servers.

4.5.2 Security Enhancements in Process and Governance

Process and governance must be an integral and transparent part of enterprise governance and complement or encompass the IT governance framework [3]. Integrated with the processes they have in place to govern other critical organisational resources. It includes monitoring and reporting processes to ensure that governance processes are effective and compliance enforcement is sufficient to reduce risk to acceptable levels. Below is the summary of the security enhancements in process and governance;

- User Access management process
- Security Patch management process
- Vulnerability management process
- Log management process
- Antivirus management process
- IT security Weekly meeting to discuss security operations and project.

- IT security reports and metrics (KPIs and KRIs).

on top of the governance and processes enhancements, all security policies reviewed and updated with management intent.

4.5.3   Security Enhancements in IT Operations

Poor configuration in IT operations can lead to cyber criminals bypassing internal policies that protect sensitive information. Setting security baselines for an organisations operational enterprise has several benefits. It standardises the minimum amount of security measures that must be employed throughout the organisation; this results in positive benefits for risk management. It also provides a convenient point of reference to measure changes to security and identify corresponding effects on risk. A lot of security enhancements has been achieved in IT operations;

- IT severs, and Client PC is full patches with the latest security updates,
- Next-generation antivirus has been installed in all client PCs and servers and monitored,
- Close all discovered critical vulnerabilities within SLA timeline,
- Physical access control to datacentres is managed and integrated with an access management tool
- Data uploading is monitored 24X7 by NOC.
- Enhancement in security incident handling,
- Upgrade or Isolate EOL/EOS systems.

In addition to all this security enhancement, new SOC function has been established to monitor security attributes for IT assets 24/7 and analysing logs and respond to a security incident to increase speed and agility in security. Preparing the workforce to protect their environments is vita.! As much as it is essential to have in place all security measures to safeguard the information systems infrastructures, hardware and software alone cannot withstand the attacks of malicious staff training, and security awareness is critical security control.

5   Current Cybersecurity Posture; Today Versus Past

The quick fix actions and short-term tactical plan are practical and affordable ways to reduce the exposure and avoid the worst to happen. The overall security program has a significant impact on information security readiness and on the capability of the staff to deter potential attackers.

Now, our enterprise has documented all of its programs and procedures, and it has a clear understanding of its risk. It is not the endpoint, in addition to strong KPIs and KRIs, continues assessment, testing and audit is an essential part for

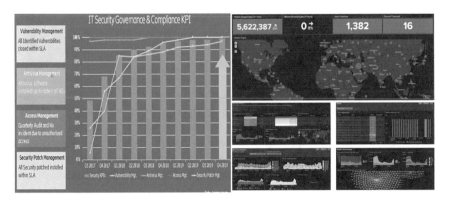

Fig. 5 Security Governance & tools (Organisation own developed tools)

continues improvement in the cybersecurity posture and ensures the security control effectiveness.

The enterprise ranked No.1 in cybersecurity posture by group risk management. Furthermore, the last security audit and penetration test report a considerable enhancement in cybersecurity capability.

A lot of security dashboards and security automation developed by a security team internally, KPIs and security metrics reflect strong security governance and risk management. Continuous monitoring for security events, incident detection, identification, handling, post-incident review and security awareness increased our ability to react to and recover from security events. Organisation 24/7 SOC increased the velocity of security event handling and detection for the intrusive/malicious/suspicious/misconfiguration/policy violation etc. events before getting a serious issue (Fig. 5).

Staff training and awareness sessions enhanced cybersecurity culture in our organisation, and this is reflected by the increased number of a reported security incident. Developing the technical staff take their skills to the next level that can be used in penetration testing and threat hunting.

6 Conclusions and Recommendations

The security status of the enterprise's assets, ability to manage defences, and the ability to react to and recover from security events are the most useful indicators for cybersecurity posture.

An organisation transformation plan, along with the best practices that have been followed, helped building a cyber-resilience strategy and improve the security posture of any organisation. It is essential to create a culture of security awareness across the organisation and among employees. This is the best way to provide a constant barrier that deters cyberattacks that take advantage of human behaviour.

A severe breach can result in data loss or potential damage to the IT infrastructure and have adverse effects on essential company operations and on the business itself through the loss of confidentiality, integrity or availability of informational assets.

Any security transformation program requires, of course, the strong commitment, direct involvement and ongoing support from senior leaders/executives. Such efforts are constant and permanent, which therefore require continuous evaluation, funding and support. Inconsistency will cancel out any steps forward and opens the organisation to increased risks.

Defending against sophisticated threats ultimately requires mature processes and competent, dedicated security professionals. Sophisticated attaches require a thoughtful process that can prevent, detect and respond to threats with speed and agility.

While cybercriminals represent a significant threat, in most cases, the critical threats to organisations are their lack of adequate defences and employees who are ignorant of cyber threats. Organisations can reduce their risks of cyber-attacks by following industry best practices and implementing key defence measures such as employee training and the use of encryption.

What cannot be measured cannot be managed. Security metrics and thresholds should be defined for specific control and process to measure the extent to which performance objectives are being achieved on an ongoing basis. Security status trends reports that are systematic and timely are a useful tool to maintain management commitment and support. Organisations had to start implementing the culture of security to their employees. Security awareness also had to be disseminated across the organisation to make sure that everyone is aligned and knows the importance of security. One of the key strategies in security is No exceptions when comes to security and important information inside the organisation [29].

References

1. Al-Mohannadi H, Awan I, Al Hamar J, Al Hamar Y, Shah M, Musa A (2018) Understanding awareness of cyber security threat among IT employees. In: Proceedings2018 IEEE 6th international conference on future internet of things and cloud workshops, W-FiCloud 2018, pp 188–192
2. Alturki A, Gable GG, Bandara W (2013) The design science research roadmap: in progress evaluation. In: PACIS 2013 proceedings
3. vila C, Chinchilla EJ, Velásquez Pérez T (2019) It governance model for state entities, as support for compliance with the information security and privacy component in the framework of the digital government policy. J Phys: Conf Ser 1409(1)
4. Bada M, Sasse A, Nurse J (2019) Cyber security awareness campaigns: why they fail to change behavior. Int Conf Cyber Secur Sustain Soc 38
5. Balbix (2020) Getting started on transforming your cybersecurity posture. Balbix. https://www. balbix.com/app/uploads/eBook-Transforming-Security-Posture.pdf. 23 Mar 2020
6. Bisandu DB (2016) Design science research methodology in computer science and information systems. Int J Inf Technol 1–7
7. Bisandu DB, Prasad R, Liman MM (2018) Clustering news articles using efficient similarity measure and N-grams. Int J Knowl Eng Data Min 5(4):333–348

8. Crumpler W, Lewis JA (2019) The cybersecurity workforce gap. Cent Strat Int Stud (CSIS) 1–10

9. Demertzis K, Tziritas N, Kikiras P, Sanchez SL, Iliadis L (2019) The next generation cognitive security operations center: adaptive analytic lambda architecture for efficient defense against adversarial attacks. Big Data Cogn Comput 3(1):6

10. de Bruijn H, Janssen M (2017) Building cybersecurity awareness: the need for evidence-based framing strategies. Govt Inf Q 34(1):1–7

11. Gartner (2020) Former NATO commander says cybersecurity most worrying threat we face. Gartner. https://www.gartner.com/smarterwithgartner/former-nato-commander-says-cyb ersecurity-most-worrying-threat-we-face/. 23 Mar 2020

12. Gundu T, Flowerday SV (2013) Ignorance to awareness: towards an information security awareness process. SAIEE Afr Res J 104(2):69–79

13. Gleasurea R (2015) When is a problem a design science problem? Syst, Signs Actions 9(1):9–25

14. Hevner R, Salvator A, Jinsoo Park T, Sudha R (2004). Design science in information science

15. Ismail N (2018) Telcos struggling to mitigate the threats of cyber attacks. InformationAge. https://www.information-age.com/telcos-cyber-attacks-123476699/. 23 Mar 2020

16. Jahankhani H, Al-Nemrat A, Hosseinian-Far A (2014) Cybercrime classification and characteristics. In: Cyber crime and cyber terrorism investigators handbook, pp 149–164

17. Kaspersky (2018) Middle-east focused threat actor Muddy Water extends attacks towards government targets in Asia, Europe and Africa. Kaspersky. https://www.kaspersky.com/about/press-releases/2018_muddy-water-final. 1 July 2020

18. Kapur R (2017) Organization and administration in adult and community Education. Int J Inf, Bus Manag 9(1):141

19. Limba T, , Agafonov K, Damkus M (2017) Cyber security management model for critical infrastructure. Int J Entrep Sustain Issues 4(4):559–573

20. Observer C (2020) 29 Must-know cybersecurity statistics for 2020. Cyber Obs https://www.cyber-observer.com/cyber-news-29-statistics-for-2020-cyber-observer/. 23 Mar 2020

21. Paul P, Bhuimali A, Aithal PS, Rajesh R (2018) Cyber security to information assurance: an overview. Int J Recent Res Sci, Eng Technol (IJRRSET) 1–9

22. Peffers K, Tuunanen T, Niehaves B (2018) Design science research genres: introduction to the special issue on exemplars and criteria for applicable design science research. Eur J Inf Syst 27(2):129–139

23. Rhodes RE, McEwan D, Rebar AL (2019) Theories of physical activity behaviour change: a history and synthesis of approaches. Psychol Sport Exerc 42(2019):100–109

24. Safa NS, Sookhak M, Von Solms R, Furnell S, Ghani NA, Herawan T (2015) Information security conscious care behaviour formation in organisations. Comput Secur 53:65–78

25. Sallos MP, Garcia-Perez A, Bedford D, Orlando B (2019) Strategy and organisational cybersecurity: a knowledge-problem perspective. J Intellect Cap. 20(4):581–597

26. Staff I (2016) CISM review manual, 15th ed. Information systems audit and control association

27. Statista (2020) Global No.1 business data platform. Statista. https://www.statista.com. 23 Mar 2020

28. SelfKey (2020) All data breaches in 2019 2020an alarming timeline. SelfKey. https://selfkey.org/data-breaches-in-2019/. 23 Mar 2020

29. Sennewald CA, Baillie C (2020) Effective security management. Butterworth-Heinemann

30. Seals T (2017) Post-breach share prices plummet below NASDAQ average. Group, InfoSecurity. https://www.infosecurity-magazine.com/news/share-prices-plummet-below-nasdaq/. 23 Mar 2020

31. SpideyMan (2020) Top 5 popular cybercrimes: how you can easily prevent them. EnigmaSoft. https://www.enigmasoftware.com/top-5-popular-cybercrimes-how-easily-prevent-them/. 9 Mar 2020

32. Vukašinovi M (2018) Cyber security measures in companies. Int J Econ Stat 6:125–128

33. Vähäkainu P, Lehto M (2019) Artificial intelligence in the cyber security environment. In: 14th International conference on cyber warfare and security, ICCWS 2019, pp 431–440

34. von Solms B, von Solms R (2018) Cybersecurity and information securitywhat goes where? Inf Comput Secur 26(1):2–9
35. Weir C, Becker I, Noble J, Blair L, Sasse MA, Rashid A (2019) Interventions for long-term software security: Creating a lightweight program of assurance techniques for developers. Softw-Pract Exp. 50(3):275–298
36. Yu Y, Kaiya H, Yoshioka N, Hu Z, Washizaki H, Xiong Y, Hosseinian-Far A (2018) Goal modelling for security problem matching and pattern enforcement. Int J Secur Softw Eng. 8(3):42–57

Cloud Computing Security Challenges: A Review

Iqra Kanwal, Hina Shafi, Shahzad Memon, and Mahmood Hussain Shah

Abstract Over the last two decades, cloud computing has gained tremendous popularity because of ever growing requirements. Organizations that are heading towards cloud-based data storage options have several benefits. These include streamlined IT infrastructure and management, remote access with a secure internet link from all over the globe, and the cost-effectiveness that cloud computing can offer. The related cloud protection and privacy issues need to be further clarified. This paper aims to discuss all possible issues that are under research and are resisting consumers to migrate from traditional IT environment to new trend of cloud computing which offers flexible and scalable environment at low-cost.

Keywords Cloud computing · Security · Confidentiality · Research challenges · Cloud security

1 Introduction

According to some researchers Cloud Computing is not completely a new technology, its roots somehow lies under "Computing as a Utility" and "Grid computing" [1, 2]. Others differ from this view and according to them it is totally independent computing [3]. At the higher level of this discussion. Cloud Computing is revolutionizing the way of computing. The vision of cloud computing is not to buy either hardware or

I. Kanwal · H. Shafi · S. Memon
Faculty of Engineering and Technology, University of Sindh, Jamshoro, Pakistan
e-mail: iqra.lakho@usindh.edu.pk

H. Shafi
e-mail: hunnyshafi@gmail.com

S. Memon
e-mail: shahzad.memon@usindh.edu.pk

M. H. Shah (✉)
Northumbria University, Newcastle upon Tyne, UK
e-mail: mahmood.shah@northumbria.ac.uk

© The Author(s), under exclusive license to Springer Nature Switzerland AG 2021
H. Jahankhani et al. (eds.), *Cybersecurity, Privacy and Freedom Protection in the Connected World*, Advanced Sciences and Technologies for Security Applications, https://doi.org/10.1007/978-3-030-68534-8_29

even software instead rent services e.g. computational power, databases, storage, and other resource one just requires a vendor for that according to pay-as-you-go model, it reduces cost and makes investment oriented to operations rather than to assets acquisition. It refers to provision of hosted services over the Internet that are scalable and dynamic. Consumers can access the services online from their web browser without knowing the underlying technical details and difficulties of the resources.

Many researchers have given various definitions but one definition by National Institute of Standards and Technology (NIST) is that "Cloud computing is a model for enabling convenient, on-demand network access to a shared pool of configurable computing resources (e.g., networks, servers, storage, applications, and services) that can be rapidly provisioned and released with minimal management effort or service provider interaction. This cloud model promotes availability and is composed of five essential characteristics, three service models, and four deployment models" is most accepted one [4].

1.1 Cloud Service Delivery Models

As the NIST definition suggests, cloud based system provides its clients with on-demand services and these can be regarded as service models which include software as a service (SaaS), platform as a service (PaaS), and infrastructure as a service (IaaS). Table 1 shows the main features of each service model, theirs users, infrastructure management, and some of the applications of these models.

Software as a Service (SaaS) It enables the customer to utilize applications running on the provider's server machine that is a cloud. Several clients can access these applications by using an interface that is a web browser, they consume the services on pay-as-you-go license subscription that significantly decreases the investment cost. SaaS is mostly used to implement business software applications at minimum charges.

Platform as a Service (PaaS) PaaS a group of software development programs and tools that are hosted on the cloud infrastructure. It is a service offered to developers which provides all the tools needed for system development. Client does not need to manage or administer computing hardware or software.

Infrastructure as a Service (IaaS) Resources can be rented as pay-per-use such resources can be storage, processing, network capacity and other computing facilities can be granted. Client can also have privilege to manage operating system and other applications. This model provides a flexibility to add or release resources and services upon requirement.

Table 1 Cloud service models

Delivery model	SaaS	PaaS	IaaS
Features	Provide software applications, used by consumers running on cloud infrastructure Examples include Software distribution model, collaboration, business processes, CRM/ERP/HR	Providing a framework for the creation, implementation and management of cloud computing solutions Examples include Web 2.0 Application, Middle-ware, Java Runtime, Tools for Development, Database	Provide hardware resources e.g. network, storage, memory, processor etc. as a virtual systems which are accessed by using Internet. Examples include servers, Networking, Data Center Fabric, Storage etc.
User	End client/End User Person or organization that subscribes a service	Developr-moderator An organization or a person that develops or deploys cloud	An organization or a person who owns cloud deployed infrastructure
Infrastructure management	Controlled by SaaS Provider	PaaS user control deployment of their individual applications and does not manage servers and storage	Controlled IaaS Provider Client is capable to launch virtual machines with any required operating systems that are managed by the clients
Applications	Google Apps (Gmail, Docs.) Salesforce CRM	Google App Engine Microsoft Azure Manjrasoft Aneka	Amazon EC2, S3 OpenNebula

1.2 Cloud Deployment Models

In literature five cloud deployment models have been discussed so far. These are public, private, hybrid, community and virtual private cloud (VPC). Among these models VPC has got less consideration by the research community [5]. Characteristics of these deployment models are summarized in Table 2.

Public Cloud In this cloud infrastructure is offered to the public on commercial basis which is own by an organization that is providing cloud services. People can rent resources and can scale up or down their utilization as desired.

Private Cloud In this model cloud infrastructure is dedicated and private system to an organization. It can be owned or rented by the organization. This may be managed by third party or organization itself.

Community Cloud This model is for the organizations having similar concerns and requirements. Cloud infrastructure is shared among more than one companies on shared cost. It is managed by the companies or a third party, located either within or outside the premises.

Hybrid Cloud Hybrid cloud is mash of multiple clouds (public, private, community), but managed and provided as a single unit by provider. The idea to use hybrid cloud mainly provide additional resources on user demand, e.g. an organization may

Table 2 Cloud deployment model characteristics

Deployment model	Characteristics
Public Cloud	Offered publically on commercial basis People rent the resources with the ability to scale them Cloud Service Provider (CSP) is responsible to own and manage a public cloud Located off-premises to the consumers Less secure than other cloud models
Private Cloud	Dedicated infrastructure for an organization for its private use Owned or rented by the organization Located either on- premises or off-premises Managed within the organization or externally More safe than the cloud that is public
Community Cloud	Shared with organizations that have similar concerns and requirements Located either on or off-premises Run by the businesses or externally
Hybrid Cloud	Mash of two or more clouds, provided as a single unit by provider Used to provide additional resources in case of demand Require both on and off-premises resources
Virtual Private Cloud	Private cloud, deployed on top of the any above mentioned cloud models over Virtual Private Network (VPN) Provide isolated resources to its consumers Characteristics are inherited by underlying cloud architectures upon which a VPC is seated

want to migrate some jobs from a private cloud to public, for this migration a hybrid cloud can be a choice.

Virtual Private Cloud (VPC) VPC is a private cloud, deployed on any above mentioned cloud models using Virtual Private Network (VPN). Its example is Amazon VPC [6]. In VPCs private and semi-private clouds are provided to the customers by VPN that provide isolated resources to its consumers. Since it is less discussed among research community, its cost, management, tenants and other characteristics are inherited from the underlying cloud models on which a VPC is seated up [5].

2 Cloud Computing Security Challenges

Although a lot of challenges are faced by emerging cloud computing technology such as interoperability, scalability, Service Level Agreement (SLA), lack of standards, continuously evolving, compliance concerns etc., security is major barrier into the adoption of cloud computing technology. Manage and maintain secure cloud computing environment is more difficult task than traditional information technology (IT) environment. As the cloud computing environment is an outsourcing of IT, along with the inherited security issues of IT environment, cloud computing also

Fig. 1   Cloud computing security challenges

come across with some additional security challenges that are focused by researcher community in the last two decades which are highlighted here [3, 5, 7–27]. These security challenges are summarised in Fig. 1.

2.1   Data Security and Confidentiality Issues

Cloud consumer need to make sure their technology save cost and never sacrifice valuable data because there are several ways of data being compromised. The amount of risk increases due to the way in which a cloud particularly increases, due to which it may demand to remove and modify record, if the encoding key is lost it's also much painful. Some of them are unique due to the nature of cloud and complex too to recover because of cloud architecture [28]. CSA's suggested solutions include strong access control API implementation, data integrity and encryption and its protection while data transfer, implementing generation of strong key, analyzing the data protection at design and also at run time. While considering security of cloud computing the data comes at top most priority. Data theft and corrupted storage are two major risks in data protection.

2.2 Data Location

One of the issues of security and confidentiality (i.e. user location, relocation of user data and services, availability and security etc.) is data location. SaaS users use these applications for the processing the business data. Users of these services have no information of the location where their data is being kept. It can be challenging in many ways, since data security and compliance rules in different states the data location is of utter important in several enterprise architecture. For instance in various South America and European countries there is some types of data that cannot cross the border due to confidentiality reasons. Besides this issue of local rules and law, when an investigation takes place, it raises a question that the data falls at which jurisdiction. SaaS service model must also provide assurance of the location of the user data.

2.3 Data Segregation

Due to multitenancy, multiple users are capable to store use the data using SaaS services. In these situations the data of multiple users is stored on same location. Data intrusion is a threat in this condition. Attackers can inject their code into the SaaS service to hijack the system. It is possible that clients may run that code accidently without proper verification which leads high potential risk to other client's data. It is obligation to SaaS application to guarantee boundary between each client data. It is also required to maintain this boundary at physical as well as at service level and an application must segregate the data from different clients.

2.4 Data Availability

It is essential for SaaS cloud applications to guaranty twenty-four-seven services to their consumers. In order to provide high-availability and scalability, it is required to take measures at both design and architectural levels. Adaption of multi-tier architecture and support for application load balancing on various servers is needed. Recovery from hardware or software crashes, and denial of service attack must be built from within the service. Simultaneously, an action plan is needs for business continuation and disaster management for future crises. It is utterly important to offer safety measures to the organizations, minimal downtime, and maximum data availability.

2.5 Regulatory Compliance

Eventually, it is the consumer who is responsible to protect their data and data integrity even though if it's on the cloud. External audits and compliance certifications are carried out on conventional service providers. Providers of cloud computing who fail to conduct the scrutiny are "signaling that clients should only use them for the most trivial functions," Gartner says.

2.6 Recovery

Data recovery is very issue concern in cloud security when considering cloud data backup. Many cloud services moves data up to 5 TB within the 12 h. However certain systems may become slower because it all depends on storage speed, the amount of time to consider, and available storage in the server that is determining and negotiating this price. Backup availability in order to keep business running is very important. Data must be backed up during the recovery process.

2.7 Investigative Support

Any conceivable object, such as processing units, storage devices or applications, is distributed as services of cloud computing. The offered services are cost-effective and expandable. Enticing benefits from cloud computing draw tremendous interest from both company owners and cyber robberies. The "computer forensic inquiry" then take measures to find evidence against criminals. As a consequence of the new technologies and approaches that are being used in cloud computing, when examining the case, forensic investigative methods come across with various types of problems. These are difficult problems to cope with multiple decisions on the variety of data stored on many servers on various locations, restricted access to cloud evidence and also the problem of seizing physical evidence for the sake of validation of credibility or presentation of evidence.

2.8   Lack of Execution Controls

A user in a cloud system requires fine-grained access control of remote execution environment which the system lacks. Therefore, memory management, access to external utilities, I/O operations, and data are some of the crucial issues which are outside the purview of the user. In many scenarios the clients require to inspect execution to make sure that no illegal operations are performed but lack of execution control restricts from it.

2.9   Long-Term Viability

In an ideal situation a well-known cloud corporation will not split or be acquired by any other organization because it is very rear. But one must be sure of data availability if such condition occur. According to Gartner "Ask potential providers how you would get your data back and if it would be in a format that you could import into a replacement application".

2.10   Malicious Insiders

Rocha et al. [29] explained ways malicious insiders are able get unauthorized access to confidential information. They have provided a demonstration of some of attacks along with videos. They showed how easily some insider can get access to cryptographic keys, passwords, and other files. It often happens that the employees are provided with restricted amount of access to the system based on company policy but with high access, it is possible for them to get sensitive and restricted data and services. CSA enforces strict check on supply chain management, transparency in information management and security practices, reporting, specification of HR requirements as part of SLA, and defining security breach notifying procedures.

2.11   Cloud Malware Injection Attack

An attacker makes an attempt to inject malicious service or virtual machines into the cloud. In such attack, the attacker uses his malicious service module (PaaS or SaaS) or an instance of virtual machine (IaaS) and tries to augment the cloud system with it.

2.12 Account High Jacking

When all authentication practices are required, the account credential details between customers/users and services and the implementation of austere authentication techniques, organizations must get as minimal details as possible to locate their users' authentication problems individually. A key role in the user authentication process should not be played by the majority of publicly accessible information. It can be inferred that, regardless of design, an Incident Response Plan is of utmost importance.

There are different ways through which attackers use to accesses cloud accounts. Some of these are the use of reused passwords, which they try to different customers account until they open them.

2.13 Issue of Multi-tenancy

Multitenancy presents a problem for the users who run their services on same physical servers. The issue is to protect user data against data theft and unauthorized access from other users. This is also a concern of current web-hosting services. This dilemma needs to be seriously reexamined with the prevalent use of cloud computing because users store their substantial data on the cloud which require proper measures of security [30].

2.14 Service-Level Agreement

Service Level Agreements (SLAs) which define minimum output standards can be anticipated by the customer, e.g. 99.99% system availability in a year. Conventionally, however, security features such as privacy and confidentiality have not been considered by SLAs. Bernsmed [31] explained how SLA of a cloud which could be expanded in order to add some security aspects that allow multiple service providers given security levels to compose cloud services.

2.15 Virtual Machine (VM) Isolation

Virtual machines running on same hardware must be separated from each other. Although logical separation is already there between VM, still physical separation need to be there since resources are shared among servers and it can lead to data leakage.

2.16 Legal Issues

Legal issues arise due to conflicting legal jurisdictions and when cloud service provider share resources in different geographical locations because sometimes different data is available in different locations with diverse digital regulations.

3 Conclusion

Traditional computing practices has evolved with time and transformed into a new trend such as cloud computing. Cloud computing provides cost-effective, innovative, flexible, and optimized computing models. It allows numerous benefits to the world of computing and sets modern trends of advance level IT. Although it offers computing as a cloud with a simple internet connection, there are numerus security challenges that are yet to be addressed. Security has always remained a major challenge in computing and IT. Along with inherent security challenges of traditional systems, cloud computing comes with additional some additional security threats, risks, and challenges. This paper presented security challenges that are focused by the research community and need to addressed to enhance security concerns of cloud computing.

References

1. Stanoevska-Slabeva K, Wozniak T, Ristol S (2010) Grid and cloud computing: a business perspective on technology and applications
2. Foster I, Zhao Y, Raicu I, Lu S (2008) Cloud computing and grid computing 360-degree compared. In: Grid computing environments workshop, GCE 2008
3. Shaikh FBF, Haider S (2011) Security threats in cloud computing. In: 2011 International conference on internet technology and secured transactions, no December, pp 214–219
4. Mell P, Grance T (2011) The NIST definition of cloud computing. In: Cloud computing and government: background, benefits, risks
5. Freire MM, Inácio PRM (2014) Security issues in cloud environments : a survey, pp 113–170
6. Chauhan S et al (2018) Amazon virtual private cloud (Amazon VPC) and networking fundamentals. In: AWS® certified advanced networking official study guide
7. Varghese B, Buyya R (2018) Next generation cloud computing: new trends and research directions. Future Gener Comput Syst 79(September):849–861
8. Birje MN, Challagidad PS, Goudar RH, Tapale MT (2017) Cloud computing review: concepts, technology, challenges and security. Int J Cloud Comput 6(1):32–57
9. Cloud computing security issues and challenges, no Jan 2011, 2015
10. Yang C, Huang Q, Li Z, Liu K, Hu F (2017) Big data and cloud computing: innovation opportunities and challenges. Int J Digit Earth 10(1):13–53
11. Khorshed T, Ali ABMS, Wasimi SA (2012) A survey on gaps, threat remediation challenges and some thoughts for proactive attack detection in cloud computing. Future Gener Comput Syst 28(6):833–851
12. Computing C (2014) A survey of cryptographic based security algorithms, vol 8, no March, pp 1–17

13. Zhang Q, Cheng L, Boutaba R (2010) Cloud computing: state-of-the-art and research challenges. J Internet Serv Appl 1(1):7–18
14. Mehraeen E, Ghazisaeedi M, Farzi J, Mirshekari S (2016) Security challenges in healthcare cloud computing: a systematic review. Glob J Health Sci 9(3):157
15. Mogos G (2019) Cloud security. Crit Anal 17(3):51–54
16. Sahmim S, Gharsellaoui H (2017) Privacy and security in internet-based computing: cloud computing, internet of things, cloud of things: a review. Procedia Comput Sci 112:1516–1522
17. Li J, Zhang Y, Chen X, Xiang Y (2018) Secure attribute-based data sharing for resource-limited users in cloud computing. Comput Secur 72:1–12
18. Paper C, Science PC (2015) State-of-the-art survey on cloud computing security challenges, approaches and solutions state-of-the-art survey on cloud computing security challenges, approaches and solutions, no June
19. Hamlen K, Kantarcioglu M, Khan L, Thuraisingham B (2010) Security issues for cloud computing. Int J Inf Secur Priv 4(2):36–48
20. Khorshed MT, Ali ABMS, Wasimi SA (2012) A survey on gaps, threat remediation challenges and some thoughts for proactive attack detection in cloud computing. Future Gener Comput Syst 28(6):833–851
21. Sengupta S, Kaulgud V, Sharma VS (2011) Cloud computing security—trends and research directions. In: Proceedings of the 2011 IEEE world congress on services, no 4, pp 524–531
22. Abdul-Jabbar SS, Aldujaili A, Mohammed SG, Saeed HS (2020) 西南交通大学学报 Integrity and security in cloud computing environment: a review 云计算环境中的完整性和安全性:回顾. J Southwest Jiaotong Univ 55(1):1–15
23. Subramanian N, Jeyaraj A (2018) Recent security challenges in cloud computing. Comput Electr Eng 71(July):28–42
24. Fatima S, Ahmad S (2019) An exhaustive review on security issues in cloud computing. KSII Trans Internet Inf Syst 13(6):3219–3237
25. Tabrizchi H, Kuchaki Rafsanjani M (2020) A survey on security challenges in cloud computing: issues, threats, and solutions. J Supercomput 24(June):133–141
26. Zissis D, Lekkas D (2012) Addressing cloud computing security issues. Future Gener Comput Syst 28(3):583–592
27. Chen D (2012) Data Security and privacy protection issues in cloud computing, no 973, pp 647–651
28. Cloud Security Alliance (2010) Top threats to cloud computing. Security
29. Rocha F, Correia M (2011) Lucy in the sky without diamonds: Stealing confidential data in the cloud. In: Proceedings of the international conference on dependable systems and networks
30. Rong C, Nguyen ST, Gilje M (2013) Beyond lightning: a survey on security challenges in cloud computing q. Comput Electr Eng 39(1):47–54
31. Bernsmed K, Jaatun MG, Undheim A (2011) Security in service level agreements for cloud computing. In: CLOSER 2011—Proceedings of the 1st international conference on cloud computing and services science

Printed in the United States
by Baker & Taylor Publisher Services